Complexities

Complexities

Women in Mathematics

Edited by
BETTYE ANNE CASE
ANNE M. LEGGETT

PRINCETON UNIVERSITY PRESS

Princeton and Oxford

Copyright © 2005 by Princeton University Press

Published by Princeton University Press, 41 William Street, Princeton, New Jersey 08540

In the United Kingdom: Princeton University Press, 3 Market Place, Woodstock, Oxfordshire OX20 1SY

LIBRARY OF CONGRESS CATALOGING-IN-PUBLICATION DATA

Complexities : women in mathematics / edited by Bettye Anne Case, Anne M. Leggett.
   p.   cm.
Includes bibliographical references and index.
ISBN 0-691-11462-5 (cloth : acid-free paper)
   1. Women in mathematics—United States—History.   2. Feminism and science—United States.   3. Women mathematicians—United States—Biography.
4. Mathematics—Vocational guidance—United States.   I. Case, Bettye Anne.
II. Leggett, Anne M.
QA27.5.C66   2005
510′82′0973—dc22   2004048843

British Library Cataloging-in-Publication Data is available

This book has been composed in Sabon and Swiss 721

Printed on acid-free paper. ∞

pup.princeton.edu

Printed in the United States of America

10 9 8 7 6 5 4 3 2 1

To Jack and Gerry, always there

CONTENTS

PREFACE

On the cusp of the new century, we began a journey through time and place. Looking back showed how women before us fared in the world of mathematics; examining the present highlighted trends and directions for today's women mathematicians, leading to a vision of an ideal future where women are full participants in all intellectual ventures. Women's journey from the margins of the world of doctoral-level mathematics to a more central position in this particular universe has been long and difficult. The struggle has led to more satisfying lives, both professional and personal. This chronicle of some women who have chosen to be professional mathematicians is intended to facilitate the journey for our sisters who follow.

Thought-provoking perspectives of special interest to those concerned with gender issues appear in many of the articles, offering encouragement and support to women in different trenches—associates in law firms, medical residents—a whole generation of women finding their paths. More directly, the stories include points of reference and parallels for women in all academic fields, especially in those where the proportions of women researchers are low, as they are, for example, in the bench sciences, computer science, and engineering.

The readings for this time travel, an eclectic anthology in five parts, have varied origins. Many of the articles in parts I through III are based on material from the newsletters of the Association for Women in Mathematics (AWM, est. 1971). This volume serves as an archival source for those selected as well as for the information in parts III and IV about the Olga Taussky Todd Celebration of Careers for Women in Mathematics. The conversational flavor of the articles that were originally talks at mathematics meetings and conferences has been preserved in their written form. Throughout the book, supplementary information from many sources augments topics introduced in the main articles, providing context and balance. Finally, writings from this new century complement the earlier works, providing deeper insight into women's place in mathematics today.

Complexities sets the stage in part I by presenting biographies, autobiographies, mathematical histories, or vignettes from the lives of fifteen women born from 1776 to 1919. Obstacles that stood in their way, often solely because they were female, appear in story after story. Part II portrays the collective efforts of women to eliminate some of these barriers or to mitigate the difficulties involved in navigating them. The narratives about AWM yield multiple examples of women's accomplishments, both as individuals and in groups. This theme is also explored in several articles illuminating women's place on the global mathematics stage.

In part III, practitioners describe multiple career paths open today to women mathematicians. The challenges explored range from how some valiant African-American women have dealt with discrimination on two fronts to how some young mothers have succeeded in doing mathematics while raising their children. Contributions from currently active women mathematicians appear throughout and form all of parts IV and V. Several of the authors in part IV describe a problem (in mathematics research or teaching), along with solutions to and perspectives on that problem. Part V is a tapestry of recent thinking from a spectrum of mathematical women—distinguished recipients of multiple honors to new doctorates—speaking freely on mathematics and life.

The book's attention to career issues parallels that of workshops held for beginning mathematicians and follows the theme of the Taussky Todd conference, which in turn took its cue from the extraordinarily varied career path of its namesake. The talks there featured not only mathematical results and descriptions of technical work but also interwoven observations on the themes of existence: overcoming the difficulties life has placed before women who want to be mathematicians, carving out a career path around the stumbling blocks, having a life and a career in today's complicated world, and working to make it easier for future generations of women to pursue mathematics.

One of our pleasures in working on this book has been noting similarities, regardless of the differences in circumstance, among the commentaries of these mathematical women. A vibrant thread links the earliest women to the most recent—the passion to do mathematics and to share it with others, despite the obstacles laid before them by lives and times. Please join our journey into the world of women mathematicians.

ACKNOWLEDGMENTS

As editors of a volume with over seventy contributors, we owe our gratitude to many individuals and organizations. We would like first of all to thank developmental editor Susan Gerstein, who provided valuable services and ideas as the manuscript began to take shape. Her sustained enthusiasm for the project encouraged us in turn. We are similarly indebted to the editorial advisors, Carolyn S. Gordon (Dartmouth College), Suzanne Lenhart (University of Tennessee), and Sylvia M. Wiegand (University of Nebraska), for volunteering their efforts whenever and for whatever we needed them. Lenhart, president of AWM during crucial phases of book production, was particularly generous.

We especially thank the volume's contributors for their responsiveness during the editing process. Many AWM members have given time to advance various stages of the editorial process—as contributors, referees, editorial advisors, and executive committee members. We owe a special debt to the organization's founders, past officers, and early newsletter editors. We have been informed and inspired by authors of earlier works that lend intellectual support for this volume, and by the comments of knowledgeable reviewers, which we have appreciated. We also thank the staff at Princeton University Press for all their efforts in producing this book.

Our colleagues at the Department of Mathematics of the Florida State University (FSU) and the Department of Mathematics and Statistics of Loyola University Chicago gave useful information and wise advice; we thank our students for their patience. We have appreciated support for project needs from our departments. The staff at AWM's Maryland office and at FSU Mathematics provided multiple types of in-kind assistance; staff members at the American Mathematical Society (AMS) and the Mathematical Association of America (MAA) provided information and advice.

A goal of the organizers and funding agencies of the 1999 Olga Taussky Todd Celebration of Careers in Mathematics for Women was the widest possible dissemination of the biographical, mathematical, and career information that was developed from that conference. Base funding for the conference and developmental stages of this volume was provided by the National Security Agency, with additional support from the Department of Energy, the Office of Naval Research, and the Mathematical Sciences Research Institute in Berkeley. "The Taussky Todd Conference," IV, lists many individuals whose work made that conference and, by extension, early efforts on this project possible. The editors gratefully acknowledge the benefits to this volume of the direct and in-kind funding received by AWM over the years; a partial listing appears in "Activities and Awards," II.

Permission to use material printed in other sources and to reproduce photographs has been graciously provided by AMS, AWM, Boston University, British Society for the History of Mathematics *Newsletter,* Computing Research Association *Newsletter,* European Women in Mathematics, Pamela Davis Kivelson, Kovalevskaia Fund *Newsletter,* Suzanne Lenhart, Deborah Lockhart, MAA, Constance Reid, Springer-Verlag Inc., St. Joseph's University, Elaine Terry, John Todd, University of Texas, Dawn Wheeler, and Sylvia Wiegand.

In addition to the above, the list of individuals that follows includes many of those who have helped or inspired this work. We apologize in advance for any omissions. The editors thank Robin Aguiar, Jean Berry, Phil Bowers, Mickey Boyd, Mimi Burbank, Barbara Deuink, Muriel Daley, Chandler Davis, Esther Diaguila, Wayne Drady, Susan Ecklund, John Ewing, Etta Falconer, Sergei Gelfand, Grace Godfrey-Brock, Mary Gray, Judy Green, Gloria Hewitt, Johnny L. Houston, Sam Huckaba, Joan Hutchinson, Allyn Jackson, Lise Jacobson, Linda Jarvis, Alison Kalett, Tanja Kazan, Beth Kegler, Pat Kenschaft, Cathy Kessel, Susan Leggett, Lee Lorch, Jim Maxwell, Joe Mayne, Susan Minnerly, Calvin Moore, Joe Mott, Dianne O'Leary, Kendra Remick Priestly, Sam Rankin, Judy Roitman, Colleen Rose, Donna Salter, Diane Saxe, Alice Schafer, Hans Schneider, Pam Schnitter, Martha Siegel, Vickie Magoon Sims, Wilma Slaight, Elaine Smith, Jackie Smith, Melissa Smith, Tina Straley, Rebekka Struik, De Witt Sumners, Jean Taylor, Laura Tedeschini-Lalli, Christine Thivierge, Priscilla Travis, and Miao Ye.

Our families have been wonderful, especially in their understanding when we have been overly busy; we thank Jack Quine and Gerry McDonald for their support and sage counsel. Finally, we give special thanks to senior editor Vickie Kearn and production editor Gail Schmitt at Princeton University Press, without whose enthusiasm, encouragement, and commitment this book would not exist. Gail cheerfully dealt with the myriad questions and changes we threw her way. For more than two years, Vickie helped us stay on target and, much more important, was our collaborator with suggestions that vastly improved the original concept. When she said, "You'll look back and say this extra work was worth it," she was always right.

Bettye Anne Case
Olga Larson Professor of Mathematics
Florida State University

Anne M. Leggett
Associate Professor of Mathematics
Loyola University Chicago

NOTE TO READERS

All bylined papers that have not previously appeared in print were refereed. Reprints have been excerpted or modified from their earlier published form. Known factual errors in reprints have been corrected, but statistics and time signifiers reflect the date of original writing. In some cases, footnotes and references have been added to provide new information. Bracketed numbers in the text indicate references at the end of the paper in which they appear.

Many of the websites in the list of abbreviations have information and bibliographies on topics discussed in this book as well as on related issues (e.g., gender issues in precollege education). There is also a website for this volume, http://www.pup.princeton.edu/titles/7915.html where additional references and updates to book material are provided. Readers are encouraged to inform us of additional relevant sources, corrections, and so forth.

The editors are coauthors of introductions and other text that is not attributed to another source.

ABBREVIATIONS

AAAS: American Association for the Advancement of Science; http://www.aaas.org/

AAUW: American Association of University Women; http://www.aauw.org/

ACM: Association for Computing Machinery; http://www.acm.org/

AIM: American Institute of Mathematics; http://www.aimath.org/

AMS: American Mathematical Society; http://www.ams.org/

AMUCWMA: African Mathematical Union Commission on Women in Mathematics in Africa; http://www.math.buffalo.edu/mad/AMU/amucwma.html

ASEE: American Society for Engineering Education; http://www.asee.org/

AWC: Association for Women in Computing; http://www.awc-hq.org/

AWIS: Association for Women in Science; http://www.awis.org/

AWM: Association for Women in Mathematics; http://www.awm-math.org/

BMSA: Board on Mathematical Sciences and Their Applications, formerly Board on Mathematical Science; http://www7.nationalacademies.org/bms/

CBMS: Conference Board of the Mathematical Sciences; http://www.cbmsweb.org/

CWS: Caucus for Women in Statistics; http://www.geocities.com/ResearchTriangle/System/2290/

EWM: European Women in Mathematics; http://www.math.helsinki.fi/EWM/

EYH™: Expanding Your Horizons, the flagship program of the Math/Science Network; http://www.expandingyourhorizons.org/

Femmes et Mathématiques; http://www.femmes-et-maths.fr.fm/

IAS: Institute for Advanced Study; http://www.ias.edu/

ICIAM: International Congress of Industrial and Applied Mathematicians; held every four years under the auspices of the International Council for Industrial and Applied Mathematicians; http://www.iciam.org/council/council_tf.html

ICM: International Congress of Mathematicians; held every four years under the auspices of the International Mathematical Union; http://www.mathunion.org/ICM/index.html

ICME: International Congress on Mathematical Education, held every four years under the auspices of the International Council on Mathematics Instruction, http://www.mathunion.org/ICMI/ICME_congress.html

ICMI: International Council on Mathematics Instruction, http://www.mathunion.org/ICMI/

IEEE: Institute of Electrical and Electronics Engineers, Inc.; http://www.ieee.org/

IMA: Institute for Mathematics and Its Applications; http://www.ima.umn.edu/

IMU: International Mathematical Union; http://www.mathunion.org/

IOWME: International Organisation of Women and Mathematics Education; http://www.stanford.edu/~joboaler/iowme/

JMM: Joint Mathematics Meetings; AWM holds its annual meeting in conjunction with this January annual conference of the AMS, MAA, and other organizations

MAA: Mathematical Association of America; http://www.maa.org/

The MacTutor History of Mathematics archive, University of St. Andrews; http://www-history.mcs.st-andrews.ac.uk/history/index.html

Mathematicians of the African Diaspora; http://www.math.buffalo.edu/mad/

MSEB: Mathematical Sciences Education Board; http://www7.nationalacademies.org/mseb/

M/SN: Math/Science Network; http://www.expandingyourhorizons.org/aboutmsn.html

MSRI: Mathematical Sciences Research Institute; http://www.msri.org/

NAM: National Association of Mathematicians; http://www.caam.rice.edu/~nated/orgs/nam/

NAS: National Academy of Sciences; http://www4.nationalacademies.org/nas/nashome.nsf/

NBS: National Bureau of Standards; predecessor agency to NIST

NCTM: National Council of Teachers of Mathematics; http://www.nctm.org/

NIH: National Institutes of Health; http://www.nih.org/

NIST: National Institute of Standards and Technology; formerly the NBS; http://www.nist.gov/

NRC: National Research Council; http://www.nas.edu/nrc/

NSA: National Security Agency; http://www.nsa.gov/

NSF: National Science Foundation; http://www.nsf.gov/

ONR: Office of Naval Research; http://www.onr.navy.mil/

Project NExT: New Experiences in Teaching; http://archives.math.utk.edu/projnext/

RAWM: Russian Association of Women Mathematicians

SACNAS: Society for the Advancement of Chicanos and Native Americans in Science; http://www.sacnas.org/

SIAM: Society for Industrial and Applied Mathematics; http://www.siam.org/

SWE: Society of Women Engineers; http://www.societyofwomenengineers.org/

TWOWS: Third World Organization for Women in Science; http://www.ictp.trieste.it/~twas/TWOWS.html

WAM: Women and Mathematics, a project of the MAA Committee on the Participation of Women; http://www.mystery.com/WAM/

WME: Women in Mathematics Education; http://www.wme-usa.org/

YMN: Young Mathematicians Network; http://www.youngmath.net/

All websites and web pages were active on April 23, 2004.

I. INSPIRATION

The history of women working in mathematics goes back at least as far as Hypatia (A.D. 370–415), but the women in part I are from the more recent past. These women persevered in the face of multiple burdens and hardships to do the mathematics they loved. Their recorded work spans just over two centuries, from 1794 to 1997, beginning with a student paper and ending with a posthumously published article written by a prolific researcher. Those whose accomplishments are archived in the first chapter earned their doctoral degrees in the twentieth century, while those in the second completed their degree requirements in the nineteenth—save for the three who worked when women were not yet permitted to pursue the advanced degree.

Clearly most of these women, now deceased, had formidable obstacles to overcome even from childhood. A poignant story shared by two of the earliest of these women has each of them hiding candles for nighttime studies of mathematics literally undercover. Severe familial discouragement of their mathematical interests led them to these extremes, their interest then characterized as "unhealthy" and—as is still too often the case more than two centuries later—as "unfeminine." The opportunities available to these women (and the successes and frustrations of their contemporaries, many of whom left no written record) were subject to the vagaries of political events and geographic location, which often had greater impact on their mathematical accomplishments and lives than did their own abilities and desires. As one author recounts, she acquired a position for which she was more than fully qualified only when "desperation again overcame prejudice."

There is a strong sense of personal immediacy to the articles in chapter 1, most of which are autobiographical or were provided by close associates or family members. The second chapter begins with an author-granddaughter; another author found a Paris street named for her subject but searched in vain for her tombstone. Read on to learn more about these remarkable women.

1. From the Twentieth Century

The women appearing in this chapter, inspirations for many currently active mathematicians, received their doctorates before 1960. Their stories look back through time, arranged in reverse chronological order of the women's births. It is appropriate to begin with reminiscences of Olga Taussky Todd, namesake of the conference that spurred the compilation of this book. Her charmingly informal remarks are augmented by excerpts from a memorial article about her. Later in the book, two articles give more formal descriptions of her life and work.

Julia Robinson's mathematics led to her election as the first woman mathematician in the National Academy of Sciences (NAS); she was also the first woman elected president of the American Mathematical Society (AMS). In 1980 she was the second woman to present the prestigious AMS Colloquium Lectures. The first woman to do so, fifty-three years earlier in 1927, was Anna Johnson Pell Wheeler.[1] Among the few women who earned a doctorate in mathematics in the first half of the twentieth century, Pell Wheeler was singular in receiving such recognition for her mathematics. She is likely best known for her active mentoring of other women. While chair of the mathematics department at Bryn Mawr College, she brought two refugees to her institution: Emmy Noether, then an established mathematician, and also, to work with Noether, the young Taussky Todd—the women whose lives bracket this chapter. Noether is indisputably among the greatest mathematicians of the twentieth century. In recognition of the importance of her work, the Association for Women in Mathematics (AWM) has sponsored since 1980 the annual Noether Lecture series at the Joint Mathematics Meetings (JMM). Robinson and Taussky Todd are two of the outstanding senior women mathematicians who have delivered these lectures.

Marjorie Lee Browne is the third African-American woman currently known to have earned a Ph.D. in mathematics; two of the women profiled in "A Dual Triumph," III.1, cite her as an inspiration. The accomplishments of Cora Ratto de Sadosky, not only an excellent mathematician but also a fiery advocate of the need for social justice, and of Mabel Barnes, a remarkable survivor with a fifty-year career in teaching, deserve also to reach a wider audience.

Familial connections abound in this chapter: both Ratto de Sadosky and Barnes had daughters who became mathematicians, while Ratto de

---

1. See AWM *Newsletter* 12(4), 1982, 4–13; and Bettye Anne Case, ed., *A Century of Mathematical Meetings* (Amer. Math. Soc., 1991), 311–319, 321.

Sadosky, Taussky Todd, and Robinson had mathematician husbands. Women with sisters will find much that resonates in Robinson's story, told by her sister Constance Reid, herself a well-known biographer of mathematicians.[2] Noether's family included several mathematicians (her father, Max, was an analyst, while she was an algebraist), so that a mention of the surname is often followed by the question, "Which Noether?" Emmy's mathematician brother Fritz was the father of Gottfried, statistician and late husband of her biographer Emiliana Pasca Noether. Browne was also influenced at home by the keen interest in mathematics of both her father and brother, though neither was a professional mathematician. Going beyond the biological family, Browne inspired and assisted in practical ways students who became close to her and to each other.

In Her Own Words

OLGA TAUSSKY-TODD

Based on a talk at the panel "Centennial Reflections on Women in American Mathematics," organized by Judy Green and Jeanne LaDuke, AMS Centennial Meeting, Providence, 1988 as reported in the AWM *Newsletter* 18(6), 1988, 10–11. Taussky Todd (1906–1995) was then a professor emerita of mathematics, California Institute of Technology.

At any time a mathematician's life is not an easy one, and mine began at a particularly critical time in history. I graduated from the university in Vienna, Austria, in 1930 and had then a variety of jobs connected with mathematical research, mostly unpaid or underpaid, at various institutions, with an assortment of bosses and pupils.

The fact that I studied and worked in several countries made me able to observe a number of facts about the behavior and treatment of women. Now we live with "Women's Lib," and it has not only changed the opportunities for women, but also their behavior toward each other. Women are now more supportive of their women colleagues. This was not always the case. Even the great and kind Emmy Noether was no exception. She was convinced that men had greater strength and that women ought not to attempt to work like men, that regular appointments ought to go to men so that they could support a family. Women ought to look for marriage.

I came to the U.S. for the first time in 1934, and this is where the story really ought to begin. I came to Bryn Mawr College, where the department

2. Reid wrote, e.g., *Hilbert* and *Courant*, both published by Springer-Verlag.

chairman was Anna Johnson Pell Wheeler. She had gone through hard times. Her first husband, Pell, became very ill, and she had to look for work. But she really made it. She was the first woman to give the AMS Colloquium Lecture series. However, at the age of about fifty her health seemed to break down. One day I discussed with her the problem of women in academic life and bemoaned their poor strength. She predicted that women would become more athletic in the future. Well, if she saw our young women at Caltech nowadays, dressed in short outfits on the coldest days, full of strength, she would see that her prediction has come true.

The next time I turned up in the U.S. was after World War II, which we (my husband and myself) had spent in Great Britain, with a truly tough time. We gave up academic life and joined the British Civil Service, and I particularly had to leave my favorite subject. This second visit was by invitation of the U.S. National Bureau of Standards, where we had hard and new work (exploitation of computers), but no more war. I think women were treated as well as the men there. We stayed there until 1957, when Caltech offered both of us very attractive positions. I myself was the first woman to teach in this real men's place, a fact that is not true any longer. So there is progress!

In 1958 I was invited to give the one-hour lecture at an AMS meeting, the first woman since Emmy Noether in 1934. The lecture, titled "Integral Matrices," was published in the AMS *Bulletin* and is cited in the book by Curtis and Reiner. At such an occasion the chairman usually says a few kind words by way of introduction. I trained myself to say "thank you for your kind words." However, he only mentioned my name and Caltech, and I almost thanked him for his "kind words."

Soon there will not be occasions for a woman to be "a first." I cannot help wondering what would happen nowadays if a woman chairman had to fill an opening and a woman and man of fairly identical qualifications applied for it. Difficult situations will, of course, still occur in many ways. Blame it all on Adam and Eve.

Some predictions for women in university departments:

universities will employ more women;
departments will employ more women;
however, in the same department some men colleagues will be more
   jealous of the achievements of women colleagues than of men
   colleagues and even persecute them if possible.

And women will always be different from men.

Reference

Linear Algebra and Its Applications 280(1), Elsevier, is a special issue honoring Olga Taussky Todd. It gives an account of Olga's life and work, with full references.

Remembering Olga Taussky Todd

CHANDLER DAVIS

Based on "Remembering Olga Taussky Todd," AWM *Newsletter* 26(1), 1996, 7–9, and "Postscript to Olga Taussky Todd," AWM *Newsletter* 26(2), 1996, 28, which were adapted for reprinting in the *Mathematical Intelligencer* 19(1), 1997, 15–17, by permission of AWM and the author. Copyright © 1997 Springer-Verlag Inc. Davis was then a professor of mathematics, University of Toronto, where he is currently a professor emeritus.

Olga Taussky is remembered by many for her lectures. One was AWM's Noether Lecture in 1981; this had a special resonance, for she had known Emmy Noether both at Göttingen and at Bryn Mawr. Others remember Olga as author of some beautiful research papers, as teacher, as collaborator, and as someone whose zest for mathematics was deeply felt and contagious. The field she is most identified with—which might be called "linear algebra and applications," though "real and complex matrix theory" would be preferred by some—did not have autonomous existence in the 1930s, despite the textbook by C. C. MacDuffee. Her stature in that field is the very highest, as was palpable in the standing ovation after her survey talk at the second Raleigh conference in 1982.[3]

It is amusing to hear the story of a job interview where a member of the committee asked her, with motivation we can imagine, "I see you have written several joint papers. Were you the senior or the junior author?" Another member of the committee was G. H. Hardy, who interjected, "That is a most improper question. Do not answer it!" At another interview she was asked, "I see you have collaborated with some men, but with no women. Why?" Olga replied that that was why she was applying for a position in a women's college![4] It is less amusing to learn that the senior woman mathematician insisted that women students not do their theses with Olga, even when male colleagues considered her the most suited to the projected research, because it would be damaging to their career to have a woman supervisor.

In 1938, while both were working at the University of London, Olga Taussky and John Todd married. Jack's scientific background was rather different—classical analysis—and his background was different—Presbyterian northern Irish. But their ensuing collaboration over fifty-seven years was close and extraordinarily fruitful. There were few joint

3. Later published as "How I Became a Torchbearer for Matrix Theory," *American Mathematical Monthly* 95(1998), 801–812.
4. John Todd, "G. H. Hardy as an Editor," *Mathematical Intelligencer* 16(2), 1994, 32–37.

papers, but they talked everything over, and everything either did was influenced by the other.

Olga went into applied work for the Ministry of Aircraft Production during the war. The problems included analysis of aircraft designs for their stability properties. The tools were the localization of eigenvalues, stability analysis (testing whether the real parts of all eigenvalues are <0, or anyway not too far above 0), and numerical computation. The Todds' war work coincided with the start of the great expansion of number-crunching technique; Jack did, but Olga did not, keep always adept at the most powerful computational methods. Don't imagine Olga uttering only abstract notions and Jack only results of machine computations. Her curiosity extended to the details of numerical examples; his encompassed the theory. A good example is the Hilbert matrix, a passion they shared.

At the end of the war they moved to the U.S. National Bureau of Standards, first in Washington and then in Los Angeles. This was the period when, stimulated by the coming of peace and the computer revolution, the new matrix theory community was being established. What now look like fundamental theorems of matrix theory—Gaussian elimination, the Cauchy interlacing theorem, the Cayley-Hamilton theorem, Sylvester's inertia theorem, the Smith and Jordan forms, Perron-Frobenius theory, the variational principle for eigenvalues—were known and had not been entirely forgotten. They weren't taught much: there is an "introductory linear algebra course" everywhere now, but then nowhere; as a consequence, when I began graduate study in physics in 1946, four different courses I took began with about six weeks on "vectors." What happened in the following decade was the recognition of matrix theory as a body of doctrine and as a necessary toolkit for the scientist. Simultaneously it recognized itself as a "field" of research; recognition by others took longer.

It had been several years since the Todds had taught. Olga had grown up in a world where women—even Emmy Noether—might be barred from university professorships. It was most welcome when Caltech invited her and Jack to join the faculty in 1957. The offer was (as was usual at the time) for the husband to become professor and the wife research associate, but their offices were adjacent and the same size, and Olga was welcome to conduct seminars and supervise theses. The anomaly in their status ceased to look ideal when, in 1971, a very young assistant professor of English was glorified by the press as the first woman ever on Caltech's faculty. The first, indeed! What about Olga? I saw no sign that Olga held this against the young woman herself, but it did rub her the wrong way; she went straight to the administration and had her rank changed to professor.

Olga Taussky always wished to ease the way of younger women in mathematics and was sorry not to have more of them to work with. She said so, and she showed it in her life. Marjorie Senechal recalls giving a paper at an AMS meeting for the first time in 1962, and feeling quite alone and far from home. Olga turned the whole experience into a pleasant one by coming up to Marjorie, all smiles, introducing herself, and saying, "It's so nice to have another woman here! Welcome to mathematics!"

Being Julia Robinson's Sister

CONSTANCE REID

Julia Robinson, 1919–1985. Based on the after-dinner talk "Being Julia Robinson's Sister," delivered at the 1996 Robinson Celebration, MSRI, Berkeley, as reported in the AWM *Newsletter* 26(5), 1996, 22–28. Reid is a noted writer on mathematics and mathematicians who resides in San Francisco.

When I was asked to speak tonight, I could not refuse. The Julia Robinson Celebration of Women in Mathematics is a truly celebratory occasion, and I feel that as Julia's sister I should be here. Yet I find myself in a very difficult position. Here I am to speak about Julia, and being spoken about is the last thing Julia would want. As a mathematician, as was done earlier in the meeting—yes. But as a person—no.

So I decided my subject would be simply "Being Julia Robinson's Sister." That is the one subject connected with Julia that I can talk freely about—because it's my life, not Julia's. But in the course of the evening, talking about our sisterhood—from not so much a personal point of view as from what one might call "a point of view pertaining somewhat to mathematics"—I can tell you something about Julia, some things that will not violate her desire for personal privacy, and something also about the feelings that she expressed to me on the subject of her other sisters—all the women here and the others who are mathematicians.

Julia was born twenty-three months after I was, essentially two years—the worst possible difference in age for siblings, in my opinion—close enough for the younger to almost catch up with the elder—who is nevertheless always just a little bit ahead. I have to confess that as children we fought almost all the time. My earliest memory of Julia is of her tearing the hair off my doll while I poked the eyes out of hers. We were not close. In addition to age and sibling rivalry separating us, there was also a serious illness that was to keep Julia away from home for a year and out of school from the time she was nine until she was thirteen. It was to affect her entire life—to prevent her from having the children she very much wanted and to make it physically impossible for her take on the rigors of a full-time professional position at Berkeley.

While I could tell you something about these early years, I prefer to concentrate on that longer period in our lives that extended up to Julia's death, when we were very close. That period began in 1950, when I married and moved to San Francisco and Julia returned to Berkeley after a year at the RAND Corporation in Santa Monica. At that time she had been married since 1941 to Raphael Robinson, who had been her number theory teacher at Berkeley; she had got her Ph.D. in 1948 under Alfred Tarski with an important result in a combination of logic and number theory, and during the year that she had just spent at RAND she had solved an important problem in game theory. She had also begun to work on Hilbert's tenth problem.

I knew practically nothing about these mathematical achievements or interests. Once, a year or two before, when Julia came home to San Diego for a visit, she had tried to explain to me what she had done in her thesis. I did not have the faintest idea what she was talking about, or why it was significant, but I remember feeling a little sorry for her because she couldn't explain something important that she had done even to her sister. Oddly enough, I didn't feel sorry for myself for not being able to understand.

Later, in the time I am talking about, when not only I but also our entire family had migrated from San Diego to the Bay Area, Julia and I saw a lot of each other. We met for lunch in San Francisco and shopped furniture stores and talked endlessly both in person and on the phone. We had many common interests. She was a housewife who did mathematics, and I was a housewife who wrote. There was also politics—this was the era of Joseph McCarthy and the infamous Loyalty Oath at Berkeley.

When we got together as a family, which we frequently did, Raphael liked to make conversation with me by telling me things about mathematics. He was a remarkable expositor, as some of you know, and he told me about Gödel's work, and Turing machines, and the theory of sets, and the pearls of number theory, and n-dimensional geometry, and knot theory—maybe even about Hilbert's problems. I was somewhat used to such "teaching" because, during a brief period in college when Julia and I shared a room, she used to tell me about things she had read in *Men of Mathematics*, which had just appeared at that time.

Well, all this effort—on both the Robinsons' part—was to bear fruit one morning in 1951 when Julia, in the course of a telephone conversation, reported to me the success of a program of Raphael's for testing the primality of very large Mersenne numbers on one of the new giant computers—this one was SWAC (the Bureau of Standards Western Automatic Computer). These computers, which were popularly called "giant brains," had been invented during the Second World War and had been known to the public for only about five years. Julia also explained to me the connection between Mersenne numbers and "perfect" numbers. This achievement of Raphael's interested me—it struck me as something I could write about that other people would be interested in, too.

Julia promptly encouraged me, in a very practical way, by inviting me to lunch with Dick Lehmer, the mathematician in charge of SWAC, so that I could find out from him what SWAC looked like and how it was operated. At that time neither Raphael nor Julia had ever actually seen one of the new computers—and it is still remarkable, even to experts, that Raphael had successfully programmed SWAC simply by studying the manual. Well, Dick was helpful, and his wife, Emma, was helpful, too—it was she who suggested that I send my article to *Scientific American*. To make a long story short, *Scientific American* published it, a publisher read it and wrote to ask if I—Constance Reid, who had left mathematics for Latin in her sophomore year in high school—would be interested in writing a little book on numbers for him.

Now what still amazes me is that Julia did not try to talk me out of this project, but actually encouraged me. Raphael did not encourage me, but neither was he negative. The publisher was thinking about a book on numbers to go with a book he had published on the alphabet called *The Twenty-six Letters*. This suggested to me a book about the ten digits, since the *Scientific American* article had been in a way a story about "six" as the first perfect number. I thought I would just treat the other digits in a similar fashion—a mixture of number theory, history, and what you might call numerology. Julia and Raphael seemed to think that I could do that. Later, though, when I got to the chapter on 9, which was to be about "casting out 9's" and other such checks, Raphael insisted that there should be some real mathematics in the book, so he explained congruences to me and the law of quadratic reciprocity.

Well, that first book, *From Zero to Infinity*, was something of a success—it has been in print now since 1955. One book led to another and another, and these I wrote more and more on my own—although Julia and Raphael always read the finished manuscripts.

While I was writing these books, handling the financial side of my husband's law practice, raising my children, and working to improve the San Francisco public schools, Julia was so absorbed in politics that she virtually gave up mathematics.

You know that Julia was a solver of mathematical problems, but do you know that she put her mind to all sorts of other problems—relatively small problems like how Marina Ratner's little daughter could learn English quickly and enjoyably—Julia's solution was to give her Nancy Drew books—and larger problems of the University of California—and it had plenty of problems during those years—the Democratic Party—the United States—the world.

I can give you an example of Julia's nonmathematical problem solving on a major scale. In 1952, when Adlai Stevenson had been badly defeated by Eisenhower and the Democratic Party was in what can be best described as disarray—Julia was concerned about the fact that the intellectual grassroots support for Stevenson was separating itself from the party and from party politics. She decided that her sister Constance should convey her ideas in a letter to the editor of the *New Republic*, since in her view I could write and she could not. This past Sunday I went down to the library and looked up that letter. There it was—a column and a third at the beginning of the Letters to the Editor column in the *New Republic* of January 26, 1953. It was odd to read it. The words were Constance Reid's, but the political passion was Julia Robinson's! The letter appeared just before an important meeting of Democratic leaders at Asilomar, to which interested citizens were also invited. At Julia's urging my husband and I went with her and Raphael. We found to our amazement that all the bigwigs at the meeting were talking about my letter and were asking, Who is this Constance Reid? I know people have sometimes suspected that Constance Reid was really Julia Robinson, and on this occasion it was so. I don't remember exactly what happened, but the end result was that Julia involved herself during those years in the nitty-gritty of Democratic Party politics—she registered voters, stuffed envelopes, rang doorbells in neighborhoods where people expected to be paid for their vote. She even served as Alan Cranston's campaign manager for Contra Costa County when he successfully ran for state controller—his first political office.

This politically active period of Julia's life concluded at the end of the 1950s when, her physical condition having become much worse, she had to undergo major heart surgery. The surgery greatly improved her general health, although she still lacked the stamina of a normal person, and when she taught a single class at Berkeley, as she frequently did, everything else had to be put on hold.

At this time I, after writing three books explaining mathematics to laymen, felt that I had exhausted, not mathematics, but the mathematics that I was capable of explaining. So I was rather at loose ends in my writing. I wanted to do something different. Well, after three successful books, Julia had begun to think of me not only as a writing asset, but also as an *asset* to mathematics. One day she came across an obituary of some mathematician who had recently died. She read it with interest and, remembering what E. T. Bell's *Men of Mathematics* had meant to her when she was a college student, she decided it would be good for students to be able to read about more modern mathematicians than those in Bell, whose names were also attached to theorems in their textbooks.

Constance Should Update E. T. Bell. To set this proposed project in the context of Julia's mathematical career, I should say that she and Martin Davis and Hilary Putnam had just published their joint paper, "The

Decision Problem for Exponential Diophantine Equations," but Julia was becoming somewhat discouraged about her ideas on the subject of Hilbert's tenth problem. A year or so before—again at Asilomar—she had explained the problem to me. By this time I had a little more understanding than I had had when she explained her thesis. She had said to me then—which had impressed me greatly—that she didn't care whether she solved the problem herself—she just had *to know* the answer, she wouldn't want to die *without knowing.*

It was during this period that she came up with the idea of my writing a collection of short biographies of modern mathematicians, and she spent a great many hours with me going through *Math. Reviews* and making out three-by-five cards for all the obituaries, memoirs, autobiographies, and biographies of mathematicians that we could find between the first issue in 1940 and the most recent one in 1964. I should mention that by 1964, although there were lots of obituaries, there were *no* full-length biographies. There were two autobiographies—Norbert Wiener's *Ex-Prodigy* and G. H. Hardy's *A Mathematician's Apology*, which was somewhat autobiographical. That was it. This situation has changed dramatically in the interim—if not in numbers, at least in percentages.

Well, Julia was very persistent, and I became interested if not excited, so we decided to go to Europe, where I could absorb local color and interview colleagues and relatives of the mathematicians on our list, all of whom had lived after the First World War—and had died.

It happened that, at the time, Julia was auditing a class of Alfred Tarski's in which the person who always arranged to sit next to her was a young Ph.D. from Göttingen, a probabilist then, named Volker Strassen. She told him that her sister was planning to write a book about *Men and Women of Modern Mathematics*, and Volker said, but of course then we must come to Göttingen and when we came he would show us around.

It was on that trip that I first realized the respect in which Julia was held by other mathematicians.

Volker's Ph.D. adviser, Konrad Jakobs, was eager to entertain us; rather, to entertain Julia. It was clear that Volker had scored a coup with his "Doktorvater." (Incidentally, Julia told me later that it was her paper on game theory, the only paper she ever wrote on that subject, which so interested Jakobs.) Volker himself, whose wife was momentarily expecting their second child, told us that if the baby was a girl—in those days people did not know before the event—he was going to name her Julia. The baby was born while we were still in Göttingen, but it turned out to be a boy.

The result of our visit to Göttingen, however, was that I abandoned the project of updating E. T. Bell and decided that I, who knew almost nothing about mathematics but what Julia and Raphael had explained to me, would write a life of David Hilbert.

I should say here that Julia had not suggested that I write about Hilbert. I came to him on my own—Hilbert simply enchanted me just as he had enchanted all the young mathematicians and physicists who had flocked to study with him in Göttingen. But if you think Julia tried to discourage her mathematically untrained sister from writing the life of the greatest mathematician of the first half of the twentieth century, you don't know Julia.

For my birthday she gave me the three volumes of Hilbert's collected works and, when her mathematical friends inquired about my qualifications to write the life of Hilbert, she told them with a perfectly straight face that I was reading all his papers.

(Incidentally, as an aside, I did read *all the words* in Hilbert's collected works—mathematicians of those days wrote more in words than they write today—and Hilbert's were quite enlightening in regard to his ideas and feelings about mathematics.)

Julia then suggested that I interview mathematicians in the area who had actually known Hilbert—Lewy, Pólya, Szegö, even Siegel, who was passing through Palo Alto on his way back to Germany. But I was hesitant about talking to real mathematicians about writing about Hilbert—Julia and Raphael, OK, they were family, but Carl Ludwig Siegel? I remember Julia's saying slyly, "You're afraid they will find out that you're a hoax, Constance"—which of course I was.

Now, even a quarter of a century after the publication of *Hilbert* and the other biographies that have followed, I still don't really understand why Julia encouraged me as she did when I might have disgraced, certainly embarrassed, both her and Raphael.

I think that perhaps at least part of the explanation lies in something Julia said to Olga Taussky after *Hilbert* was published and was an unexpected success. Olga was complaining that there were other important things that she would have told me about her mathematical relationship to Hilbert if she had known "that *everybody* was going to read the book," but many people had come in the past to talk to her about her days in Göttingen, and then nothing had ever happened, so she had thought it would be the same with me.

"Olga," Julia said, "you should have known that the Bowman girls always finish what they start."

At that time Julia had not been a Bowman for thirty years, and I had not been a Bowman for twenty, but I think that the strong sense our parents conveyed to us that being a Bowman was something special—although in actuality the Bowmans were quite ordinary people—was at the foundation of Julia's sense of herself—and of course she knew it had rubbed off on me too. I might write as Constance Reid, but at bottom I was *Constance Bowman*.

After *Hilbert* I wrote a life of Richard Courant at the suggestion of K. O. Friedrichs, who became my mathematical collaborator in that project.

I can't say that Julia and Raphael were exactly "miffed" to see me going off on my own, but they did feel a little out of it—although, as I have said, both of them always read my manuscripts before they were published. Naturally, after I had written three biographies—one shortly after Julia's death—and Julia had become famous and Saunders Mac Lane had proposed her for membership in the NAS, and Alfred Tarski and Jerzy Neyman, who were old and not well and didn't much care for each other, had both made the trip back to Washington, D.C., just so that they would be present to help explain the importance of Julia's work—people began to make what they always thought was an original suggestion—*why don't you write a life of your sister?*

The truth of the matter is that I never considered doing so.

I knew Julia—and I knew myself—and neither of us would want our biographies written—by anyone. I did think, however, that Julia should let herself be interviewed for *More Mathematical People*, which I was helping to edit, because—and this was a telling point—she had objected in regard to the earlier book, *Mathematical People*, that it had contained interviews with three women—me, Mina Rees, and Olga Taussky-Todd—*people*, not *mathematicians*, being the operative word in the title—but only one of the three was a research mathematician.

"Julia," I said, "how can you object when you yourself refused to be interviewed?"

She of course had no answer to that.

Well, after her election to the NAS in 1976—you have all heard, I am sure, the great story about Julia's being identified as "Professor Robinson's wife" when the university press office called the mathematics department to find out just who Julia Robinson was—Berkeley started to think about how to get this new Academician into its stable. There was the problem that Julia because of her health—although it was much improved—did not want and could not handle the rigors of a full professorship.

(Incidentally, Julia once told Cathleen Morawetz—this must have been in the early 1970s, when she and Raphael began to talk about his retiring early so he could devote more of his time to mathematics—that what she would really like was to share a half-time job with him, but I am sure she had never suggested this to anybody in the department. Certainly I had never heard anything about it nor, according to Raphael, had he, but it is a kind of "Julia solution" to a problem.)

Well, after she was elected to the Academy, the Berkeley mathematics department came up with the idea of offering her a full professorship with the duty of teaching just one-fourth time—which was just about exactly what she had been doing for a number of years. The department seems to have been a little concerned about the appropriateness of such an offer because the chairman consulted University of Chicago mathematics professor Saunders Mac Lane, who recently sent me a copy of his reply:

"In my opinion it would be eminently appropriate that Dr. Robinson receive a professorial appointment, under such part time arrangement as may be mutually agreeable," Mac Lane wrote. "Her accomplishments in mathematical logic and related topics are, in my considered opinion, outstanding and would justify her appointment as a Distinguished Service Professor, or its equivalent, at any leading American university, but most appropriately at the University of California at Berkeley."

Julia accepted Berkeley's offer. But that was not the end. She was showered with more and more honors. I can still hear her, telephoning me about some new award and saying, almost in despair, anyway in mock despair, "Constance, what next?"

This may, in fact, have been when she was asked if she was willing to have her name put up as the unopposed candidate for president of the AMS.

Raphael did not think that she should accept but should save her energy for mathematics, as he would have done. He did not try to impose his view on her—he simply stated his opinion. But when she consulted me, I said that I felt there was no way she could not accept, and she agreed—not because that was my opinion, but because it was the same as her own. It might be a long time before another woman mathematician was offered the position. In fact, of course, it was almost ten years.

I should tell you, however, that Raphael accepted Julia's decision and her many absences, learning to cook and take care of himself—skills which were to stand him in good stead after Julia's death.

So here my sister was, famous for her mathematical work and famous for her firsts, steadfastly refusing to be written about.

"Dear So and So," she wrote to someone who wanted to include her in a book about women scientists, "I am of course very flattered to be considered for your book but I must ask you not to write about me. I am appalled at the prospect of details of my life and beliefs appearing in print. (I don't even want to be written about after I'm dead but that is difficult to manage.) This has nothing to do with your abilities and qualifications, as I will continue in the future to discourage any account of my life."

In her view a mathematician was his or her work; personality/personal details could do nothing to illuminate that and so were of no importance. She detested what she saw as the *cult* of personality, the prying into every aspect of what was private, which was and is still prevalent in biographical—and for that matter, autobiographical—writing.

Although I felt very much the same, I thought that her position in relation to *any* writing about her life and views was logically untenable. She, however, stubbornly maintained that position until it was clear to her and to me that she was going to die.

Then I brought forth my most telling argument. Given her achievements, somebody was bound to write a biography of her. How much better if her sister wrote it, and she herself had the opportunity to approve it. She finally agreed.

On June 30, 1985—as it turned out, just thirty days before she died—we had an interview about what she recalled as significant about her life. She was lying on the couch in her living room, and Raphael was present, although he never said a word, or even made a sound, except to agree with a chuckle that Julia was indeed very stubborn.

Almost immediately I got the idea of writing her life, in imitation of Gertrude Stein, as "The Autobiography of Julia Robinson." I think this was because Julia had told me at this time how struck she had been by something she had read to the effect that the only reason for writing one's autobiography was to give credit where credit was due. There were people to whom Julia very much wanted to give credit. Beyond our parents and others from her early days, these were all men. A young assistant professor at San Diego State College who, in opposition to the head of his department, told her to go—and to go to Berkeley. Her husband, Raphael Robinson—of whom she said she did not think she would have become a mathematician had it not been for him. Alfred Tarski, her thesis adviser—he and his mathematics were so completely right for Julia that it is hard to imagine her career if he had not come to Berkeley. Jerzy Neyman, who by providing financial support made it possible for her to continue graduate study at Berkeley after she got her A.B. Yuri Matijasevich, who provided that last thing that was needed to prove that the solution of the tenth problem is indeed negative and whose friendship and collaboration over the barriers of age, sex, and geography were so satisfying to her during the last years of her life. I have to tell you that when Julia was in the hospital the nurses marveled at the number of phone calls from men that she received—they had never had as a patient such a woman!

I worked very hard on the "Autobiography," knowing I was working against time, and each day read to Julia, who was back in the hospital, what I had written. She listened attentively, making suggestions or deletions, and today when I reread the "Autobiography" I feel that I am reading something that Julia herself wrote—it is an eerie sensation. "The Autobiography of Julia Robinson" was published in the *College Mathematics Journal* in 1986 and reprinted in 1990 in *More Mathematical People*.

Julia and Raphael always felt that mathematics and the University of California *had been good to them*—these were Julia's own words to me on one occasion—and they intended to leave whatever they had to the university for the benefit of mathematics at Berkeley. After Julia's death Raphael decided that the bulk of his (really *their*) quite substantial estate should go to endow the Julia B. Robinson Fellowship Fund. Raphael died in 1995 and named me executor. Since he had not disposed of Julia's papers, photographs, and memorabilia after her death, I became in a sense her executor as well. I gave her mathematical letters, including her long correspondence with Matijasevich, to the Bancroft Library with the proviso that nothing personal was to be quoted without my permission. I cooperated with the AMS's wish to publish Julia's collected papers.[5] But there were still many photographs and much memorabilia that I couldn't help wishing I had had to illustrate the "Autobiography"—particularly things that, although not strictly mathematical, were relevant to Julia's mathematical career. What was I to do with the script of a University Explorer program on "Mathematics by Machinery" that Julia had heard and sent for when she was fourteen,[6] or a theme on mathematics that Julia wrote as a freshman in college, or this statement of Julia's made in response to a question as to whether she had ever experienced discrimination as a woman mathematician:

"No," she wrote, "—except for a semester or two when the nepotism rule was enforced.[7] Also, there was one case when both my husband and I were invited to a conference and the committee decided it would be unfair to pay expenses for both of us because the other families would have to pay for the wives. We didn't particularly care, and perhaps they were right."

It seemed to me that something more about Julia was wanted and needed—a book that could be placed in the hands, not only of professional mathematicians, but of mathematics teachers and students and even non-mathematicians. Perhaps the "Autobiography" should be reprinted as a little book and expanded with some of the material I had found among Julia's things. But the book should include as well something about Julia's mathematical work to give a sense of the character of her thought and the personal warmth that she brought to mathematical collaboration. So I asked Lisl Gaal, Martin Davis, and Yuri Matijasevich for permission to reprint articles they had earlier written, which had been published in widely separated places. The result of our "collaboration," which brings all of these writings together, is *Julia, a Life in Mathematics*.[8]

Two years ago I established an award in Julia's name.[9] This was to be made each year at the high school from which she graduated—exactly

---

5. *The Collected Works of Julia Robinson*, with an introduction by Constance Reid, edited and with a foreword by Solomon Feferman, Collected Works 6 (Providence, R.I.: AMS, 1996).

6. The program was based on an interview with mathematician Dick Lehmer and his father.

7. This was at the beginning of her career when, as she says in the "Autobiography," she was more interested in having a family than having a job at the university.

8. Constance Reid, *Julia, a Life in Mathematics* (Washington, D.C.: MAA, 1996).

9. My share of royalties from the *Collected Works of Julia Robinson* and all royalties from *Julia* fund a Julia Bowman Robinson Prize in Mathematics at San Diego High School.

sixty years ago last month—and where at the time she was the only girl taking mathematics after plane geometry. When famous alumni of the high school were written about, they were always movie stars, athletes, authors, politicians, you name it, anybody but a mathematician. My idea was that the students should know that an outstanding mathematician had come from their high school, and that the prize should be large enough to impress them with the respect in which mathematics is held. My prize was to go to the best mathematics student, female or male, in the high school graduating class. I understand that the ratio of females to males in the advanced mathematics classes at the high school is now not 1 in 30, as it was sixty years ago in Julia's day, but 50–50.

Julia firmly believed that there is no reason that women cannot be mathematicians, and she just as firmly believed that there should be affirmative action to bring women onto mathematical faculties at colleges and universities. "If we don't change anything," she said to me in that last interview, "then nothing will change." She didn't expect that the percentages would be 50–50, but she did say that affirmative action for women mathematicians should continue until men mathematicians no longer considered women mathematicians unusual.

Julia thought of mathematicians—these were her words once to a group of young people—"as forming a nation of our own without distinctions of geographical origins, race, creed, sex, age, or even time (the mathematicians of the past and you of the future are our colleagues too)—all dedicated to the most beautiful of the arts and sciences."

Euphemia Lofton Haynes

When the next story originally appeared, it referred to Marjorie Lee Browne as one of the first two African-American women Ph.D.'s in mathematics. Although Browne's degree from the University of Michigan was not awarded until February 1950, she had completed the requirements in 1949; Evelyn Boyd Granville received her degree from Yale in 1949.[10] Because they both finished their theses in 1949, they were long regarded as more or less tied for the honor of being first.

Euphemia Lofton Haynes, Catholic University of America (CUA) 1943, is now the first known African-American woman mathematics Ph.D. Scott Williams (State University of New York at Buffalo), at his website "Mathematicians of the African Diaspora," credits Robert Fikes Jr. of San Diego State University for bringing Haynes to his attention. Haynes's degree was

10. See part IV for the story of her life and career.

earned eighteen years after Elbert Cox (Cornell) became the first mathematics Ph.D. of African descent now known.[11]

According to the CUA library (where thirty feet of shelving archive her family records), Haynes received her bachelor's degree from Smith College in
1914 and a master's in education from the University of Chicago in 1930. She
taught in the public schools of Washington, D.C., for forty-seven years, during
which time she completed her master's and Ph.D. degrees. She was the first
woman to chair the D.C. school board. She figured prominently in the integration of the D.C. public schools and also of the Archdiocesan Council of
Catholic Women. A fourth-generation Washingtonian, Haynes was active in
many community functions. Upon her death in 1980, she bequeathed
$700,000 to CUA in a trust fund established to support a professorial chair
and student loan fund in the School of Education.[12] An annual lecture series
is named in her honor.[13]

Marjorie Lee Browne

PATRICIA CLARK KENSCHAFT

Marjorie Lee Browne, 1914–1979. Based on "Marjorie Lee Browne: In Memoriam,"
AWM *Newsletter* 10(5), 1980, 8–11. Kenschaft was then an assistant professor of
Mathematics, Montclair State College, where she is currently a professor at the institu-
tion now called Montclair State University.

Until 1949, only one American Black woman had earned a Ph.D. in mathematics, Euphemia Lofton Haynes in 1943. That year there were two who
fulfilled the requirements: Evelyn Boyd Granville from Yale University
and Marjorie Lee Browne from the University of Michigan. By 1979,
there had been about seventeen more, and on October 19, 1979, Marjorie
Lee Browne became the first of them to die.

Browne had several publications and had ambitious plans for her
retirement, just begun at the time of her sudden death, but most of her

11. Scott W. Williams, "Martha Euphemia Lofton Haynes, First African American Woman Mathe-
matican," Mathematicians of the African Diaspora [online]. The State University of New York at Buffalo,
2001 [accessed 25 June 2002]. Available from World Wide Web: http://www.math.buffalo.edu/mad/
PEEPS/haynes.euphemia.lofton.html.

12. "Haynes-Lofton Papers," Manuscript Collections, American Catholic History Research Center
and University Archives, CUA Library [online; accessed 25 June 2002]. Available from World Wide Web:
http://libraries.cua.edu/achrcua/manuA-K.html.

13. Press release, "African Americans in Catholic Education" [online]. CUA Office of Public Affairs,
Washington, D.C., 1999 [accessed 25 June 2002]. Available from World Wide Web: http://publicaffairs.
cua.edu/news/99haynes.htm.

career was devoted tirelessly and effectively to helping Black students share the joy and creativity of studying mathematics and to enabling them to use mathematics for a rewarding career. Less than two weeks before her death, she told me, "If I had my life to live again, I wouldn't do anything else. I love mathematics."

She taught from 1949 to 1979 at North Carolina College at Durham (renamed North Carolina Central University [NCCU] in 1969); she was department head from 1951 to 1970. For twenty-five years, she was the only mathematics Ph.D. in the department. She taught fifteen hours a week, both undergraduate and graduate courses, and she served as graduate advisor for ten master's theses. No wonder she had little time for research!

Under her leadership, NCCU became the first predominantly Black institution in the United States to be awarded a grant for an NSF Institute for secondary teachers of mathematics; she directed the mathematics section of these institutes for thirteen years. Her summers were filled with teaching secondary school teachers, and she wrote four sets of lecture notes for their use: "Sets, Logic, and Mathematical Thought" (1957), "Introduction to Linear Algebra" (1959), "Algebraic Structures" (1964), and "Elementary Matrix Algebra" (1969).

William T. Fletcher, one of Browne's many protégés, earned a doctorate in mathematics from the University of Idaho and succeeded her as department head at NCCU. When he recommended her for the position of professor emeritus, he wrote:

> Her manifestations of conspicuous attainment and scholarship, coupled with her dynamic academic leadership, inspired many high school teachers to receive graduate degrees or advanced training and, thereby, she contributed significantly to the improvement of the quality of Mathematics Education in schools and colleges throughout North Carolina and the South. . . .
>
> Her thoroughness, demands for excellence and rigor, wisdom, vision, and productive powers in the classroom have profoundly influenced not only the academic growth and development of countless students but also their aspirations to achieve and succeed in the field of Mathematics. She helped students—many of whom came to her with less than adequate preparation—discover that mathematics was a challenging creative pursuit, and her encouragement and instruction equipped many to pursue the study of mathematics to the completion of the Ph.D. degree. Graduates of this department during her tenure have made significant achievements in the professions . . . and have performed in the graduate schools of other universities with a high degree of success.

In 1975, Browne became the first recipient of the W. W. Rankin Memorial Award for Excellence in Mathematics Education, given by the North Carolina Council of Teachers of Mathematics. The announcement says, "She pioneered in the Mathematics Section of the North Carolina Teachers Association, helping to pave the way for integrated organizations."[14]

During the academic years 1966, 1967, and 1973, she served on the Advisory Panel of the NSF Undergraduate Scientific Equipment Program. In the 1960s, she obtained a grant from IBM for the first computer at NCCU, served as a faculty consultant in mathematics for the Ford Foundation, and obtained the first Shell grant for awards to outstanding students in her department.

Marjorie Lee Browne was born on September 9, 1914, in Memphis, Tennessee, to Lawrence Johnson Lee, a railway postal clerk, and Mary Taylor Lee. Her only sibling was a brother, two years older than she. After his undergraduate mathematics degree and a master's in physical education, he taught physical education and coached at Southern University in Louisiana. Their father had also attended college for two years, unusual for that time, and was known as a "whiz" in mental arithmetic. He taught his children about the fun of mathematics and kept up with their mathematical studies as long as he could.

Marjorie herself told me, "I always, always, always liked mathematics. As a child I was rather introverted, and as far back as I can remember I liked mathematics because it was a lonely subject. I could do it alone." Her family sent her to LeMoyne High School, a private school started after the Civil War by the American Missionary Association to educate Negroes. Her father took very seriously his responsibility as one of the few steady earners in the community (with a civil service job) and borrowed money not only for his children's education but also to help others. He would invite the high school football team to his home for nourishing meals because they could get them no other way. (From then on, she didn't like spaghetti and meatballs because she had so much at that time!)

The students at LeMoyne were all Black, but the faculty was interracial, and Browne felt they were excellent teachers. She credited much of her later success to her excellent preparation there. She was graduated in three and a half years; during this time she also won the Memphis city championship for women's singles in tennis.

College funding was difficult during the depression, but some combination of scholarships, working, and borrowing took her through Howard University. Browne, who had a fine voice, sang in the Howard University choir. In 1935, she graduated cum laude.

14. "W. W. Rankin Memorial Awards," *Math Newsletter*, North Carolina Department of Public Instruction, no date or volume number.

Her teaching career began at Gilbert Academy in New Orleans, a Methodist secondary school for Blacks. During this time, she lived with an uncle. A cousin who lived in the same house still vividly remembers taking a course from her and at the end receiving one of the few Fs of his career. "She was completely honest. It didn't matter what relationship you were to her. Once you portrayed an interest in mathematics, she stuck right with you. Otherwise she had no time for you. That was her life—mathematics and physics." He tempered these remarks by remembering also how they bought a record player together. The payments were fifty cents a week, so they each contributed twenty-five cents. Browne had a lifelong interest in music; her cousin described her as "a tremendous listener with an ear for the classics."

When Browne began to think of graduate education, she talked with a neighbor who had gone to the University of Michigan and who reported that the fees there were not too high, a most important consideration. In 1939, she received her M.S. from that institution, going there during the summers.

She then joined the faculty of Wiley College in Marshall, Texas, and began working toward her doctorate. Eventually, she took a leave from Wiley College; in 1947–1948 she was a teaching fellow at the University of Michigan. Her dissertation, written under the supervision of G. Y. Rainich, was "On the One Parameter Subgroups in Certain Topological and Matrix Groups." In 1948, she was elected to Sigma Xi and became an institutional AMS nominee. The following year she received her Ph.D. in mathematics. She joined the MAA in 1950. In 1955, her article "A Note on the Classical Groups" was published in the *Monthly*.[15]

Browne obviously believed in continuing education for herself as well as for her many students. She attended many conferences.[16] In the academic year 1952–53, she was a Ford Foundation Fellow, sponsored by the Fund for Advancement of Education; she studied combinatorial topology at Cambridge University in England. She traveled throughout Western Europe during that year. In 1958–59, she was an NSF Faculty Fellow studying numerical analysis and computing at UCLA. While there, she seized the opportunity to travel in Mexico. The academic year 1965–66 found her again an NSF Faculty Fellow, this time studying differential topology, especially Lie groups and Lie algebras, at Columbia University in New York City.

My only conversation with Marjorie Lee Browne was the first time I telephoned a stranger to ask about her life. I was full of apprehension about

---

15. *American Mathematical Monthly* 62(1955), 424–427.

16. For example, in 1957 she attended the Conference on Mathematics in the Behavioral Sciences at Stanford University cosponsored by the Social Science Research Council and the MAA and in 1973, the Conference on Applications of Mathematics in Behavioral, Engineering, Medical, and Management Sciences at the Georgia Institute of Technology.

my reception, but she had not responded to my written request for information about her to use in my upcoming talk, "Black Women in Mathematics." Since she was one of the first to earn a Ph.D., I did not want to leave her out if I could help it. She was very kind to me and gave me courage to telephone others and thus collect much varied information. However, she kept asking, "Why are you doing this *now*?" I couldn't bring myself to say that she was a pioneer and no longer young and that I wanted to obtain primary source material on her life while she was still alive and could check it. I simply told her I was preparing a speech. She offered to send me her complete résumé, and it arrived promptly. The following week she died.

Her sonorous voice was kindly, but firm and businesslike, and impressed me even over the telephone. It seemed to me she could have made a career on the stage with that voice, and I'm sure it was an asset in the classroom. She told me repeatedly, however, how much she liked working alone. "I do have plenty of friends, and I talk with them for hours at a time, but I also like to be alone, and mathematics is something I can do completely alone." Her résumé says she was divorced with no children.

The conflicting demands of teaching and administration for the sake of others, pitted against the desire to develop fully her own intellectual gifts, is clear in this account of her life. There is little time for research and writing when one feels an urgent obligation to share one's own achievement—to the point of teaching fifteen hours a week, chairing a department, and teaching and administering a program for secondary school teachers in the summers. Browne planned a monograph on the real number system for her retirement, but never found the time.

During the last years of her life, she often gave personal financial aid to gifted younger people so they could pursue their education. To continue these efforts, the Marjorie Lee Browne Trust Fund was established in her honor to give scholarships each year to able students in the NCCU Department of Mathematics and to support student-oriented programs in mathematics.

References

1. Telephone conversation on October 6, 1979, with Browne, and the three-page vita she subsequently sent me.
2. Telephone conversations and correspondence with William Fletcher in 1980.
3. Telephone conversation on July 11, 1980, with Lavern Taylor Pierce, a first cousin.
4. Telephone conversation on July 12, 1980, with Thaddeus Taylor, another first cousin.

Cora Ratto de Sadosky

CORA SADOSKY

Expanded version of "Cora Ratto de Sadosky (1912–1980)," from the article "Sadosky Prize Established," in the Kovalevskaia Fund *Newsletter* 12(1), January 1997, 5–6, reprinted in the AWM *Newsletter* 27(2), 1997, 20–21. Sadosky is a professor of mathematics at Howard University.

Cora Ratto de Sadosky was an Argentinean mathematician, an inspiring teacher who devoted her life to fighting against oppression, discrimination, and racism and to instilling her passion for mathematics in the many students she taught and mentored.

Born in 1912 to a middle-class family of Italian origin, Cora Ratto graduated with a degree in mathematics from the University of Buenos Aires. Toward the end of her studies, she headed the Federación Universitaria Argentina (FUA), the Argentinean Student Union. Actively opposed to Nazism and Fascism, Cora was active also in the solidarity effort in favor of the Spanish Republic and in the denunciation of the Chaco War forced on neighboring Bolivia and Paraguay by the United States and Great Britain in the early 1930s.

In 1937 Cora Ratto married Manuel Sadosky, her lifelong companion both in mathematics and in political activities. Manuel was a bookish and intense young man from a poor Jewish family whose parents had fled the pogroms in tsarist Russia in 1905. Since the thirties were times of rampant anti-Semitism, Cora chose to use her married name, Cora Ratto de Sadosky, from that time on.

Until her sudden death on January 2, 1980, Cora and Manuel shared a long and happy marriage. Their house was always full of friends, attracted by a hostess who was not only a superb cook, but also a great conversationalist, mixing witticism with wisdom and genuine caring. Cora and Manuel had one daughter—who also became a mathematician—and one granddaughter, who became the greatest joy of her life. Both of them are called Cora!

During World War II, immediately after the Nazi invasion of the Soviet Union, Cora Ratto de Sadosky created La Junta de la Victoria (The Victory Union), a women's organization devoted to helping the anti-Nazi war effort. In 1945 La Junta, with Cora at its helm, had 50,000 volunteers—in a country with fewer than 12 million people—and made a substantial contribution to the Allies' war efforts with money, clothing, and food for the fighting troops. Not only was La Junta significant for its solidarity against the Nazis, but it was the first women's mass organization of its size in Latin American history. In December 1945, as one of two representatives of her organization, Cora was a founding member of the International Women's Union at its first meeting in Paris.

Soon after the war was over, Cora and Manuel, with their then-little daughter, went to Europe to continue their mathematical studies. In Paris, Cora worked under the direction of Maurice Frechet. She did not complete a doctoral dissertation at that time because the family left for Italy, where her husband received postdoctoral training in applied mathematics. Later on he was to introduce scientific computing in Latin America, serving as the director of the first Computer Center at the University of Buenos Aires in the sixties and helping develop similar projects in Uruguay, Chile, and Paraguay. The Sadoskys returned to Argentina during a period of turmoil and political repression. They were banned from teaching at public universities, as were all those not affiliated with the party in power, and Cora went to work at a commercial enterprise to support the family.

When the Argentinean universities regained their autonomy in 1956, Cora and Manuel were part of the team that built a modern school of sciences at the University of Buenos Aires. During a time of strenuous activity in teaching, doing research, and organizing the new school, Cora received her doctorate, with a thesis on hyperbolic singular integrals written under Mischa Cotlar, at the age of forty-six. From 1958 to 1966 she was an associate professor of mathematics. (She never became a full professor, nor did she achieve due recognition for her research. While she did not complain about having a harder time in her profession for being a woman, she was acutely aware that this was the case.)

During her decade at the university, Cora initiated and edited a celebrated series of research publications (the first volume, *Mathematics and Quantum Physics*, was written by Laurent Schwartz, Fields Medal, 1950, the third, *Singular Integrals and Hyperbolic Partial Differential Equations*, by the Argentinean-born mathematician Alberto Calderón, National Medal of Science, 1991); this series became an important resource for library interchanges. She also organized a host of advanced courses for several generations of mathematicians, physicists, and chemists, many of whom later became leading scientists in the United States, Latin America, Canada, and Europe.

Cora was most remarkable as a teacher. She devoted a part of her immense energy to teaching the first courses in calculus and algebra. She was idolized by her students, who clung to her every word and used to say that they "were willing to go to the barricades for modern algebra!" With her husband, she co-organized an entire teaching system for several hundred students, where thanks to the use of scores of senior and regular teaching assistants and student mentors, they gave a personalized education to each and every one.

Culminating her experiences of teaching first-year algebra, Cora coauthored, with Mischa Cotlar, *Introducción al Algebra Lineal* (Buenos Aires: Eudeba, 1966). This book was a remarkably modern and rigorous text, the first of its kind in Spanish (for a comprehensive review, see *Mathematical Reviews* 35#4061). Cora also developed, in collaboration with others, university textbooks for training high school teachers.

Possibly Cora's most dramatic contribution while at the University of Buenos Aires was the creation of the Fundación Alberto Einstein, a foundation aimed at supporting talented mathematics and science students in need of financial help. Her fellowship and mentoring programs helped hundreds; the work of the Albert Einstein Foundation was the first step in establishing a university-wide scholarship system.

In 1965 Cora created and codirected the monthly magazine *Columna 10*, with the primary aim of raising the awareness of the Argentinean public on foreign affairs, especially the Vietnam War, which at the time barely reached the pages of the local newspapers. *Columna 10* was a landmark publication, exposing the secret war waged by the U.S. Green Berets, the effects of defoliation, and the human rights atrocities committed during the Vietnam War.

In 1966 the first of a string of increasingly repressive military dictatorships took control of Argentina. Following a violent assault on the School of Sciences by the military and the police, 400 faculty members resigned their positions at the University of Buenos Aires. That event, remembered as the Night of the Big Sticks, ended normal scientific life in Argentina for several decades. Cora retired from mathematics research and teaching and concentrated her efforts on denouncing the increasing human rights abuses perpetrated by the military.

Under threats on their lives, Cora and her family left Argentina at the end of 1974, exiled, first in Caracas, Venezuela and then in Barcelona, Spain. Her constant dream of returning to her country was never realized, and she died in exile.

In 1996, the Cora Ratto de Sadosky Prize was instituted in Vietnam to encourage the girls performing best in their national mathematics olympiads. The prize, sponsored by her family, the Kovalevskaia Fund, the Vietnam Ministry of Education and Training, and the Vietnam Women's Union, honors her enthusiasm for knowledge, her passion for justice, her human warmth, and her love for mathematics.

Her legacy resides in the hundreds of students she inspired and helped and in the many friends around the world who saw in her an example of integrity, farsightedness and commitment to change.

Fifty Years in Mathematics

MABEL S. BARNES

Based on a talk at the panel "Centennial Reflections on Women in American Mathematics," organized by Judy Green and Jeanne LaDuke, AMS Centennial Meeting, Providence, August 1988, as reported in the AWM *Newsletter* 18(6), 1988, 6–8. Barnes (1905–1993) was then a professor emerita of mathematics, Occidental College.

My reflections on women in American mathematics will begin with my student days in the Midwest in the 1920s and end with my retirement from Occidental College in Los Angeles in 1971. The account of ups and downs eventually has a happy ending, with my finding just what I was looking for at Occidental. I shall try to give you some impression of the flavor of those times.

I always enjoyed mathematics, beginning with arithmetic in a one-room country school in Iowa. In high school and college an interest in mathematics was considered unusual for a girl but certainly was not discouraged. I entered college with a firm intention of majoring in Latin. Along with Latin I took mathematics because I liked it too, and taking calculus changed my mind about majoring in Latin. (In those days calculus was a college sophomore subject.) In the mathematics department my interest was encouraged, but more generally the attitude was that pursuing a Ph.D. in mathematics was something for a man to do, not a woman.

In the 1920s women were received cordially in the Midwest graduate schools. There were even a few women on the university mathematics faculties. There was an underlying assumption, though, that of course men would go into better jobs than women after finishing graduate work.

When I was completing my graduate work in 1930, I enrolled in a teachers' employment agency and through it found a job filling in a leave of absence at Nebraska State Teachers College, Wayne. At that time the job scarcity of the Depression had not yet quite hit. There was no prejudice against women on the faculty at Wayne, possibly because they could be had for less money. In view of the number of them in the education department, I think we were actually a substantial majority. There were two and a half positions in the mathematics department, and the other full-time person was a woman. Mathematics was not an especially popular major for girls, though. In the small high schools of northeastern Nebraska it was traditional that the coach teach mathematics. Consequently, quite a few athletes turned up in my classes, among them some very good students.

The curriculum at that time would have been very poor preparation for graduate work. Among our advanced courses were subjects like theory of equations and higher Euclidean geometry. Our students didn't learn advanced calculus or modern algebra, but they did learn why you couldn't trisect angles, square circles, or duplicate cubes with ruler and compasses, and they learned about some remarkable sets of collinear points and tangent circles connected with triangles, and about the nine-point circle.

It was a vastly different world in the early thirties. Transportation and communication were, from the present point of view, extremely slow. Consequently distances seemed very great. For most of the young people of northeast Nebraska, going to college meant going to Wayne, whether or not they were interested in teaching as a career. Wayne was at least accessible. The total student enrollment was something only in the hundreds.

The leave of absence that resulted in my job stretched to three years, at the end of which my predecessor returned, having acquired three years of graduate work but no Ph.D. (at that time the two were considered equivalent for accreditation purposes for the college). There was a prospective vacancy at Nebraska State Teachers College, Chadron, in the northwestern corner of the state, and the president at Wayne was prepared to give me a strong recommendation to his friend, the president at Chadron, but the vacancy did not materialize. Had I gone to Chadron, my life would have been very different.

By this time the Depression was upon us. Anyone who did not experience it can hardly imagine what it was like from hearing descriptions of it and seeing pictures. My experience then was decidedly less traumatic than that of many people. There was plenty of uncertainty and anxiety along the way, but as a last resort I could always have gone home to my parents. They did not lose their farm as many of their neighbors did, and they ate well. But in 1933 there were no jobs at all for men or women. What to do now?

Even in remote Nebraska I had heard about a place called the Institute for Advanced Study opening in faraway Princeton. I applied for admission and was accepted. For some years the School of Mathematics was the only school of the Institute and was housed with the mathematics department in Fine Hall at Princeton. Soon after I arrived the director of the School of Mathematics took me aside and warned me that Princeton was not accustomed to women in its halls of learning and I should make myself as inconspicuous as possible. Otherwise I found a very friendly atmosphere and spent a valuable and enjoyable year there. Had I not gone east, I would not have met Olga Taussky as early as I fortunately did.

At the end of that year there were, of course, still no jobs. Six men and I from the Institute and from Princeton University took a special qualifying exam to be taken on as substitutes for mathematics teachers on leave from New York City high schools. Our duties were to teach light schedules in the high schools to which we were assigned and to give jointly an alertness course, as it was called, for high school teachers (it qualified them for raises). At the end of the year I was asked to give an alertness course by myself the next year. There had been many requests for it. No prejudice here! However, that summer I married and went to live in Massachusetts, where my husband had a *job* as an assistant professor at Tufts College, as it was called then.

I kept my hand in somewhat by marking papers and substituting for him when he was away and by helping him edit the mathematics section of Eschbach's *Handbook of Engineering Fundamentals*, second edition. I remember there was a question as to whether my name should appear on the list of contributors as just M. S. Barnes instead of Mabel S., but the first name did go in. Fortunately, from our location we could attend the Harvard Mathematics Colloquia.

In 1942, during World War II, I was back in Princeton. My husband, on leave from Tufts College, was doing war work for Bell Labs in New York and commuting. Somehow I encountered a Princeton physics professor who asked if I would read physics papers—they needed help desperately. Although I had done only a bare undergraduate physics major along with my mathematics major, I agreed—you can always study texts and mark papers, and they paid the handsome rate of one dollar per hour. After a while the situation was really desperate—the physicists were off trying to make an atom bomb. The exigencies of war swept prejudice aside, and I was asked to teach some classes. With my sketchy background in physics I wouldn't have considered trying to teach it, but just at that time we were preparing to move to Red Bank, New Jersey, because my husband had been transferred to the Bell Lab there. There wasn't the remotest chance of my finding mathematical employment in Red Bank. I didn't even think of trying—but I did grow some beautiful tomatoes.

After the war, in 1946 my husband went back to Tufts to his job as chairman of the Department of Applied Mathematics. By that time there was a heavy influx of war veterans. The afternoon before classes were to begin he came home and said, "Mabel, you'll be teaching in the morning." It seemed that no man could be found, the last lead having just failed. Nepotism and my being a woman were overlooked. Desperation again overcame prejudice. I was rehired for the next year; however, the next summer my husband accepted a job at UCLA, and we moved to California.

Note that in making moves it never once occurred to either my husband or me that the possibilities of my getting a job should be given any consideration at all.

I found that jobs for women in mathematics were in very short supply in California. Santa Monica City College would have hired me gladly, but I wasn't "qualified." I didn't have California junior college credentials. To remedy that deficiency I'd have to take some education courses at UCLA. My three years of teaching at a teachers' college might help some. I decided to pass this opportunity up. Then a little later, when the moving van was at the door to move us into a house we had just had built, the phone rang. It was an offer of a job to teach one advanced calculus class at USC. At that time I had hopelessly too much to do to accept.

In the summer of 1950 we were sitting with a friend from Caltech at a luncheon at a mathematics meeting. Across the table from our friend there was a young man who started to tell him that at Occidental College they were looking for a man with a Ph.D. in mathematics, preferably not from nearby Caltech—Oxy already had many Caltech men on its faculty—who was interested in undergraduate teaching. I nudged my husband and said in a low voice that I thought I was the man he was looking for. We approached him after the luncheon, and he said he would arrange for an interview with the dean. This vacancy had occurred because of a serious illness. Although they were somewhat dubious, I was hired as an instructor by special appointment for one semester. (As an aside, that September there came to Occidental also a junior college transfer student, a track team shot-putter, attracted by the reputation Occidental had at that time in track. The shot-putter's name was Ken Hoffman, now for many years at MIT and well known as the voice of mathematics in Washington.)

I was rehired for the second semester and afterward promoted fairly rapidly, although the administration at that time was not particularly friendly to women. There were very few on the faculty. I found Occidental a very congenial place and one that I enjoyed for twenty-one years.

My salary wasn't exactly handsome, and I was given the, at that time, perfectly reasonable and acceptable explanation that I had a husband to support me, while the men on the faculty had to support families. By the time I retired in 1971, that kind of argument had lost some of its weight and effect, and there were quite a few women on the faculty at Occidental.

In the early 1970s when my daughter was looking for a job in mathematics, times had really changed. In looking for a job it had become almost an advantage to be a woman.

Emmy Noether

EMILIANA PASCA NOETHER

Emmy Noether, 1882–1935. Based on "Emmy Noether: Twentieth Century Mathematician and Woman," a talk at the panel "History of Women in Mathematics," moderated by Lenore Blum, 1976 Summer JMM, as reported in the AWM *Newsletter* 6(7), 1976, 1–6. Emiliana Pasca Noether was then a professor of modern European history at the University of Connecticut, where she is now a professor emerita. The Emiliana Pasca Noether Chair in Modern Italian History has been established in her honor.

> Within the past few days a distinguished mathematician, Professor Emmy Noether, formerly connected with the University of Göttingen and for the past two years at Bryn Mawr College, died in her fifty-third year. In the judgment of the most competent living mathematicians, Fräulein Noether was the most significant creative mathematical genius thus far produced since the higher education of women began.
> —*Albert Einstein* [3, 12]

Born in 1882, Emmy Noether was the daughter of the mathematician Max Noether, professor at the University of Erlangen. The Noethers belonged to the intellectual middle class in Wilhelminian, Germany, a class marked by orderly customs and great stability. The family occupied the same apartment for forty-five years, until the death of Max Noether in 1921. The Noethers respected education for its own sake and were interested in intellectual pursuits. Their home life must have been warm and companionable. Emmy was the firstborn to Max and his wife, Ida Amelia Kaufman, who came from a rather wealthy family. Emmy and her brother Fritz grew to share their father's dedication to mathematics.

As a child, however, Emmy gave no signs of precociousness or extraordinary ability in mathematics. From 1889 to 1897 she attended the State Girls' School in Erlangen. There, with other daughters of the bourgeoisie, she learned the elements of language and arithmetic. She also seems to have studied the piano and to have learned French and English. As she grew to young womanhood she developed a great love of dancing and spent many evenings at family parties. Her schoolmates and friends of this period remember her as a clever, friendly, and rather endearing child. In 1900 she took the Bavarian state examinations to become certified as a teacher of English and French. These demanding examinations lasted four days, and she did well in them. It would seem that now her education was complete. She had gone through the schooling deemed necessary for a young woman of her class and breeding. Moreover, she had acquired a certificate which would enable her to earn her livelihood, should the need ever arise.

It is at this point that the story of Emmy Noether becomes interesting. For it was shortly after 1900 that she decided that she would like to attend the university to study mathematics, following in her father's footsteps. What is intriguing about this decision is why at this point in her life—after having given no outward sign until then of being anything but a conformist to the path outlined for her by her family and social class—did she suddenly decide that she did not want to continue as a dutiful "fille de famille" or become a teacher of French and English to other well brought up young ladies? Was it the example of her younger brother Fritz, now beginning his university studies, and undoubtedly brimming with enthusiasm over his mathematical studies, which he probably discussed at dinner with his father, Max? Was it the cumulative effect of years of having been exposed to discussions by the Erlangen mathematicians who frequented the Noether home? The reasons will probably never be fully known, for none of Emmy Noether's personal papers have survived her exile from Germany, her death, and the dispersal of her possessions. What matters for the history of mathematics is that she did take the step, she did persist, despite all the odds against women, and did go on to become one of the most distinguished algebraists of the twentieth century.

In the early years of the twentieth century, it was not easy for a woman in Germany to do what Emmy Noether set out to accomplish. Women could enroll as auditors in courses at the university, but only with the professor's permission, and they were not allowed to take examinations, except again by special permission of the instructor. Only in 1908 did the German Ministry of Education permit coeducation at the universities. One of the difficulties facing any woman who wished to qualify for admission to a German university was that she was barred from attending the gymnasium, which provided the rigorous education required for university admission. The various girls' schools were little better than finishing schools for the daughters of the bourgeoisie and trade schools for those of workers. There was, however, an escape clause to the limitations placed upon women. They could take matriculation examinations (a type of entrance exam) to demonstrate their mastery of the course of study prescribed in the gymnasium.

After attending the University of Erlangen from 1900 to 1902 as an auditor, Emmy Noether, on July 14, 1903, took and passed the matriculation examination at the Royal Gymnasium in Nürnberg for admittance to the university. In the following school year, 1903–1904, we find her enrolled for the first semester at the University of Göttingen, more liberal, perhaps, toward women than other German universities at this time: in 1906 Vera Lebedoff-Myller was to get her degree, working under the mathematician David Hilbert, who later was to invite Emmy Noether to join his group at Göttingen. The following year Emmy returned to continue her studies at the University of Erlangen, and on October 24, 1904, she was duly inscribed as student number 586 in this university. She was the only woman enrolled in the faculty of mathematics with forty-six male fellow students. On December 13, 1907, she successfully defended her thesis, and on July 2, 1908, was awarded her degree summa cum laude.

Undoubtedly, the first two influences on her development as a mathematician were her father and her mentor, Paul Gordan, under whom she wrote her dissertation. She remained devoted to him, even after his death in 1912, and his picture hung in her study in Göttingen all the years she was there. Max Noether was "a very intelligent, warm-hearted, harmonious man of many-sided interests and sterling education" [5, 55]. Paul Gordan was a very different sort of person. Weyl described him as "a queer fellow, impulsive and one-sided. A great walker and talker . . . either with friends, and then accompanying his discussions with violent gesticulations . . . or alone, and then murmuring to himself and pondering over mathematical problems; or . . . carrying out long numerical calculations" in his head [5, 55].

Having received her degree, in 1907–1908 Emmy worked without any formal appointment or pay at the Mathematical Institute in Erlangen, partly helping her father and partly working on her own research. Slowly, recognition began to come her way. In 1908 she was elected to membership in the Circolo mathematico of Palermo, Italy, and she attended her first International Mathematical Congress in Rome with her father. Then an unknown young mathematician, she must have played an inconspicuous role during the proceedings, listening to her elders, and probably touring the Roman ruins, Baedecker in hand, before returning home to Erlangen. In the following year she was invited to join the German Mathematical Union and began to lecture publicly.[17]

In 1913, while continuing her research and publishing its results, she began to substitute for her father at the University of Erlangen. Max Noether, who had suffered polio as a child, was feeling the effects of age. Two years later, David Hilbert and Felix Klein seem to have invited her to go to Göttingen. By then Emmy had published some half dozen papers and had an impressive knowledge of certain aspects of mathematics, which Hilbert and Klein felt would complement their work on relativity theory. Emmy accepted the invitation and moved to Göttingen, where she remained until forced out by the Nazis in 1933. But, while in 1915 Göttingen had a more liberal policy toward women students, it was not ready to admit women to the faculty. The mathematicians wanted some sort of appointment for Emmy Noether to regularize her position at the university. At the very least they wanted the *Habilitation*[18] for her, but the entire Philosophical Faculty, which included philosophers, philologists, and historians as well as natural scientists and mathematicians, had to vote on the acceptance of the *Habilitation* thesis. Particular opposition came from the nonmathematical members of the faculty.

17. Her first lecture was at the Salzburg section of the German Mathematical Union in 1909; in 1913 she was invited to read a paper at the Vienna branch. These lectures were interrupted by the war in 1914, but resumed during the Weimar period, when she spoke in such diverse cities as Bad Nauheim (1920), Jena (1921), Leipzig (1922), Marburg an der Lahn (1923), Innsbruck (1924), Danzig (1925), and Prague (1929).

18. The *Habilitation* is a certification that is required in some European countries for professional appointments; it signifies independent research beyond the doctorate.

They argued formally: "How can it be allowed that a woman become a *Privatdozent*? Having become a *Privatdozent*, she can then become a professor and a member of the University Senate. Is it permitted that a woman enter the Senate?" They argued informally: "What will our soldiers think when they return to the University and find that they are expected to learn at the feet of a woman?!" ... Hilbert answered their formal argument ... with ... directness: "Gentlemen, I do not see that the sex of the candidate is an argument against her admission as a *Privatdozent*. After all, the Senate is not a bathhouse."

When, in spite of this rejoinder, he still could not obtain her *Habilitation* he solved the problem of keeping her at Göttingen in his own way. Lectures would be announced under the name of Professor Hilbert, but delivered by Fräulein Noether [4, 143].

November 1918 saw the end of both the war and the monarchy in Germany, and the Weimar Constitution was adopted on August 11, 1919. Emmy was a woman in her late thirties by now, whose scholarly reputation was growing, but she still had no secure position in the academic world. Politically, she had great hopes for the future of the new German republic and she was a convinced pacifist.

The changes in the political structure of Germany represented by the Weimar Republic did not really change the conservative outlook of the mandarins of academia.[19] But they were forced to make some few concessions, and Emmy Noether was finally allowed to take her *Habilitation* on June 4, 1919. Now she could lecture openly under her own name. However, it should be pointed out that the title *Privatdozent*, which she was able to use after her *Habilitation*, was a purely honorific one, carrying no pay with it. Three years later, she was appointed *ausserorderntlicher Professor*, again a largely honorific title without remuneration, since she was not, as we would say in the United States, on the tenure track. The early twenties were years of extraordinary inflation for Germany. Whatever private means Emmy may have possessed that had enabled her to sustain herself were decimated by the inflation, so in the spring of 1923, she was given a *Lehrauftrag*, or lectureship, for algebra, to which was attached a stipend. Finally, she could teach officially, give examinations, direct dissertations, and be paid, even if very little. The first student to complete a doctorate under her direction was a woman, Grete Hermann, who finished in February 1925.

Many of her younger colleagues—including Hermann Weyl, who became a permanent member of the Göttingen faculty in 1930—recognized Emmy's outstanding abilities and tried without success to get a better

---

19. For an analysis of German academia, see Fritz K. Ringer, *The Decline of the German Mandarins: The German Academic Community, 1890-1933* (Cambridge, Mass.: Harvard University Press, 1969), especially 213–252, 435–449.

position from the ministry for her. Weyl also failed to get her elected to membership in the Göttingen Society of Science. Traditional hostility and prejudice toward women overbalanced her scientific contributions and growing reputation [5, 60]. But Emmy Noether, so long as she could continue her work and disseminate her ideas among willing students, did not seem embittered by the shabby treatment officialdom gave her. She was, according to Weyl and other mathematicians, the strongest center of mathematical activity in Göttingen from 1930 to 1933, both because of her research and because of her influence upon a wide circle of students.

Among them were two mathematicians who came to Göttingen not as aspiring Ph.D. candidates but as already recognized young scholars. One was the Dutch mathematician B. L. van der Waerden and the other was the Russian P. S. Alexandroff, who had been professor at the University of Moscow since 1926 and came to Göttingen as a guest professor. Both became members of the group known as the "Noether boys," who joined Emmy in long walks during which the only topic of conversation was mathematics. Many years later, after the ravages of Nazism and the Second World War had passed over Germany, van der Waerden singled out the years from 1920 to 1934 as the period during which Emmy Noether, Emil Artin, and Alexander Ostrowski had given algebra a new turn.

Emmy's friendship with Alexandroff endured during these years. Her work became known and gained recognition in Russia, and during the academic year 1928–1929, she was invited to Moscow, where she gave a course in abstract algebra at the University of Moscow and conducted a seminar in algebraic geometry at the Communist Academy.

International recognition also came to Emmy. At the 1928 International Congress of Mathematicians held at Bologna, Italy, she delivered a major paper at one of the section meetings. Then in 1932 at the International Congress in Zürich, she addressed a plenary session.

The rise of Nazism to power in January 1933 shattered the life and work of many German scholars in universities throughout Germany. Göttingen was no exception. National Socialism had been gaining support among students. The predominantly conservative faculty members, though having reservations about some of the extreme aspects of National Socialism, were not averse to its coming to power. By and large, academia had only tolerated the Weimar Republic; many of its members thought of themselves as apolitical intellectuals and felt that they could accept the changes that Nazism seemed to offer for the benefit of the fatherland.

By a series of anti-Semitic laws enacted shortly after they came to power, the Nazis sought to purge the civil service of Jews and persons of partially Jewish ancestry [2, 130]. The impact of these laws was immediately felt in the Mathematical Institute of Göttingen, by then the foremost center for mathematical studies in Germany. The director of the Institute, Richard Courant, as well as Emmy Noether and Max Born, the theoretical physicist and later Nobel Prize laureate, left Germany as a result.

Forbidden to lecture at the university, Emmy Noether continued to receive some of her students, eager to continue their work under her, in her home; one of them repeatedly turned up in a Nazi uniform. The summer of 1933 was a difficult and troubled one, but she seemed never to lose her serenity. Weyl recalled that "her courage, her frankness, her unconcern about her own fate, her conciliatory spirit, were, in the midst of all the hatred and meanness, despair and sorrow . . . a moral solace" [5, 62–63].

Alexandroff initiated talks with the People's Commissariat for Education in Russia about appointing her to a chair at the University of Moscow. But, as he reported after her death, it was slow in making a decision. Emmy, deprived of her modest stipend, unable to participate officially in the academic life of Göttingen, could not wait for the Russian People's Commissariat for Education to make up its mind, and she accepted the offer of a one year's guest professorship from Bryn Mawr [1, 147].

At the end of October 1933 Emmy Noether left Germany for her year at this unknown American girls' college, in a strange country, whose customs were unfamiliar to her. But with her usual buoyancy she adjusted to her new surroundings. She was pleased to learn that Bryn Mawr had offered a graduate fellowship in her name for the academic year 1933–1934 to a qualified student. Her seminar drew few students, but in February 1934 she started to give weekly lectures at the Institute for Advanced Study in Princeton. At Bryn Mawr she slowly began to form a group around her; this time they were the "Noether girls" rather than the "Noether boys" of Göttingen. Unwilling to give up her love of walking and talking mathematics, she would take her American students for hikes on Saturday afternoons. Often she would become so absorbed in her discussion of some fine mathematical point that she would forget about traffic. One of her American colleagues at Bryn Mawr, Grace Shover Quinn, remembered her as "sincere, straightforward, kindly, thoughtful, and considerate" [1, 148]. [See also note 26, p. 92.]

In the summer of 1934 she returned to Germany to see her brother Fritz and his family. He, too, had lost his position at the Technische Hochschule in Breslau; he had accepted a professorship at the University of Tomsk in Siberia, where he was moving with his wife and two sons. This was the last time that Fritz and his family were to see Emmy. She also returned to her beloved Göttingen, visited other friends in Germany, and then went back to Bryn Mawr. During her second year there she acquired a few more graduate students, and her first Ph.D. candidate, Ruth Stauffer, finished her dissertation under Emmy's direction. In the spring of 1935 she entered Bryn Mawr Hospital to be operated on for the removal of a tumor; she seemed to be recovering well, when she died suddenly on April 14 at the age of fifty-three. Thus ended the life and career of this remarkable woman, whose death was mourned by her friends and colleagues not only as a loss to the work of mathematics, but also on a deep, personal level.

Her qualities as a human being left a lasting impression on those who had known her. Weyl said that she was "warm like a loaf of bread. There irradiated from her a broad, comforting, vital warmth" [5, 57]. While she did not have the outward attributes usually associated with femininity, for she has been variously described as having "a loud disagreeable voice," as looking like "an energetic and very nearsighted washerwoman" in "baggy clothes" [4, 143], Alexandroff remembered that "she loved people, science, life, with all the warmth, all the cheerfulness, all the unselfishness, and all the tenderness of which a deeply sensitive . . . soul is capable" [1, 147].

References

1. P. S. Alexandroff, Address to the Moscow Mathematical Society, 1935, quoted in C. H. Kimberling, "Emmy Noether," *American Mathematical Monthly* 79(1972), 136–149.
2. Karl Dietrich Bracher, *Die Auflösung der Weimarer Republik* (Stuttgart: Ring, 1957), quoted in *From the Middle Ages to Nazi Rule*, Year Book XVIII, Publications of the Leo Baeck Institute (London: Secker and Warburg, 1973).
3. Albert Einstein, "Letter to the Editor," *New York Times*, May 4, 1935.
4. Constance Reid, *Hilbert* (New York: Springer-Verlag, 1970).
5. Hermann Weyl, "Emmy Noether," *Gessamelten Abhandlungen Hermann Weyl*, vol. 3, ed. K. Chandrasekharan (Heidelberg: Springer-Verlag, 1968).

2. From Earlier Times

The stories here continue the journey back in time begun in the first chapter. These women were born in Europe over roughly a century, the first in France in 1776. The theme of mathematical families extends into this chapter, which introduces the matriarch of the largest of these families. Several mathematicians are found among the children of Grace Chisholm Young and her collaborator and husband, William. Their granddaughter Sylvia Wiegand (also married to a mathematician, as well as the daughter of one) writes on the life they shared, both in the home and in the world of mathematics. Chisholm Young's story is followed by remarks on the career and mathematics of her daughter Cecily Young Tanner.

Four of the women profiled here found it necessary to relocate in order to pursue the doctoral degree or to find appropriate employment. Both Chisholm and Charlotte Angas Scott received their undergraduate degrees from Girton College, associated with Cambridge University, but at that time Cambridge did not award the doctorate to women. Chisholm traveled to Göttingen for her degree, while Scott completed the work for her degree at Cambridge but received her D.Sc. from the University of London via "external examination." For appropriate academic employment, Scott moved to the United States and joined the faculty of Bryn Mawr College, where she greatly influenced the shaping of that institution, still today a leader in women's education.

Elizaveta Litvinova, one of numerous Russian women who left their homeland in those years for advanced study, received her doctoral degree in 1878 from Bern University. She was the second woman to receive a doctorate in mathematics, the first having been her countrywoman Sonia Kovalevsky, who received her degree in 1874 from Göttingen, as did Chisholm in 1895 and Noether in 1908. When Litvinova returned to Russia, the government prevented her from obtaining a position that allowed time for mathematics research. She nevertheless became a well-respected expert in mathematical pedagogy. After earning her doctorate, Kovalevsky also had difficulty finding appropriate employment. She found work in Sweden, her career ending there with her untimely death at the age of forty-one.

Ada Byron Lovelace worked with Charles Babbage on his protocomputers. Although her work was unheralded during her brief lifetime, many accounts of her life have appeared in the press over the past twenty-five years. When Dana Angluin wrote her article in 1975, the programming

language commissioned by the U.S. Department of Defense to help software designers and programmers develop large, reliable applications had
not yet been named Ada in honor of Lovelace's accomplishments. According to the Ada Information Clearinghouse, "The Ada programming language was originally christened 'DoD-1' by the general press. . . . In 1979,
Jack Cooper of the Navy Materiel Command thought of naming the language Ada, which was widely accepted by the [DoD's High Order Language Working Group]."[20] This surely pleased the feminist scholars working
to bring to light the "forgotten" work of women from earlier times.

Sophie Germain likewise did not have the opportunity to pursue a career
as a mathematician but worked as an independent scholar in both mathematics and philosophy. Her work on the theory of elasticity was worthy of a
prize in 1816. There are interesting relationships and parallels among the
lives of Germain, Lovelace, and other women of those times, most notably
Mary Fairfax Somerville. This chapter concludes by illustrating some of these
with vivid quotations from multiple sources. Throughout this book, the biog-
raphers are themselves often mathematicians with fascinating life stories—
none more so than Christine Ladd Franklin, a biographer of Germain. Ladd
Franklin completed the work for her doctorate in 1882, the first woman in the
United States to do so. But although Johns Hopkins University had permitted
her to study and then later to work there, it would not award her the degree
she had earned. She nevertheless had a productive career in teaching and
research, the latter mostly in the optics of color vision. After Johns Hopkins
belatedly awarded her the degree in 1926, the *New York Times* reported that
"one of the outstanding features of the day was an ovation accorded [her]
when the names of candidates for degrees were announced."[21]

My Grandmother, Grace Chisholm Young

SYLVIA M. WIEGAND

Grace Chisholm Young, 1868–1944. Based on the talk "Grace Chisholm Young," at
the AWM session "British Women Mathematicians," chaired by Lenore Blum, 1977
JMM, as reported in the AWM *Newsletter* 7(3), 1977, 5–10. Wiegand was then an
assistant professor of mathematics, University of Nebraska–Lincoln, where she is now
a professor of mathematics.

20. "The Naming of Ada," AdaIC Flyer [online]. IIT Research Institute, all rights assigned to the U.S.
Government (Ada Joint Program Office) 1998 [updated 11 August 1998; accessed 16 January 2003].
Available from World Wide Web: http://archive.adaic.com/docs/flyers/naming.html.
21. Judy Green, "Christine Ladd-Franklin," *A Biobibliographic Sourcebook for Women Mathematicians*, Grinstein and Campbell, eds. "To restore ideal at Johns Hopkins," *New York Times*, 23 February
1926, 12, as cited in Green.

Grace Chisholm was one of the first women to receive a Ph.D. in mathematics. She studied under Felix Klein and received the Ph.D. magna cum laude from Göttingen in 1895 at age twenty-seven. When Grace passed her final oral exam, she became the first woman to earn a Ph.D. in any field in all of Germany through the normal procedure. Sonia Kovalevsky had been granted a Ph.D. in absentia, for which she submitted a paper, but Grace took all the necessary exams and fulfilled the necessary requirements for the degree; she had to go herself to get official permission from the German government.

Grace had been educated at home by her mother and a governess; nevertheless, she managed to pass the Cambridge Senior Examination in 1885 at the age of seventeen. Grace's brother went to Oxford, but at that time women rarely went to a university. Girton College, associated with Cambridge University, was the first institution in England dedicated to educating women at the university level, and it had only been in existence since 1869. At first, Grace was encouraged to spend her time helping the poor and otherwise making herself "useful." However, she had great initiative and desire for learning, and she entered Girton in 1889. She was awarded the Sir Francis Goldschmid Scholarship, which her father matched in value. In the Cambridge Tripos Part I examination in 1892, Grace scored the equivalent of a first class, and then, in response to a challenge, she took the Oxford examination unofficially and obtained the highest mark for all students at Oxford that year. Grace went to Göttingen to continue her studies because to do so would have been impossible in England.

A letter written by Grace to her old college friends at Girton describes the situation at Göttingen:

> Professor Klein's attitude is this, he will not countenance the admission of any woman who has not already done good work, and can bring him proof of the same in the form of degrees or their equivalent, or letters from professors of standing and further he will not take any steps till he has assured himself by a personal interview of the solidity of her claims. Prof. Klein's view is moderate. There are members of the Faculty here who are more eagerly in favor of the admission of women and others who disapprove altogether.

On his fiftieth birthday, Klein was honored in Turin, where Grace was then studying. At dinner, he was seated next to Grace, said to be his favorite pupil, and he whispered to her: "Ah, I envy you. You are in the happy age of productivity. When everyone begins to speak well of you, you are on the downward road."

The year after Grace earned the Ph.D., she married William Henry Young, an Englishman who had been her tutor at Girton before she went to Göttingen. They both became internationally known mathematicians in their time, and their results are still widely quoted today. Together they produced 220 mathematical articles and several books.

For several years after their marriage, Will continued to tutor at Cambridge, where he earned an Sc.D. in 1903 at the age of forty. Later he held professorships at Liverpool, Calcutta, and Aberystwyth, Wales. Will served as president of the London Mathematical Society from 1922 to 1924, and he was president of the International Union of Mathematicians in 1929. Among other things, he is famous for his discovery (independent of Lebesgue) of the integral and for his work on Fourier series and cluster sets.

The pattern of their life together was that Will traveled to and from their family home to earn a living, while Grace brought up their six children and followed her other interests. At the same time, both of them worked intensely on mathematics. Among Grace's other interests were medicine, languages, and music. She completed all the requirements for a medical degree except the internship (her medical practice was therefore limited to the family). Grace knew six languages and taught them to the children when they were young. To the children she also communicated her love of music; each of the children played some instrument, and the family gave informal concerts together.

A few years after the marriage, Grace and Will changed their roles. Previously Grace had been the researcher of the pair, but she became so impressed by Will's creativity that she unofficially became his scribe. She wrote up his papers for publication, often filling in proofs, correcting mistakes, and so on. When they were together, and in their correspondence, they discussed these papers extensively. In short, Grace helped a great deal with papers signed by Will, but, more than that, he would probably have accomplished very little without her, and he realized it. Will certainly had a profusion of ideas and a great intelligence, but he would have had neither the time nor the temperament to carry them through. An illustration of the kind of help she gave him is given by this footnote to his paper "On Integration with Respect to a Function of Bounded Variation" in the *Proceedings of the London Mathematical Society* (1914):

> Various circumstances have prevented me from composing the present paper myself. The substance of it only was given to my wife, who has kindly put it into form. The careful elaboration of the argument is due to her.

Also a letter written to Professor Lida Barrett by their daughter, Rosalind Cecily,[22] demonstrates Grace's assistance to Will:

> Another famous partnership, that of George Eliot and Lewes, can be taken in many respects as the counterpart of that of my parents. There

---

22. See following story for information on Cecily Young Tanner's mathematics.

it was the man who took the brunt of life off the woman's shoulders and spent his creative energies in fostering her genius. This, my mother clearly appreciated.

When all is said, it remains that my father had ideas and a wide grasp of subjects, but was by nature undecided; his mind worked only when stimulated by the reactions of a sympathetic audience. My mother had decision and initiative and the stamina to carry an undertaking to its conclusion. Her skill in understanding and in responding, and her pleasure in exercising this skill led her naturally into the position she filled so uniquely. If she had not had that skill, my father's genius would probably have been abortive, and would not have eclipsed hers and the name she had already made for herself.

When Will traveled to work in India, Grace began publishing under her own name again, and she did her best work.

Parts of Grace's dissertation, "The Algebraic Groups of Spherical Trigonometry," are mentioned in Klein's book, *Elementary Mathematics from an Advanced Standpoint*, Volume I, page 179. To each spherical triangle on the unit sphere, Klein associates a point of 12-space as follows:

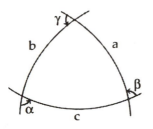

corresponds to $(\cos a, \cos b, \cos c,$
$\cos \alpha, \cos \beta, \cos \gamma, \sin a, \ldots, \sin \gamma) \in \mathbb{R}_{12}.$

If M_3 is the image of the set of all spherical triangles, then M_3 is a three-dimensional set of $\mathbb{R}_{12}$. Klein found nine identities satisfied by M_3, but they did not determine the set. He asked several questions about M_3, and then went on to say:

One could gain familiarity with these things by consulting investigations which have been made in exactly the same direction but in which the questions have been put somewhat differently. These appear in the Göttingen dissertation, 1894, of Miss Chisholm (now Mrs. Young), who by the way was the first woman to pass the normal examination in Prussia for the doctor's degree.

Klein describes Grace's work briefly: she considered an image of the spherical triangles in $\mathbb{R}_6$ and completely characterized this version of M_3.

Grace's most important independent work appears in a group of papers appearing from 1914 to 1916, in which she studied derivates of real functions. Recall that the upper right Dini derivate is defined by

$$D^+f(t_0) = \limsup_{h \to 0^+} \frac{f(t_0 + h) - f(t_0)}{h} \text{ and } D_+ f(t_0), D^- f(t_0), D_- f(t_0) \text{ are defined}$$

similarly. The theorem now known as the Denjoy-Saks-Young theorem gives a complete description of the possible behaviors of these derivates:

Theorem: Except at a set of measure zero, there are three possible dispositions of the derivates of a function $f(x)$, either (i) they are all equal and the function is differentiable, or (ii) the upper derivates on each side are $+\infty$, or (iii) the upper derivate on one side is $+\infty$, the lower derivate on the other side is $-\infty$, and the two remaining extreme derivates are finite and equal.

According to Riesz and Nagy, the theorem was proved for continuous $f(x)$ by Denjoy and Grace independently, Grace proved it for measurable $f(x)$, and then Saks generalized it to all $f(x)$. Lloyd Jackson at the University of Nebraska told me that this theorem was one of his favorites, and that it was extremely useful to him in his work because eliminating options (ii) and (iii) is sufficient to show that a function is differentiable.

In 1905 a geometry book was published in both Will's and Grace's names, but it was probably written by Grace, as the German edition was published under her name alone. In those days, the study of geometry was limited to theorems for the plane, but Grace's book was intended for elementary school children and, in fact, her seven-year-old son Frank helped her with some parts.

Will and Grace jointly published *The Theory of Sets of Points* in 1906, the first book of its kind. Set theory was not popular with most mathematicians at that time, but Georg Cantor was very enthusiastic about the book:

> It was with great joy that I received the day before yesterday the copy you have most kindly sent me of your joint work with your husband, *The Theory of Sets of Points*. My sincerest thanks to you both.
>
> It is a pleasure for me to see with what diligence, skill and success you have worked and I wish you, in your further researches in this field as well, the finest results, which, with such depth and acuteness of mind on both your parts, you cannot fail to attain.[23]

The authors commented on the importance of set theory in the original preface to the book:

> In subjects as wide apart as Projective Geometry, Theory of Functions of a Complex Variable, the Expansions of Astronomy, Calculus of

---

23. From the preface to the second edition, by R. C. H. Tanner.

Variations, Differential Equations, mistakes have in fact been made by mathematicians of standing, which even a slender grasp of the Theory of Sets would have enabled them to avoid.

Two joint papers with Will are referred to in Collingwood and Lohwater's book *The Theory of Cluster Sets*. These papers and some signed by Will were important to the development of the theory of cluster sets and prime ends.

In addition to Grace's mathematical works, she wrote two books for children that were lessons on elementary biology (they included cell structure seen under a microscope and contained the story of a family like the Youngs). She worked for five years on a sixteenth-century historical novel (unpublished) called *The Crown of England*.

While discussing Grace's works, it seems appropriate to mention her six children. Unfortunately, the oldest child, Frank, apparently the most precocious as a child, died in World War I in 1917 at age twenty. As I write this, the second child, Rosalind (Cecily) Tanner, is a mathematician and historian. Another daughter, Janet Michael, fulfilled her mother's dream of becoming a medical doctor, and Helen Marion Canu, the last daughter, studied mathematics at the graduate level. The last two sons, Laurence and Pat, became a mathematician and a chemist, respectively. Cecily, Janet, Laurence, and Pat are all actively involved in their work and with their own children and grandchildren.

Grace Chisholm Young died in 1944, before I was born, so unfortunately we never met. Everything I have heard about her indicates she was a wonderful person, extremely talented, but also kind and loved by everybody.

My father is Laurence Chisholm Young, the fifth child of Grace Chisholm and William Henry Young. In 1976, my father retired from the University of Wisconsin, where he had been on the faculty of the Mathematics Department for twenty-eight years.

The presence of so many mathematicians in my family (and particularly women mathematicians) has been inspirational for me. Growing up with a tradition of mathematics for women made it seem a natural thing to do and provided helpful role models that most women don't have. Now as I work on my own theorems, I imagine that Grace is looking over my shoulders with helpful suggestions!

To close, I would like to quote a letter from Will to Grace that describes their situation as he saw it:

> I hope you enjoy this working for me. On the whole I think it is, at present at any rate, quite as it should be, seeing that we are responsible only to ourselves as to division of laurels. The work is not of a character to cause conflicting claims. I am very happy that you are getting on with the ideas. I feel partly as if I were teaching you, and

setting you problems which I could not quite do myself but could enable you to. Then again I think of myself as like Klein, furnishing the steam required—the initiative, the guidance. But I feel confident too that we are rising together to new heights. You do need a good deal of criticism when you are at your best, and in your best working vein.

The fact is that our papers ought to be published under our joint names, but if this were done neither of us get the benefit of it. No. Mine the laurels now and the knowledge. Yours the knowledge only. Everything under my name now, and later when the loaves and fishes are no more procurable in that way, everything or much under your name.

There is my programme. At present you can't undertake a public career. You have your children. I can and do. Every post which brings an answer from you to my last request or suggestion gives me a pleasurable excitement. Life here is more interesting with such stimulants. I am kept working and thinking, too, myself. Everything seems to say we are on the right track just now. But we must flood the societies with papers. They need not all of them be up to the continental standard, but they must show knowledge which others have not got and they must be numerous.

Notes

1. The main source for my talk was an article by I. Grattan-Guinness in the *Annals of Science* 29(1972), 105–186. This article includes many letters written by Grace, as well as selections from some notes she wrote as a start on an autobiography. There are very appealing descriptions by Grace of her final oral, of getting official permission for the Ph.D., of her childhood, and of her children.

2. Grattan-Guinness also compiled a joint bibliography of W. H. and G. C. Young in *Historia Mathematica* 2(1975), 43–58. He says he combines their publications because Grace collaborated so closely with Will on papers signed by Will. Another interesting article is M. L. Cartwright's obituary notice for Grace in the *Journal of the London Mathematical Society* 19(1944), 185–192.

3. These titles of some other papers by Grace show her diversity of interests:

   a. "On the Curve $y = \left(\dfrac{1}{x^2 + \sin^2 \psi} \right)^{3/2}$ and Its Connection with an Astronomical Problem," *Monthly Notices, Royal Astronomical Society* 57(1897), 379–387.

   b. "On the Form of a Certain Jordan Curve," *Quarterly Journal of Pure and Applied Mathematics* 37(1905), 87–91.

   c. "On the Solution of a Pair of Simultaneous Diophantine Equations Connected with the Nuptial Number of Plato," *Proceedings of the London Mathematical Society* (2)23(1924), 27–44.

   d. "Pythagore, comment a-t-il trouvé son théorème?" *L'Enseignement mathématique* (1)25(1926), 248–255.

Like Mother, Like Daughter

IVOR GRATTAN-GUINNESS

Based on "Obituary: Cecily Tanner 1900–1992," from the British Society for the History of Mathematics *Newsletter* 23(1993), Spring 1993, 10–15, and reprinted in the AWM *Newsletter* 23(6), 1993, 21–24. Grattan-Guinness is a professor of mathematics and statistics, Middlesex University Business School.

In 1925 Cecily entered her mother's college, Girton (Cambridge), for postgraduate work under the direction of E. W. Hobson. Soon afterward her younger brother Laurence came up, and he wrote a Cambridge Tract on *The Theory of Integration*, which was published in 1927 thanks to Cecily typing out the text. During this time she translated into English Konrad Knopp's beautiful book on infinite series [1], still one of the best sources for the classical theory. In addition, she completed her thesis [2]; but the British system allowed her to gain only the title of Ph.D., in contrast to the full *Dissertation* that Grace had gained under Klein's direction in 1895. But she did obtain a fellowship at Girton, for her research work had begun to flower with a flurry of papers in 1928 and 1929.

Cecily worked in the family field of mathematical analysis, especially Riemann-Stieltjes integration [3]. In the course of her doctoral research she came to her best mathematical idea, elaborated as "The Algebra of Many-Valued Quantities." Her motivation for this theory was to simplify the intricacies involved in developing Riemann-Stieltjes integration. This lovely theory never gained the attention that it deserved, although "a certain Mr. von Neumann" of Budapest wrote to ask for an offprint. However, it was used in a branch of numerical analysis called "interval analysis" and later in the extension of the membership function of fuzzy sets [4].

Cecily's house was in Boundary Road in Wallington: an appropriate name, for in a few years the balance of her concerns changed quite sharply from mathematical research to its education and history. In the early 1950s she wrote some papers of an educational character and came to recognize the central importance of inequalities in mathematics. She spoke on the matter in an unpublished lecture on "Equal and Unequal" at the 1958 International Congress of Mathematicians in Edinburgh.

Around that time Cecily increased her interest in the history of mathematics; with the support of G. J. Whitrow she founded a seminar in history and education at Imperial College in 1961, which ran until her retirement six years later. Its inauguration was part of Klein's legacy to her family; he had stressed the importance of the research seminar, and hers imitated the private one run after the Second World War in Frankfurt am Main by Klein's disciple, the mathematician and historian Wilhelm Lorey, whom she knew well.

Concern with education and inequalities united in a study of the history of inequalities, upon which she published several papers [5; 6; 7]. By this means she came to the principal inventor of the notations ">" and "<" the Elizabethan mathematician Thomas Harriot. At that time virtually forgotten, he was to be rediscovered and popularized largely by her efforts. In 1967 she financed at Oxford (his hometown) the Thomas Harriot Seminar, to discuss the great man and his context. It ran intermittently until 1983, and now an annual lecture is held there. Meanwhile, activity had switched to the energetic leadership of Gordon Batho at the University of Durham, where in 1979 she also financed a seminar that now meets alternately there and in Cambridge. Two years later the university granted her an honorary M.A.

Cecily had the surviving Harriot manuscripts photocopied and worked extensively on them, in the hope that the circle around the Oxford Seminar would prepare a collected edition of these difficult materials. To her great irritation this project lapsed, but she published a number of short papers on him and also three long ones [8; 9; 10]. In 1971 she participated in a Harriot Seminar held at the University of Delaware, contributing to the proceedings with a study of his associates [11]. Her main interest in Harriot's mathematics, which centered upon his algebra, led to a paper on Pythagorean triples [12] and then a thought-provoking pair [13; 14] contrasting his understanding of algebra with that of his Renaissance predecessor G. Cardano.

References

1. Konrad Knopp, *Theory and Application of Infinite Series*, trans. Cecily Tanner (London and Glasgow: Blackie, 1928).
2. Cecily Tanner, "Foundations for the Generalisation of the Theory of Stieltjes Integration etc. An n-Dimensional Treatment," Cambridge University Ph.D., 1929.
3. Fyodor A. Medvedev, *Razvitie ponyatya integrala* (Moscow: Nauka, 1974).
4. Ivor Grattan-Guinness, "Fuzzy Membership Mapped onto Intervals and Many-Valued Quantities," *Zeitschrift für mathematische Logik und Grundlagen der Mathematik* 22(1976), 149–160.
5. Cecily Tanner, "On the Role of Equality and Inequality in the History of Mathematics," *British Journal for the History of Science* 1(1962), 159–169.
6. ———, "Mathematics Begins with Inequality," *Mathematical Gazette* 45(1961), 292–294.
7. ———, "La symétrie locale des fonctions et ensembles arbitraires," *L'Enseignement mathématique* (2)8(1962), 192–194.
8. ———, "The Study of Thomas Harriot's Manuscripts," *History of Science* 6(1967), 1–16.
9. ———, "Thomas Harriot as Mathematician: A Legacy of Hearsay," *Physis* 9(1967), 235–292.

10. ———, "Thomas Harriot (1560?–1621)," in *»Geometrija i algebra po v̂etkom XVII stoljeća« povodom 400-godišnjice rodenya Marina Getaldića* (Zagreb; Jugoslavian Academy of Sciences, 1969), 161–170. [In French]

11. ———, "Henry Stevens and the Associates of Thomas Harriot," in *Thomas Harriot, Renaissance Scientist*, ed. J. Shirley (Oxford: Clarendon Press, 1974), 91–106.

12. ———, "Nathaniel Torporley's 'Congestor analyticus' and Thomas Harriot's 'De triangulus laterum rationalum,'" *Annals of Science* 34(1977), 393–428.

13. ———, "The Ordered Regiment of the Minus Sign: Off-beat Mathematics in Harriot's Manuscripts," *Annals of Science* 37(1980), 127–158.

14. ———, "The Alien Realm of the Minus: Deviatory Mathematics in Cardano's Writings," *Annals of Science* 37(1980), 159–178.

Charlotte Angas Scott

PATRICIA CLARK KENSCHAFT

Based on "Charlotte Angas Scott: 1858–1931," which appeared in two parts in the AWM *Newsletter* 7(6), 1977, 9–10, and 8(1), 1978, 11–12. Kenschaft was then an assistant professor of mathematics, Montclair State College, where she is currently a professor at the institution now called Montclair State University.

In 1885 it was an act of daring for any mathematician, especially a woman, to head for the United States with a doctorate from a European university. Then primarily concerned with expansionistic activities such as taming the "Wild West," Americans had little interest in intellectual pursuits and practically none in creative mathematics. Yet Charlotte Scott, possibly spurred by the lack of job opportunities in Europe, left her native England at the age of twenty-seven to become the first mathematics professor of a new college called Bryn Mawr. She kept up her European contacts by making frequent summer trips abroad.

For Bryn Mawr's first forty years, Scott was the chair of its "mathematical department." She helped it become a noted center of learning, not only as an organizer and as a teacher (many of her students became noted mathematicians), but also as a textbook author and as a researcher in analytical geometry. Soon after her arrival, she became active in the fledgling New York Mathematical Society. In 1892, she became one of the founding members of the AMS. In 1906, she became its first female vice president, almost seventy years before Mary Gray became the second.

Scott came from a line of innovators. Both her father (the Rev. Caleb Scott) and her father's father (the Rev. Walter Scott) were Congregational ministers, called "nonconformist" because they did not belong to the established Church of England. Education for women was an active concern of the British Congregational Church in the latter nineteenth century. When Caleb Scott's gifted daughter displayed a desire to study mathematics seriously, he did his best to provide her with competent home tutors, since there were few secondary schools open to women. In 1876, Charlotte won a scholarship to Girton College[24] in Cambridge, the first women's college in England and then only seven years old.

Scott first received widespread public attention in 1880 when she obtained special permission to take the Tripos Exams, the final undergraduate exams at the University of Cambridge. She tied for eighth place in mathematics and thus, had she been a man, would have received the title of Eighth Wrangler. As it was, she was not allowed to be present, or even officially mentioned, at the commencement ceremony. The young men were so incensed at this injustice that when the name of the official Eighth Wrangler was called, the hall rang with shouts of "Scott of Girton! Scott of Girton!"

This was a welcome change from former publicity for the college. A few years earlier Cambridge society had been horrified by a rumor that two Girton students had been riding unchaperoned in a dogcart with a clergyman. Actually their chaperone, dressed in mourning with a stiff black hat, had been mistaken for a clergyman. Nonetheless, the rumor did much damage to the status of Girton College women.

Thus the Girton community did its best to honor Scott: she was led in to a celebration to the tune of "See the Conquering Hero Comes" and crowned with laurels after an ode in her honor was read. The February 7, 1880, issue of *Punch* pleaded her case. Also, a petition was circulated throughout England asking that women be admitted to the Cambridge examinations as a right, not a special privilege, and that they be permitted to receive Cambridge degrees corresponding to their performance on the exams. Over 8000 signatures were collected in three months. As a result, the following year women were granted the right to be admitted to the Cambridge honors exams and to have their names announced along with those of their male competitors. Thus they would have public recognition of their intellectual achievements, an important asset in seeking employment.

Nevertheless, Cambridge did not grant degrees to women until 1948. Meanwhile, women could study there and receive degrees from the University of London by taking "external examinations." In this fashion, Scott received her bachelor's degree in 1882 and her doctorate in 1885 from the University of London for work done at Girton. In 1882 she joined the Girton College faculty as a resident lecturer in mathematics,

---

24. Girton College was founded as Hitchin College, which opened in Benslow House, Hitchin, in 1869. In 1872 Hitchin College was renamed Girton College, the present site in Cambridge was purchased, and an association was formed to increase ties with the University of Cambridge. In 1873 the college moved to its current premises in Cambridge. In 1948 it finally received the status of college of the university.

teaching undergraduates as she pursued her graduate studies. Girton was still very small; she was the third member of the staff.

In 1885, the first faculty of Bryn Mawr was not much larger. It had seven members, including another recent doctoral recipient—Woodrow Wilson. But Bryn Mawr had a graduate school and was the first women's college in the United States to grant the Ph.D. degree. Scott was the first chair of the mathematics department at Bryn Mawr College and held the position from 1885 until she retired in 1925. She accepted the challenge of training young, superior mathematicians with zest. Every two or three years from 1894 until her retirement, a young woman earned a doctorate under Scott's supervision. Each of these students spent at least one year of her graduate program at some mathematical center in Europe due to Scott's skill in obtaining grants for them and because of Scott's dedication to making the U.S. mathematical community equal to any in the world.

Isabel Maddison, who received her doctorate under Scott's supervision in 1896 and later joined the mathematics faculty at Bryn Mawr, wrote: "Professor Scott was an extraordinarily good teacher. She has the rare gift of lucid explanation combined with an intuitive perception of just what the student could grasp so that she never bored by being too easy or discouraged by being too difficult. Nor did she spare any effort to help a stupid student who really tried, though she was ruthless with the lazy or casual."[25]

There were, of course, students with reservations. A copybook contains the following ditty: "S is for Scott/Superior Scott/She is kind in the main/If you have any brain/But when you have not—/Superior Scott." In contrast, Scott's letters reveal sympathy for the problems of students. One, lamenting that she cannot recommend a student for further graduate study says, ". . . in every particular save the one essential, that of capacity, she is all that I could possibly wish." Another pleads for seven pages on behalf of a student abruptly dismissed because of a serious illness, arguing that her physical infirmity makes intellectual work all the more important and that it is cruel to heap one tragedy upon another.

She also produced a continual stream of research papers. Her specialty was the study of specific algebraic curves of degree higher than two. According to F. S. Macaulay, "A favorite topic of Miss Scott's was higher singularities, on which she wrote several papers. The subject is abstruse, and not widely and completely known. . . ."[26]

We may guess that Charlotte Scott found her experience in the New World rather lonely. She retained her church membership in England for

25. Isabel Maddison, "Charlotte Angas Scott: An Appreciation," *Bryn Mawr Alumni Bulletin* 12(1932), 10.

26. Readers who want further information about her research may find Macaulay's paper on pages 230–240 in the July 1932 issue of the *Journal of the London Mathematical Society*.

over a decade after she came to America. When Scott moved into a house in 1897, she persuaded a cousin to cross the Atlantic and take care of the house for her; the cousin presumably also provided companionship. When Scott retired, she immediately bought a house in Cambridge, never to return to the United States. A biography of Martha Carey Thomas, first dean and second president of Bryn Mawr College, and only one year older than Scott, reports that Scott found it "a source of bewilderment and some resentment" that she was not encouraged to drop in at Thomas's office for "friendly gossip."[27] Their letters reflect a noticeable lack of warmth. Scott's last doctoral student, Marguerite Lehr, who also had a distinguished career on the Bryn Mawr mathematics faculty, reported that "she was always extremely kind to me, but I never would have dreamed of asking her anything personal."

In 1904 Scott had a severe attack of rheumatoid arthritis, and from then on her unpredictable health and increasing deafness were extreme frustrations. However, Lehr remembers that in her last years at Bryn Mawr she could lecture "perfectly well" despite being almost totally deaf. She depended on graduate students to answer the undergraduates' questions.

Her doctor recommended increased out-of-door exercise. This was a novel idea to women of her generation, but even at Girton she had helped introduce lawn tennis for women. Now in her middle years, she took golf lessons and became a "very fair player."[28] She also turned to the traditional British activity of gardening. Lehr wrote, "For many . . . Scott's name will bring a vivid picture of a garden, brought year after year, unbelievably, to greater beauty. . . ."[29] She even won a prize for a new variety of chrysanthemum she developed.

Some historians of Bryn Mawr College have said that her influence on that institution is second only to that of Thomas. Scott served on innumerable committees and boards. She was involved in such varied matters as faculty selection, student admission standards, arrangements for the entrance exams, mathematical prerequisites, course offerings, allotment and size of library holdings, decisions regarding public health and student social behavior, and a heated debate as to exactly where a foot path should cross the campus. She argued especially firmly for funding an ample mathematics library with a steady supply of current journals.

She also played an important role in the growth of the American mathematical community. She made pilgrimages to Europe almost yearly, attending international mathematical meetings, placing her graduate students, and persuading leading mathematicians to visit (and sometimes

27. Edith Finch, *Carey Thomas of Bryn Mawr* (New York: Harper, 1947), 174.

28. Maddison, "Scott," 9.

29. Marguerite Lehr, "Charlotte Angas Scott: An Appreciation," *Bryn Mawr Alumni Bulletin* 12(1932), 12.

move to) the United States. She recruited a second mathematician from Cambridge for the Bryn Mawr faculty in 1887 and a third in 1903. In 1900 she attended the International Congress of Mathematicians in Paris and wrote a twenty-two-page report that appeared in the AMS *Bulletin*.[30] She persuaded a young couple, Bertrand and Alys Russell, then not widely known, to visit the United States in 1896 and to spend a few weeks lecturing at Bryn Mawr. "With their markedly unconventional statements, radical theories, and probing questions, they set the college, and in particular, the trustees, by the ears."[31]

In 1922 her former students arranged a dinner in her honor on the Bryn Mawr campus. Over 200 people attended, many traveling long distances to give a brief speech of recognition. Alfred North Whitehead made a special trip to the United States to attend. A short quote from his featured speech provides an appropriate conclusion to this story:

> A friendship of peoples is the outcome of personal relations. A life's work such as that of Professor Charlotte Angas Scott is worth more to the world than many anxious efforts of diplomatists. She is a great example of the universal brotherhood of civilization.[32]

Sonia Kovalevsky

LINDA KEEN

Sonia Kovalevsky, 1850–1891. Based on "Sonya Kowaleskaya—Her Life and Work," a talk at the panel "History of Women in Mathematics," moderated by Lenore Blum, 1976 Summer JMM, as reported in the AWM *Newsletter* 7(2), 1977, 2–6. Keen was then a professor of mathematics, Lehman College and Graduate Center, City University of New York, where she is now a professor of mathematics and computer science.

Sonia Kovalevsky was born in Russia in 1850. During her teens she began to develop various intellectual interests, but the idea of a learned woman was abhorrent to her father. When he was told she was particularly good in algebra, her algebra book was taken away. However, she was able to reacquire the book and continued to read it secretly at night.

Both she and her older sister very much desired to obtain university educations for themselves but felt it was impossible; Russian universities

---

30. Charlotte Angas Scott, "The International Congress of Mathematicians in Paris," *Bulletin of the American Mathematical Society* 7(1900), 57–79.

31. Charlotte Angas Scott, letter to M. Carey Thomas, "M. Carey Thomas Papers," Bryn Mawr College Archives, Bryn Mawr, Pennsylvania.

32. Emily James Putnam, "Celebration in Honor of Professor Scott," *Bryn Mawr Alumni Bulletin* 2(5), 1922, 13.

were completely closed to women, and the only way a woman could study was to go abroad. On a visit to Petersburg, Sonia's sister Anyuta heard of a way to do so. They would try to arrange, with the help of some intellectual friends, "fictitious marriages," that is, marriages in name only that would, however, free them from legal ties to their parents. In 1868, Anyuta was introduced to a young geology student, Vladimir Kovalevsky, who was willing to marry her. However, that year he married Sonia instead.

When they arrived in Heidelberg, Sonia was shocked to discover that it was not easy for women to study at the university and that, in fact, they were barred from matriculating. After enormous effort, she received permission to audit lectures without matriculating. The next three semesters were among the happiest of her life. During this period Sonia decided that she wanted to devote herself entirely to mathematics, and that there was only one man with whom she wanted to study, Karl Weierstrass at the University of Berlin. Weierstrass took her as a private pupil. In 1874 Sonia finished three articles, any one of which, Weierstrass claimed, would make a fine doctoral dissertation. The problem was to obtain a degree for her. Eventually Göttingen agreed to award her a doctorate in absentia.

After receiving her degree, Sonia and her husband decided to go back to Russia to live and to turn their fictitious marriage into a real one. Her daughter was born in October 1878. Sonia became unsatisfied and unhappy and decided to return to mathematics. Her husband committed suicide, after which Mittag-Leffler helped her obtain a position at the University of Stockholm in 1883. She was made professor for life in 1889. She died in 1891 and was buried in Sweden. She was finally honored by the Russian government on the centenary of her birth.

References

1. K. R. Biermann, "Karl Weierstrass," *Journal für die reine und angewandte Mathematik* 223(1966), 208–210.
2. R. Cooke, *The Mathematics of Sonya Kovalevskaya* (New York: Springer-Verlag, 1984).
3. A. Hibner Koblitz, *A Convergence of Lives. Sofia Kovalevskaia: Scientist, Writer, Revolutionary* (Boston: Birkhäuser, 1983; 2nd ed., New Brunswick: Rutgers University Press, 1993).
4. D. H. Kennedy, *Little Sparrow* (Athens: Ohio University Press, 1983).
5. S. Kovalevsky, "Zur Theorie der partiellen Differentialgleichungen," *Journal für die reine und angewandte Mathematik* 80(1875), 1–32.
6. ———, "Über die Reduction einer bestimmten Klasse von Abel'scher Integrale 3-en Ranges auf elliptische Integrale," *Acta Mathematica* 4(1884), 393–414.
7. ———, "Sur le problème de la rotation d'un corps solide autour d'un point fixe," *Acta Mathematica* 12(1889), 177–232.
8. ———, *A Russian Childhood*, translated, edited, and introduced by Beatrice Stillman (New York: Springer-Verlag, 1978).

9. P. Ja. Kochina, ed., *Briefe von Karl Weierstrass an Sofie Kowalewskaja 1871–1891* (Moscow: Nauka, 1973).

10. ———, *Love and Mathematics: Sofya Kovalevskaya*, translated from the Russian by Michael Burov (Moscow: Mir Publishers, 1985), revised from the 1981 Russian edition.

11. G. Mittag-Leffler, "Sophie Kovalevsky, notice biographique," *Acta Mathematica* 16(1893), 385–392.

12. ———, "Weierstrass et Sonja Kowalewsky," *Acta Mathematica* 39(1923), 133–198.

13. J. Spicci, *Beyond the Limit: The Dream of Sofya Kovalevskaya* (New York: Forge Books, 2002). Historical novel.

14. B. Stillman, "Sofya Kovalevskaya: Growing up in the Sixties," *Russian Literature Triquarterly* 9(1974), 276–302.

15. V. Volterra, "Sur les vibrations lumineuses dans les milieux biréfringents," *Acta Mathetica* 16(1892), 153–206.

Elizaveta Fedorovna Litvinova

ANN HIBNER KOBLITZ

Based on "Elizaveta Fedorovna Litvinova (1845–1919)—Russian Mathematician and Pedagogue," AWM *Newsletter* 14(1), 1984, 13–17. Then a lecturer in the Honors Program, University of Washington, Ann Hibner Koblitz is now a professor of women's studies, Arizona State University.

Elizaveta Fedorovna Litvinova was a member of a group of women who were inspired by nihilism, a radical sociopolitical movement that sprang up in the 1860s in St. Petersburg and other large cities of the Russian Empire. The early nihilist philosophy (not to be confused with later more extreme movements bearing the same name) taught that the natural sciences were a progressive force in society. The sciences improved people's lives and by their very nature contributed to the fight against backwardness and superstition. The early nihilists believed that intensive study of the natural sciences would not only help the material lot of the Russian peasant masses, but also hasten along the day of (peaceful) social revolution, which they were convinced was not far off.

In addition to this faith in the power of the natural sciences, the nihilists were staunch advocates of the equality of women. They felt that women of the educated classes and the nobility had a right, and even a duty, to develop their minds and strive toward a career in the natural sciences or medicine. Moreover, nihilist men believed that they had a moral obligation to help women realize their full intellectual potential.

Given this sociopolitical climate, it is not surprising that Russian women in significant numbers, including Litvinova, wanted to enter the sciences and medicine. She became, after Sofia Kovalevskaia, the second woman in Europe to obtain her doctorate in mathematics. Her life unfolded along different lines from that of Kovalevskaia, however. Litvinova was forced by circumstances and the tsarist government to work in positions that did not use her graduate mathematical training for the whole of her life. Her story is interesting for the glimpses it affords us into the problems faced by the pioneer women scholars of the last century.

Elizaveta Fedorovna Ivashkina was born in 1845 to a landowning family in the Tula region of Russia. She was lucky enough to be sent to one of the few women's gymnasia in Russia, the Marinskaia in St. Petersburg. These institutions were on a far lower level than boys' gymnasia, but at least provided some education other than sewing and deportment. Moreover, the teachers in the Marinskaia were known for their competence and (some of them) for their progressive views on women's education.

In St. Petersburg, Elizaveta soon fell under the influence of nihilism. She became involved in discussion circles of revolutionary young people, wrote radical poetry, and decided to pursue advanced studies. She encountered many obstacles to her desire: the strenuous objections of her parents, her poor preparation, and the difficulty of being admitted to the university. She was not alone, however; fortunately, the informal nihilist women's education network helped her during this time.

It is necessary to explain here that universities in Russia, as in all of Europe, did not allow women to enroll. But from the early 1860s, Russian women had hopes that their universities would soon allow them to matriculate and prepared themselves for university study as if this hope were a certainty. They formed groups of prospective university students, enlisted the aid of sympathetic professors, and held preparatory classes in the apartments of wealthy supporters.

By the mid-1860s, some of the more educationally advanced women decided they could not wait for the Russian universities to open and resolved to try their luck abroad. To a large extent, it was this generation of Russian women who pioneered higher education for women in continental Europe. They were among the first official students in Zürich, Geneva, Bern, Heidelberg, and Paris. They were particularly successful in Zürich, where a Russian, Nadezhda Suslova, was the first woman in Europe to obtain her medical degree. Zürich became the mecca for Russian women interested in higher education, and in the late 1860s and early 1870s there was a considerable colony of women there.

Elizaveta looked toward this colony with yearning, as she writes in her memoirs of her student years, but at first she could do nothing. In 1856, she had married a doctor Litvinov. Apparently he was willing for her to continue her studies in the capital, but unlike some husbands of the sixties (including Kovalevskaia's), he could not see his way clear to permit Litvinova to go abroad. Litvinova worked as best she could in St. Petersburg and made considerable progress toward preparing herself for the university. Yet without a formal curriculum, and without the financial resources to make private full-time study possible, her progress, and that of her friends, was not nearly so steady or so quick as they would have liked.

Litvinova's description of her student years in St. Petersburg and then abroad is fascinating and poignant. Most of the women did not have much money. For the most part, they were able to convince nihilist-oriented professors to donate their time; the professors, especially in the natural sciences, were not much older than their students and were firm believers in the equality of women. Still, there were other expenses: books, laboratory supplies, and so on. Transportation, too, was a problem, since the physics lectures might be located in the center of town, while the biology laboratory might be two hours' walk away. Underlying the women's thoughts was the uncomfortable knowledge that they had so much to make up, so much to learn before they could hope to benefit from the university. Litvinova remarked, "Quick success in our studies gave us joy. But solving a problem that would have been easy for a pupil of the fifth class of a boy's gymnasium made us realize how far we still were from the study of real science."[33]

For Litvinova, the thought of Sofia Kovalevskaia, at that time studying mathematics and the sciences at Heidelberg, ". . . was for me as for every young woman who wanted to study, a bright point toward which our eyes turned." Rumors of Kovalevskaia's success focused Litvinova's attention on mathematics. Another factor in her decision to specialize in mathematics over the other sciences was mathematician A. N. Strannoliubskii, one of the most faithful and interested tutors for the young women.

Strannoliubskii did not make any particular mark on the history of mathematics. Rather, his forte lay in teaching, in instilling a love for mathematics in his pupils, and in propagandizing the discipline. Strannoliubskii was a typical "man of the sixties," as the early nihilists sometimes called themselves. He sacrificed much of his spare time to prepare young women for the university and established free schools for workers in St. Petersburg as well. Moreover, he used the story of his most successful pupil, Sofia Kovalevskaia, to encourage his women students. Strannoliubskii convinced Litvinova that with enough work, she too would be able to earn a degree in higher mathematics.

By 1870 or so, Litvinova was sufficiently well-prepared to enter university and had acquired a certificate of competency equivalent to that

33. "E. El" [Litvinova's pseudonym], "Iz vremen moego studenchestva," *Zhenskoe delo* 4(1899), 34–35.

received by graduates of a boys' gymnasium. But she seems to have viewed her marriage as an obstacle to her further education. It is unclear what happened next: perhaps her husband died, or, less likely, perhaps her husband left her. In any case, she declares in her memoirs that "fate itself" freed her from her marital duties and made it possible for her to leave St. Petersburg in 1872 for the women's student colony in Zürich.

Life in Zürich was wonderfully exciting and intellectually stimulating for Litvinova and the other Russian women, but there were many problems. As always, financial difficulties loomed in the students' minds. They tried to minimize these by forming cooperatives for lodging, eating, and studying. Some of the poorer ones are even reputed to have shared winter coats and shoes, by arranging their class schedules so that only one of them had to be outside at any given time!

Another problem was the attitude of the Swiss citizens toward the Russian women. The Swiss could not understand the eager desire of the Russians for education, especially for education in such "unfeminine" fields as the natural sciences and medicine. Moreover, the Swiss disliked the politics of the majority of the Russian students. The amount of time most of them spent discussing the form of the revolution to come could not win favor with the Swiss burghers, and the comradely, casual relations of the Russian women with their student countrymen scandalized them. For them, the Russians' tendency to talk in each other's lodgings late into the night, go about in mixed-sex groups, and treat each other with informality meant that the Russian women were little better than prostitutes. The women consequently found it difficult to obtain rooms, and they were discriminated against in the shops and markets as well.[34]

Litvinova's lot was even harder than most. She chose to pursue her studies at the Polytechnic Institute rather than at the university, where there were far more women students. In many of her lectures, she was the only woman in a class of 150, and she was afraid to raise her eyes or look in anyone's direction too long for fear they would get the wrong impression of her.[35] (Because the Swiss had such prejudices against the Russian women, it was all too easy to give them "the wrong impression.")

Fortunately, Litvinova's instructors were cordial to her. She studied under the French professor Méquet and the well-known analyst Hermann Schwarz. Schwarz gave her occasional tutoring sessions at his home and sometimes invited her to tea or for the evening with his family. One summer, he even repeated just for her a course of lectures he had given before she arrived in Zürich.

34. For descriptions of the trials of the Russian women students see J. M. Meijer, *Knowledge and Revolution* (Assen: Van Gorcum, 1955); and Koblitz, *Science, Women and Revolution in Russia* (Amsterdam: Harwood, 2000).

35. "E. El," "Studenchestva," 40.

Litvinova settled down happily in Zürich and looked forward to completing her studies within four years. But the tsarist government was beginning to become uneasy about the Zürich student colony, considering it to be a hotbed of revolutionary ideas with the women contributing more than their share to the political climate. In June 1873, the Russian authorities issued a decree stating that all Russian women studying in Zürich had to return to Russia by the first of the following year whether or not they had finished their training. If they refused, the decree warned, they ran the risk of being forbidden entry to any Russian institutions of higher education that might open to women in the future, and they would be banned from all licensing exams and civil service posts.

For most of the Russian women, this decree marked the end of their plans for a scientific career. They were afraid to disobey the ban and took hope from the promise of women's universities implied in the document. Litvinova and several others, however, could not bear to have their studies interrupted. With the encouragement of her professors, Litvinova defied the ban and remained in Zürich until she received her baccalaureate degree in 1876. She stayed in Switzerland two more years to complete her graduate work, receiving her doctoral degree in function theory from Bern University in 1878.

Litvinova returned to Russia in 1878 and found, to her dismay, that the tsarist government had meant every word of its threat. She was not allowed to sit for the licensing exams for teachers in the higher grades of the gymnasium or to be hired as a full-time instructor in any state-licensed institution. Moreover, she was forbidden to take the exam for the Russian *magister* degree, which would have enabled her to teach at the university level. She was effectively banned from any position commensurate with her training. To complete her humiliation, Litvinova had to watch less qualified women who had obeyed the decree succeed where she could not. The Higher Women's Courses, which opened in St. Petersburg in 1878, hired several women baccalaureates from Zürich as laboratory instructors and classroom supervisors, although none were given professorships.[36]

Litvinova took a post as a teacher in the lower classes of a gymnasium. She was paid only by the hour, however, and did not have the rights to pension and vacations to which she would have been entitled had the government allowed her to take the licensing exam. Not until nine years later, after repeated pleas on the part of herself and her superiors, was she permitted by the Ministry of Education to teach in the older classes of the gymnasium. She thus became the first woman in Russia to teach

---

36. L. N. Gratsianskaia, "Elizaveta Fedorovna Litvinova" [in Russian], *Matematika v shkole* 4(1953), 64–67.

mathematics at that level, although because of her earlier defiance she was never given pension rights.

Litvinova did not pursue her research once she returned to Russia. Indeed, as a partially blacklisted teacher who had to earn her salary by the hour, she could not have had energy for original mathematical work. Yet, during her thirty-five years of teaching, she wrote biographies of several mathematicians, including Kovalevskaia and Lobachevskii, and her contributions to Russian mathematical pedagogy were considerable. She published over seventy articles on the philosophy and practice of teaching mathematics and was respected as one of the foremost pedagogues in Russia. Her methods were surprisingly modern; she emphasized alternative approaches to proofs and the use of word problems to stimulate clear thinking.[37]

In addition, Litvinova was a constant source of inspiration and encouragement to her women students. Several of them went on to become scientists in their own right. The crystallographer and applied mathematician Varvara Tarnovskaia, for example, remembered with gratitude the support and training she received from Litvinova.[38]

Some of Litvinova's students began mathematical studies under her direction, but were sidetracked along the way. The influential educator Nadezhda Krupskaia, for example, majored in mathematics at the Higher Women's Courses before her arrests and activities as an agitator cut short her academic career. Krupskaia remembered Litvinova with great affection and respect. She traced her interest in pedagogy to the influence of Litvinova and cited the latter in her own pedagogical writings.

Litvinova was also active in the European women's movement. In the 1890s and 1900s, the feminist *Bulletin de l'Union universelle des Femmes* contained articles with the initials "E.L." Since the pieces deal with Russian affairs, and one of them is an obituary of Kovalevskaia, it is safe to assume Litvinova wrote them. Certainly we know that Litvinova was one of four Russian delegates to the 1897 International Women's Congress in Brussels. And her biography of Kovalevskaia contains some beautifully sensitive lines about woman's internalization of feelings of inferiority which prove that she was supremely conscious of the multitude of problems faced by pioneering women.[39]

Not much is known about Litvinova's life after she retired from teaching. She apparently lived with her sister in the country during the turmoil of the Russian Revolution. She died in 1919, at the age of seventy-four.[40]

37. The journal *Zhenskoe obrazovanie*, later *Obrazovanie*, contains most of these articles.
38. Gratsianskaia, "Litvinova," 66.
39. Elizaveta Litvinova, *S. V. Kovalevskaia* (St. Petersburg: P. P. Soikin, 1894), 58.
40. Gratsianskaia, "Litvinova," 67.

Ada Byron Lovelace

DANA ANGLUIN

Ada Byron Lovelace, 1815–1852. Based on "Lady Lovelace and the Analytical Engine," AWM *Newsletter* 6(1), January 1976, 5–10, and 6(2), February 1976, 6–8. Then a graduate student in the Department of Computer Science, University of California at Berkeley, Angluin is now a professor of computer science, Yale University.

In 1843, about a century before the earliest digital computers were built, a paper that described in some detail the design of a general-purpose digital computer and gave several examples of programs for it was published in Taylor's *Scientific Memoirs*. The design proposed is not very different from today's digital computers, and it demonstrated a rather sophisticated grasp of the concepts and problems of computing. One of the principal characters in the story of this development was an Englishwoman, Augusta Ada Byron King, Countess of Lovelace, daughter of the poet Byron. Lady Lovelace has been called "the first computer programmer."

Early Years

Our story begins when Lord Byron proposed marriage to Annabella Millbanke, a somewhat spoiled and very moralistic young woman. During their courtship, one of Byron's nicknames for her was "The Princess of Parallelograms" [2]. She was studying plane geometry, but the name is suggestive of her rectitude as well. The two were married in 1814, and their daughter Augusta Ada was born the following year.

Soon after Ada's birth, her parents separated. When Ada was about five years old, Byron sent some of his hair in a locket for the child to wear and received a miniature of her from her mother. Also at about that age, Ada asked whether a Papa was the same thing as a Grandpapa, and whether God be a man or a lady.

In 1823, Byron went to Greece to join the battle for Greek independence. From Greece he wrote asking that Annabella send "some account of Ada's disposition, habits, studies, moral tendencies, and temper, as well as her personal appearance. . . . I hope the Gods have made her anything save *poetical*—it is enough to have one such fool in the family." In reply, her mother sent a description of the eight-year-old Ada: "Her prevailing characteristic is cheerfulness and good temper. Observation. Not devoid of imagination, but it is chiefly exercised in connection with her mechanical ingenuity—the manufacture of ships, boats, etc. . . . Tall and robust" and enclosed a profile of the child. In 1824, Byron died in Greece of a fever. On his desk lay an unfinished letter thanking Annabella for the description. Apparently the child was immediately told of her father's death, for her mother writes: "Ada shed large tears. . . . It is a great comfort to me that I have never had to give her a painful impression of her father."

Ada's education was not neglected. She and her mother spent a year on the Continent after Byron's death. When Ada was fourteen, her mother engaged a tutor at 300 pounds a year to teach Ada mathematics. Annabella described Ada to the tutor: "There are no weeds in her mind; it has to be planted. Her greatest defect is want of order, which mathematics will remedy. She has taught herself part of Paisley's *Geometry*, which she likes particularly." Ada also studied music. There is an anecdote to the effect that she took to walking round and round the billiard table while she practiced the violin, because her constant practicing had deprived her of necessary physical exercise. As was the convention, Ada was presented at Court when she was seventeen, and participated in two London "seasons," going to balls, parties, dinners, the opera, the theater.

It was about this time that Ada went to one of Charles Babbage's Saturday evening parties with Sophia De Morgan, wife of the English mathematician Augustus De Morgan. Charles Babbage had invented a calculating machine that he called the Difference Engine. He kept a small model of it in his house and exhibited it to guests.

Sophia De Morgan writes of seeing the Difference Engine: "While other visitors gazed at the working of this beautiful instrument with the sort of expression, and I dare say the sort of feeling, that some savages are said to have shown on first seeing a looking glass or hearing a gun . . . Miss Byron, young as she was, understood its workings, and saw the great beauty of the invention." In 1834, Ada was delighted to go to the Mechanics' Institute to hear Dionysius Lardner's lectures on Babbage's Difference Engine. She paid frequent visits to Babbage's house in the company of Sophia De Morgan or of Mary Fairfax Somerville, a famous mathematician in her own right.

Meanwhile, Ada's foray into the marriage market had succeeded. Her marriage when she was nineteen to Sir William King (who three years later became Lord Lovelace) and the bearing of three children put a temporary stop to Ada's active interest in Babbage's work, though she and her husband had Babbage to visit as often as they could tear him from his work.

Introduction to Babbage and His Work

Babbage was born in 1792, and so was nearly of an age with Ada's father, Lord Byron. Babbage studied mathematics at Cambridge [1]. In 1820, he and his lifelong friend John Herschel were appointed to carry out some calculations for the Royal Astronomical Society. During the tedious process of verifying the error-ridden calculations of the (human) computers[41] they had engaged for the numerical work, Babbage exclaimed, "I wish to God these calculations had been executed by steam!" to which Herschel replied, "It is quite possible." Babbage became excited about the idea of making a machine to calculate mathematical tables by the method of differences. He designed and constructed a small model, and eventually got a model with six figures and two orders of differences to calculate a few simple mathematical tables.

The makers of mathematical tables had long known the mathematical principle involved. Suppose we take a polynomial, say x^2, and begin to tabulate its values for a stretch, say beginning at 17. So we have 289, 324, 361, 400. If we look at the sequence we have constructed a little more closely, we might calculate how much each value of x^2 exceeds its predecessor, which would yield the sequence Δx^2: 35, 37, 39. These numbers in turn seem to be increasing by 2 at each step, so we might conjecture that the next value of Δx^2 is 41, and, in order for that to be true, the next value of x^2 must be 441, which is indeed the correct value of 21^2. This is not fortuitous, as it is an instance of a general theorem on polynomials, that the nth difference of a polynomial sequence of degree n is a constant sequence. This gives us a method of extending our sequence indefinitely at a cost of two additions per new entry.

Mechanical calculating machines capable of at least addition and subtraction, sometimes multiplication and division, had been built before Babbage's time—the mathematicians Pascal and Leibnitz had each built one in the seventeenth century. Babbage used his idea of combining the mechanisms found in these adding machines with the principle of constant differences to produce a design for what he called the Difference Engine.

For example, we might use three sets of four discs each, where each disc may independently have one of ten positions (representing the digits 0 to 9). These are used to hold the three values that we need to calculate the next value of x^2:

A:	0	4	4	1		0	4	4	1		0	4	8	4
B:	0	0	4	1	$\Rightarrow$	0	0	4	3	$\Rightarrow$	0	0	4	3
C:	0	0	0	2		0	0	0	2		0	0	0	2

---

41. Originally this term referred to people whose job was to do computations, not to mechanical devices.

These are to be connected by adding machinery and sequencing mechanisms so that at the turning of a crank (say), the number recorded in register C above would be added to that in register B; then the contents of register B would be added to register A. Babbage arranged for the machine to print the result and then automatically to cycle back to produce the next value of x^2.

In 1822, at the age of thirty, Babbage read a paper to the Royal Astronomical Society describing his idea and the experiments he had made. The government decided to advance him money for the construction of a full-scale version of his Difference Engine. The construction of the machine was a formidable task. The art of precision machining of metal parts (as for the steam engine) was less than half a century old, and Babbage required thousands of identical, precisely machined parts in the construction of his machine. Even worse, he constantly scrapped parts as he improved his design. His chief engineer eventually quit and could not be induced to return. The government began to doubt that the machine would ever be completed and, encouraged by Babbage's rather powerful scientific enemies, withdrew its support about twenty years after Babbage's first paper on the idea.

Even as work on the Difference Engine was coming to a halt, Babbage began to conceive of the Analytical Engine, a generalization of his original idea that would not merely tabulate polynomials of some fixed maximum degree, but would be able to execute *any* analytical calculation. As an example, he considered a sequence developed according to the rule $\Delta^2 u_n = $ last digit u_{n+1}, namely:

u	2		2		4		10		16		28		48		76		110
Δu		0	$\downarrow$	2	$\downarrow$	6	$\downarrow$	6	$\downarrow$	12	$\downarrow$	20	$\downarrow$	28	$\downarrow$	34	$\downarrow$
$\Delta^2 u$		2		4		0		6		8		8		6		0

His design used a cascade of three registers as before, provided that the last digit of the result formed by addition in register A is transferred to register C in preparation for the next cycle. He called this cyclical action of the registers "the Engine eating its own tail."

The Collaboration

After the birth of her third child, Lovelace was becoming impatient to pursue her mathematical education again. In one of her notes to Babbage, she asked him to propose a tutor for her; he knew of none. She gradually came to the idea of helping Babbage in his work on the Analytical Engine and began studying the method of finite differences because she knew that it was related to the work. She wrote expressing her anticipation of being of use to Babbage in his work, saying: "I intend to make such arrangements in Town as will secure me a couple of hours daily (with very few exceptions) for my studies."

Even that "couple of hours daily" was probably difficult for the mother of three small children to arrange. From Mary, Countess of Lovelace, the wife of her younger son, we have this account: "To these children she was always a very dear and sacred memory, and in their hearts a halo of romance forever hung about her name. . . . Little Annabella and Ralph King clearly found nothing wanting. . . . But she led a curiously detached existence, caring little for ordinary society, or even for the more satisfying pleasures and duties of country life. She was always absorbed in study."

The Analytical Engine was to be divided into a collection of registers called the "Store" and a separate mechanism capable of performing an arithmetic operation called the "Mill." An addressing mechanism under external control was to connect the inputs and output of the Mill with any of the registers in the Store, enabling the machine to combine the contents of any two registers in the Store and to deposit the result in any third such register.

The means adopted for directing the successive steps of the machine—for giving directions like "add the contents of registers 3 and 4 and place the result in register 10; now multiply the contents of registers 5 and 10 and place the result in register 11 . . ."—was a sequence of punched pasteboard cards. Such sequences of punched cards had already been used with great success (around 1800) in the Jacquard loom, to control which warp threads were to be lifted for each passage of the woof thread.

A feature of the Difference Engine allowed it to stop and ring a bell whenever it found a zero of the function being tabulated. In the Analytical Engine, this was expanded to the idea of a full conditional—alternative sets of controlling cards which were to come into play depending upon whether or not some result was zero. The necessity of repeating some block of cards over and over led to the idea of a "for-loop," that is, a block of instructions to be repeated under the control of a program-variable.

The English government scorned the new idea. However, Babbage was invited to Italy to present his idea on the Analytical Engine before a group of distinguished mathematicians in Turin. One of them, Luigi F. Menabrea, later to become the first prime minister of Italy, published "a lucid and admirable description" of Babbage's ideas in the *Bibliothèque Universelle de Genève* in October of 1842.

By the end of June 1843, Lovelace's studies had progressed to the point where she had translated into English the "Sketch of the Analytical Engine Invented by Charles Babbage" by Menabrea and had begun annotating it. Numerous letters passed between Lovelace and Babbage regarding revisions and corrections of her commentary, which came to be called the Notes and grew to be longer than Menabrea's original memoir [6].

Since the prevailing conventions made it "unfeminine" for any woman, least of all a countess, to sign any literary production, the question arose of how the translation and Notes were to be identified when published. Lovelace's husband suggested that she should at least affix her initials to the document. Babbage considered the Notes worthy of an original paper and pressed Lady Lovelace to publish them separately, but she would not default on what she considered to be her obligation to the editor of the journal that was to publish Menebrea's memoir. After some months of feverish activity, the translation and Notes were finally published, initialed only "A.A.L." [4].

The Notes

Menabrea's memoir and Lovelace's Notes are a remarkable document. Therein, a century before its time, is the concept of a general-purpose digital computer, developed to an amazing degree of sophistication. For many concepts from computer design or programming—memory, central processor, switch register, program library, for-loops, indexing, comments, program trace, coder, keypuncher, analysis of algorithms—it is possible to quote a passage of the memoir or the Notes to demonstrate their grasp of the idea. Menabrea and, to a greater extent, Lovelace wrote out some rather complex programs for the machine; these may have led to the development of some of the more subtle and powerful concepts, for example, looping and indexing.

Menabrea and Lovelace understood the capabilities of the machine in quite general terms. From the memoir: "   the cards are able to reproduce all the operations which intellect performs in order to attain a determinate result, if these operations are themselves capable of being precisely defined." From the Notes: "The engine can arrange and combine its numerical quantities exactly as if they were letters or any other general symbols; and in fact it might bring out its results in algebraical notation were provisions made accordingly." Elsewhere, Lovelace suggests that the Engine might be made to compose music if the fundamental relations of pitched sounds in the science of harmony and musical composition could be formulated precisely enough. The two thereby establish some claim to priority in formulating Church's Thesis, which posits a similar universality for another type of computing device.[42]

Shortly after the publication of the memoir and Notes, when Lovelace was twenty-eight, Augustus De Morgan (who had taught mathematics to Ada's mother when she was a girl and afterward gave Ada much help in mathematics) wrote a confidential letter to Lady Byron regarding her daughter, saying that: "The tract about Babbage's machine is a pretty thing enough, but I think I could produce a series of extracts, out of Lady Lovelace's first queries upon new subjects, which would make a mathematician see that it was no criterion of what might be expected from her." He said that she displayed the makings of "an original mathematical investigator, perhaps of first-rate eminence." Nonetheless, he had never told Lovelace his true opinion of her abilities, saying: "I always feared that it might promote an application (to mathematics) which might be injurious to a person whose bodily health is not strong" [3].

Later in the letter, De Morgan goes on to expound his idea of the connection of strength and mathematics:

> All women who have published mathematics hitherto have shown knowledge, and power of getting it, but no one, except perhaps (I speak doubtfully) Maria Agnesi, has wrestled with difficulties and shown a man's strength in getting over them. The reason is obvious: the very great tension of mind, which they require is beyond the strength of a woman's power of application. Lady L. has unquestionably as much power as would require all the strength of a man's constitution to bear the fatigue of thought to which it will unquestionably lead her.

Thus he restrained his enthusiasm for her efforts to comments like "very good" and "quite right."

Encouraged or not, Lovelace was full of ambitions and plans for the future. She wanted to see the Analytical Engine built, though by that time the English government had withdrawn all support for the Difference Engine and was not favorably disposed toward its inventor.

Babbage had at one time done work in probability and statistics, and it seems that he, and Lovelace and her husband, developed what they thought was an infallible scheme for betting on the horses and began trying it out. This enterprise was almost certainly undertaken with the intention of raising enough money to construct the Analytical Engine (Babbage had

---

42. Some have suggested that Lovelace was merely Babbage's scribe. John Walker, founder of Autodesk, Inc. and coauthor of AutoCAD, at his website The Analytical Engine emphatically disagrees: "This 1842 document is the definitive exposition of the Analytical Engine, which described many aspects of computer architecture and programming more than a hundred years before they were 'discovered' in the twentieth century. If you have ever doubted, even for a nanosecond, that Lady Ada was, indeed, the First Hacker, perusal of this document will demonstrate her primacy beyond a shadow of a doubt." John Walker, "The Analytical Engine: Table of Contents," Fourmilab Switzerland [online] [accessed 16 January 2003]. Available from World Wide Web: http://www.fourmilab.ch/babbage/contents.html.

considered and discarded other harebrained schemes for financing the Engine). The three lost heavily in their betting; Lovelace's continued gambling after the men stopped led to estrangements between her and both her mother and her husband.

To these difficulties was added a marked and progressive degeneration in Lovelace's health, eventually determined to be caused by cancer. She suffered considerable agony and died in November 1852 at the age of thirty-six. At her request, she was buried beside her father, Lord Byron.

The Analytical Engine was never built,[43] though Charles Babbage lived nearly twenty more years. Some parts of it were assembled under the direction of his son, Henry, who wrote a paper describing some details of its construction [6]. The Menabrea/Lovelace paper remains as almost the sole witness to the power and scope of the ideas of Babbage's Analytical Engine.

References

1. Charles Babbage, *Passages from the Life of a Philosopher* (London: Longmans, 1864). Babbage's autobiography and source for much of the information here.
2. *Byron's Letters and Journals,* vol. 2, *Famous in My Time (1810–1812),* ed. Leslie A. Marchand (London: John Murray, 1973–1982). U.S. publication: Cambridge, MA: Belknap/Harvard, 1994.
3. Ethel Colburn Mayne, *The Life of Lady Byron* (London: Constable, 1929). Contains material on Ada's childhood and De Morgan's letter.
4. L. F. Menabrea, of Turin, Officer of the Military Engineers, "Sketch of the Analytical Engine invented by Charles Babbage, Esq." translated with notes by Ada Byron Lovelace, in *Scientific Memoirs, Selections from The Transactions of Foreign Academies and Learned Societies and from Foreign Journals,* ed. Richard Taylor, Volume III, Article XXIX, 666–731 (London: 1843). Also available from the World Wide Web: http://www.fourmilab.ch/babbage/sketch.html [accessed 16 January 2003].
5. Philip and Emily Morrison, ed., *Charles Babbage and His Calculating Engines* (New York: Dover, 1961), with an introduction by the editors. Contains the Menabrea-Lovelace paper on the Analytical Engine, Dionysius Lardner's description of the Difference Engine, selections from Babbage's autobiography, and some other writings of Babbage and his son.
6. Maboth Moseley, *Irascible Genius: A Life of Charles Babbage, Inventor* (London: Hutchinson, 1964). A fifth of the book is devoted to Lovelace and was the source of much of the material in this note. Parts of the Lovelace-Babbage correspondence on the Notes are included.

43. At the Science Museum in London, it is now possible to see a Difference Engine and a computer printer built to Babbage's designs by a team of engineers, beginning in 1985. The Engine, weighing about two tons, was completed in 1991; the printer, designed to print tables computed by the Engine and weighing about two and a half tons, in 2000. No Analytical Engine has been built, but an emulator of such a machine may be found at John Walker's website, http://www.fourmilab.ch/babbage/contents.html [accessed 16 January 2003].

Sophie Germain

MARY W. GRAY

Sophie Germain, 1776–1831. Based on "Sophie Germain, A Bicentennial Appreciation," a talk at the panel "History of Women in Mathematics," moderated by Lenore Blum, 1976 Summer JMM, as reported in the AWM *Newsletter* 6(6), 1976, 10–14. Gray is chair and professor of mathematics and statistics, American University.

I want to speak of the bicentennial of Sophie Germain, mathematician and philosopher. The poignant story that brought Germain to my attention was in an 1879 edition of her life and works intended as a centennial remembrance. Her biographer, Hippolyte Stupuy, tells of going to the Père Lachaise cemetery and finding, near the well-frequented graves of philosopher Auguste Comte and a popular actress, that of Sophie Germain, the simple tombstone crumbling and in ruins, but a chestnut tree flourishing, perhaps indicating a higher power's making up for human neglect. A later edition of Stupuy's work, 1896, reports that the tombstone was restored, but I spent most of one wet and cold January afternoon on my knees among decaying leaves and moss-covered crumbling gravestones unable to locate it. As a result of the centennial edition of her life and works, both a street and a school for girls were named after her; the Paris City Council specially commissioned a bust of Germain for the school. Also, a plaque, which can still be seen, was placed on the house at 13 rue de Savoie in which she died on June 27, 1831.

Marie-Sophie Germain was born April 1, 1776, in a house on rue St. Denis in Paris. Of Sophie's mother, we know little beyond her name, Marie-Madeleine Gruguelin. Her father, Ambroise-François, was, as his father before him, a goldsmith, and a prosperous member of the bourgeoisie he was elected to represent in the États-Généraux in 1789. It is assumed that the Germain home was a meeting place for those interested in liberal reforms and that the young Sophie must have heard exciting political and philosophical discussions swirling around her.

M. Germain's career in what became the Constitutional Assembly was not particularly distinguished. A record cites only two speeches by him, October 8, 1790, and May 5, 1791. His concerns were those of a merchant; he complained against banks and speculators. After the Constitution, his name disappears from the Assembly records. Apparently the radical trend passed him by. Notwithstanding his earlier railings, he seems later to have become a director of a bank and adroitly to have survived the shifting trends of French politics to die at the age of ninety-five in 1821. Clearly, the family continued to be sufficiently prosperous that

Sophie could devote herself to research and writing without worrying about her means of support.

Retreating from the turmoil in the Parisian streets to her father's library during the Reign of Terror, Sophie resolved to find a serious occupation for herself. In spite of the barbs of Molière's play *La femme savante*, there was in France something of a tradition of women of a certain class devoting themselves to serious pursuits—after all, there must have been targets for the satire. In fact, there is probably more scope and less encumbrance even today in France than elsewhere for women to follow intellectual pursuits. We need only compare the current concentration of women mathematicians at the Université de Paris with the situation at leading American universities.

Sophie's conversion occurred as she read the story of the death of Archimedes at the siege of Syracuse in Montucla's *l'Histoire des mathématiques*; she became determined to be a geometer. Her biographer terms this a heroic decision—for what use could a woman make of geometry? Her family was apparently none too thrilled, for they took away her fire, her light, and her clothes to force her from her books to the needed sleep. Sophie, however, waited until everyone else slept, wrapping herself in her covers to study by the light of contraband candles.

Studying by herself presented severe difficulties—other than courting pneumonia and eyestrain. For example, in order to read the seminal mathematics of Newton and Euler, she had to teach herself Latin. Moreover, she had no benefit of discussions or explanations from a teacher. Her interests were not confined to mathematics; she explored "all the domain of knowledge." The early work in psychology remained a particular interest of hers throughout her life.

This was an exciting period in French mathematics, with great developments soon to come. Instrumental in this flowering was the establishment of the École central des travaux publiques, later to become the École polytechnique. In the heat of the Revolution, the French Academy of Sciences was abolished, but later such excesses were regretted. In 1794 a moving speech by the chemist Fourcroy on the importance of science and of public education led to the founding of the École. Among the first professors were the French mathematicians and scientists Lagrange, Monge, Fourcroy, and Berthollet.

What an opportunity for an eighteen-year-old eager for knowledge! But alas, women were not allowed to attend lectures at the École polytechnique. However, one of the innovations in this new scheme of education was to make lecture notes available to all who asked, so Germain was able to obtain these, in particular the analysis lectures of Lagrange. Another innovation was the practice of having students submit written observations, which Sophie did under the name of M. LeBlanc. This was the beginning of extensive communication, both written and direct, with many of the well-known mathematicians of her day, in particular Gauss, Legendre, Lagrange, Poisson, and Fourier.

There is an oft-repeated tale of her concern for the safety of Gauss. After the *Disquisitiones arithmeticae* appeared in 1801, Sophie started to correspond with him as M. LeBlanc, an "enthusiastic amateur." In 1806, the French occupied Brunswick, where Gauss was living. Remembering the fate of Archimedes that had so moved her earlier, she feared for Gauss's life. She sent a message to a family friend, General Pernety, chief of the artillery. He, in turn, sent a messenger to Gauss to learn whether he was safe and to express Mlle. Germain's concern. The messenger did so and invited Gauss to dine with the French governor. It seems, however, that in her anxiety Sophie had forgotten that she was known to him only as LeBlanc, so he replied that he knew no Mlle. Germain, nor any general, and anyway he would rather do mathematics. Eventually the situation was explained and her anonymity removed.

In this grand epoch of mathematics, rigor was not a prime consideration. Great advances were made, but the detailed justifications took many years to develop. Freedoms were taken which undoubtedly had Bishop Berkeley turning in his grave.[44] The governing philosophy was that of the French mathematician d'Alembert, whose advice to mathematicians was "go ahead and faith will come to you." In her work on elasticity, Sophie was an ardent disciple of this method that, especially because she had no real formal training in topics such as the basics of integration and the calculus of variations, led to certain difficulties.

Pythagoras and Aristotle had some notion of the physics of the propagation of sound, but like so much ancient knowledge, this was lost until the time of Bacon and Galileo. Newton did some early work on vibrating strings, and the English mathematician Taylor solved a special case of the problem. Building on the work of Daniel Bernoulli, one of the Swiss family of mathematicians, in 1747 d'Alembert solved the problem of the equation for a vibrating string. After that, many mathematicians worked on the problem for a vibrating surface.

The success of theories for thin rods, founded on special hypotheses, gave rise to hopes that a theory might be developed in the same way for plates and shells so that the modes of vibration of a bell might be deduced from its form and the manner in which it is supported. In 1766, Euler was first to attack the problem. He proposed to regard a bell as divided into thin annuli, each of which behaves as a curved bar. This method ignores the change of curvature in sections through the axis of the bell. In his

---

44. Bishop Berkeley disputed the foundations of calculus, objecting to the use of "ghosts of departed quantities" in calculating derivatives.

work of 1789, Jakob Bernoulli (the younger) assumed the shell to consist of a double sheet of curved bars, the bars in one sheet being at right angles to those in the other. Reducing the shell to a plane plate, he found an equation of vibration that we now know to be incorrect.

In Germany, the scientist Chladni had done some experimental work on surfaces; he sprinkled sand on a sheet and transferred the nodal figures formed during vibration to wet sheets of paper. Attempts were made to discover a theoretical basis for the experimental results. In 1808, Chladni came to Paris and repeated his experiments. By some accounts, one of his demonstrations was before Napoleon, who was fascinated and persuaded the Institut de France to offer a special prize for "giving the mathematical theory of elastic surfaces and comparing it to experimental results."

In spite of being advised by Lagrange of the great difficulties of the problem, Sophie studied the phenomenon and in September 1811 sent an anonymous memoir to the prize competition. She probably consulted with Legendre while preparing the paper. He and the mathematicians Laplace, Lagrange, Lacroix, and Malus were appointed as a commission to examine the submissions. They pointed out the inexactitude of her equation, which did point the way, however, for Lagrange to come up with an equation that is correct under special assumptions. No prize was awarded.

Another competition on the same question was announced, and in September 1813 Sophie sent a second anonymous memoir, this time receiving honorable mention in spite of an error in the calculus of variations involved.

A third competition was announced, to which Sophie, after consulting with the mathematician Poisson, submitted a memoir in her own name. This time she won the prize in spite of the fact that the equation was not rigorously demonstrated and that the agreement with experimental data was not too close because of her employment of an incorrect equation of Euler. Germain's equation was:

$$N^2 \left[\frac{\partial^4 z}{\partial x^4} + 2 \frac{\partial^4 z}{\partial x^2 \partial y^2} + \frac{\partial^4 z}{\partial y^4} \right] + \frac{\partial^2 z}{\partial t^2} = 0$$

(N^2 a constant).

A journal of the time relates the public disappointment when Germain did not appear at the award ceremony in January 1816. She preferred to work without such public recognition. However, she was no longer denied admission to the École polytechnique lectures; Fourier, secretary of the Academy, wrote a letter saying that, in light of her accomplishments, a central seat would be reserved for her at its meetings.

After winning the prize, she continued to work on the theory of elastic surfaces. In 1821, she published her 1816 prize memoir, "Recherches sur la théorie des surfaces élastique," pointing out some of its errors, and in 1826 produced another memoir on related ideas. She also did some work on cylinders and annuli, suffering again from deficiencies in the calculus of variations.

In her work, Germain used the concept of *mean curvature* of a surface in 3-space: if k_1, k_2 are the maximum and minimum values of the normal curvature, then $\dfrac{k_1 + k_2}{2}$ is the mean curvature. This is not as useful a concept as Gaussian curvature $k_1 k_2$ with which she seemed not to have been familiar, although she should have been from her correspondence with Gauss. In *Crelle's Journal* (volume 7) for 1831 appears her "Memoire sur la courbure des surfaces."

Many mathematicians have occupied themselves with trying to prove Fermat's Last Theorem. Germain was not immune to its fascination. In her first letter to Gauss, November 21, 1804, she reports her solution for $n = p - 1$, where p is a prime of the form $8k + 7$.

In the supplement to the second edition of Legendre's *Théorie des Nombres*, Sophie Germain is credited with the proof of Fermat's Last Theorem for n an odd prime less than 100. This result follows from the theorem:

> If there exists an odd prime p such that $\xi^n + \eta^n + \zeta^n \equiv 0 \pmod{p}$ has no set of integral solutions ξ, η, ζ, each not divisible by p, and such that n is not the residue of the nth power of any integer modulo p, then $x^n + y^n + z^n = 0$ has no integral solutions in which x, y, z are all coprime to n.

She had a simple but ingenious proof of this. The result was reproved by E. Wendt in 1894 and used by Dickson in 1908 to prove Fermat's Last Theorem for n an odd prime less than 1700.

In 1910, E. Dubouis defined a "sophien" of a prime n to be a prime θ, of the form $kn + 1$, for which $x^n + y^n \equiv 1 \pmod{\theta}$ is impossible to solve in integers prime to θ.

Another work of Germain in number theory, a note about finding y and z in $\dfrac{4(x^p - 1)}{x - 1} = y^2 \pm pz^2$, appeared in *Crelle's Journal* in 1831. This paper and the paper in the same volume on curvature were composed under most trying conditions. In 1829, Sophie learned that she had cancer, and in July of 1830 she finished her work on the *Crelle's Journal* papers to the sounds of the cannons of another revolution. She also published in

Annales de chimie et de physique an examination of principles which led to the discovery of the laws of equilibrium and movement of elastic solids.

However, her biographer says that Germain's real recognition was as a philosopher for her posthumously published "Considérations générales sur l'état des sciences et des lettres aux différentes époques de leur culture." This reflected the optimism of the period. While the school of the philosopher-mathematician Descartes believed that laws governed the inorganic world, the new philosophy believed that the living world also has governing principles that could be discovered and used to calculate events. Clairau, Euler, d'Alembert, the Bernoullis, Lagrange, and Laplace had presided at the end of the great period of celestial discoveries and systemization; Cavendish, Priestley, Lavoisier, and Berthollet had discovered the composition of earth, air, and water; Jussieu, Linné, Buffon, Goethe, and others developed botanical classifications, so there was optimism about similar progress in the study of human behavior.

Germain observed that science results from the classification of observed facts, the abstraction and concomitant simplification from these, and finally idealization into a system as the German mathematician Leibniz had done for mathematics, Laplace for astronomy, and Jussieu for biology. She hoped to do the same in what we now call psychology and sociology. Her philosophy is a precursor of Comte's positivism and was highly praised by him. One might even read into her discussion of nonstable systems an early inkling of catastrophe theory.

Works by Sophie Germain

"Remarques sur le mémoire d'Euler: 'Investigatio motuum quibus laminae et virgae elasticae constremiscunt,' *Acta Acad. Petrop. Ann (1779)*: 103 et seq." Manuscript, Bibliothèque Nationale, MS Fr. 9114 f. 155–194.

Recherches sur la théorie des surfaces élastiques. Paris: Huzard-Courcier, 1821.

Remarques sur la nature, les bornes et l'étendue de la question des surfaces élastiques. Paris: Huzard-Courcier, 1826.

"Examen des principes qui peuvent conduire à la connaissance des lois de l'équilibre et du mouvement des solides élastiques." *Annales de chimie et de physique* 8(1828): 123–131.

"Mémoire sur la courbure des surfaces." *Journal für die reine und angewandte Mathematik* 7(1831): 1–29.

"Note sur la manière dont se composent les valeurs de y et z dans l'équation $4(x^p - 1)/(x - 1) = y^2 \pm pz^2$ et celles de Y' et Z' dans l'équation $4(x^p - 1)/(x - 1) = Y'^2 \pm pZ'^2$." *Journal für die reine und angewandte Mathematik* 7(1831): 201–204.

"Considérations générales sur l'état des sciences et des lettres aux différentes époques de leur culture." Edited by Armand-Jacques Lherbette. Paris: 1833.

"Mémoire sur l'emploi de l'épaisseur dans la théorie des surfaces élastiques." *Journal de Mathématiques pures et appliqués* 6(1880): Supplement S5–S64.

Works about Sophie Germain

Bucciarelli, Louis L., and Nancy Dworsky. *Sophie Germain: An Essay in the History of the Theory of Elasticity*. Dordrecht, Holland: D. Reidel Publishing Company, 1980.

Oeuvres philosophiques de Sophie Germain (suiviés de pensées et de lettres inédités et précédées d'une notice sur sa vie et ses oeuvres). Edited by H. Stupuy. Paris: Paul Ritti, 1879. New ed. Paris: Firmin-Didot, 1896.

Raphael, Ellen. "Sophie Germain, Mathematician: A Biographical Sketch." Senior thesis, Brown University, 1978. Contains the text of the Germain-Gauss correspondence.

Christine Ladd Franklin and Mary Fairfax Somerville

[Sophie Germain] had no teachers, she had few books, but she had an unlimited store of energy. She studied by day and by night. Her family were alarmed at so much ardor, and endeavored to turn her attention to more ladylike pursuits. They tried the plan of putting out her fire and taking away her clothes at night, but she was found in the morning wrapped up in blankets, absorbed in her studies in a room so cold that the ink was frozen in the inkstand. It is a curious coincidence that Mrs. [Mary Fairfax] Somerville, at that very same time, in her little village in Scotland, was obliged to wrap herself up in blankets to pursue her studies before breakfast, because her whole day had to be devoted to the practice of music and painting, and to her lessons at the shop of a pastry-cook.[45]

Thus wrote Christine Ladd Franklin in 1894 about Germain, the "forgotten mathematician." It is sad irony that for many years Ladd Franklin herself was also almost forgotten, as was recounted in the introduction to this chapter. Other sources on these eighteenth- and nineteenth-century women augment

---

45. Christine Ladd Franklin, "Sophie Germain: An Unknown Mathematician," *Century* 48(1894), 946–949, reprinted in AWM *Newsletter* 11(3), 1981, 7–11, at the suggestion of Patricia Clark Kenschaft.

the details above, including one reporting that Somerville's candles were "confiscated by her father, who maintained that 'we must put a stop to this, or we shall have Mary in a straitjacket one of these days.' "[46] Sonia Kovalevsky's story differs only in the source of her illumination; after her father attempted to curtail her study of mathematics, "at bedtime [she] used to put the book under [her] pillow and then, when everyone was asleep, [she] would read the night through under the dim light of the icon-lamp or the night lamp." When her father was convinced of his daughter's talent, however, he relented and hired tutors for her.[47] In contrast, from the beginning Lovelace's mother "strongly believed in mathematics as a discipline of the mind and saw to it that Ada was well grounded in this subject. She felt that it would be a way to provide a stable mental state and a good antidote to the 'heedlessness, imprudence, vanity, prevarication and conceit' that Ada was bound to have inherited from her immoral father."[48] Some other short quotations follow; together they form a patchwork quilt of glimpses into the lives of these women from a very different time.

As proof that women may be pure mathematicians, Mrs. Somerville has had, outside of Italy and Russia, to stand alone. This is unfortunate, for the detractors of her sex have maintained that her work, though exceedingly profound, was not remarkable for originality. That charge cannot be brought against Sophie Germain. She showed great boldness in attacking a physical question which was at that time entirely outside the range of mathematical treatment, and the more complicated cases of which have not yet submitted themselves to analysis. The equation of elastic laminae, which is still called Germain's equation, formed the starting-point of a new branch of the theory of elasticity.[49]

Sir David Brewster, inventor of the kaleidoscope, wrote in 1829 . . . that Mary Somerville was "certainly the most extraordinary woman in Europe— a mathematician of the very first rank. . . ."[50]

In 1830, when she was 15, Ada [Byron] met Mary Fairfax Somerville, a well-known female mathematician from Scotland. Mary had two daughters

46. Simon Singh, *Fermat's Enigma: The Quest to Solve the World's Greatest Mathematical Problem* (New York: Walker and Company, 1997), 103.

47. Sofya Kovalevskaia, "An Autobiographical Sketch," in *A Russian Childhood*, ed. Beatrice Stillman (New York: Springer Verlag, 1978), 217–218.

48. Clement Falbo, "Augusta Ada Byron, Countess of Lovelace," in *Math Odyssey* (Champaign, Ill.: Stipe Publishing, 2000). Also available from the World Wide Web [accessed 17 July 2002]: http://www.sonoma.edu/Math/faculty/falbo/AdaByron.html.

49. Ladd Franklin, "Sophie Germain," 8.

50. J. J. O'Conner and E. F. Robertson, "Mary Fairfax Greig Somerville" [online]. School of Mathematics and Statistics, University of St. Andrews, Scotland, November 1999 [accessed 17 July 2002]. Available from the World Wide Web: http://www-groups.dsc.st-and.ac.uk/~history/Mathematicians/Somerville.html.

the same age as Ada, and the four women, Ada, Mary, and her daughters, attended geography lectures at the University of London. (It seems that the mathematician Charles Babbage had persuaded the university to allow women to attend lectures in 1830, a privilege which was rescinded within a year.) Ada corresponded with Mary Somerville on mathematical topics for the next twenty years, until Ada's death.[51]

Ada herself found Mrs. Somerville and her daughters delightful and they liked her. . . . Mary Somerville lent Ada books, advised her on her studies, set mathematical problems for her and helped with their difficulties, and above all talked mathematics to her. . . .

Lovelace turned to Babbage and Augustus De Morgan for mathematical instruction. . . . [De Morgan] concluded that Ada showed more mathematical genius [than Somerville]. Mary Somerville would rejoice in the genius and deplore the lack of steady application, but her affection for her friend would remain undiminished.

In the coming decades there would be steady pressure for women's education and Mary Somerville's pioneering accomplishments would be widely hailed. In her own time she did whatever she could to promote learning among women and open opportunities to them.[52]

51. Clement Falbo, "Augusta Ada Byron."
52. Elizabeth Chambers Patterson, *Mary Somerville and the Cultivation of Science, 1815–1840* (Boston: Martinus Nijhoff, 1983), 149–150.

II. JOINING TOGETHER

In this new century, despite the concerted efforts of women around the world, the question still arises: are women as capable as men at doing mathematics? In a letter to the editor in the Association for Women in Mathematics (AWM) *Newsletter*, an eminent male mathematician (and AWM member) answered that question with an unequivocal *yes!* In part II, stories are told by a cast of mathematicians spanning the globe; they have joined together to assure women the opportunity to demonstrate this capability.

Around 1970, encouraged by the vitality of feminism at that time yet often discouraged by their experiences as faculty and students, groups of women began meeting to look for ways to become full participants in the collegial and intellectual life of the mathematical community. Some of them decided to establish a new mathematical professional society, the AWM. By doing so, they hoped to encourage women and girls to learn more about mathematical careers, as well as to improve work environments for themselves and for those women who would come after them, their goal being nothing short of eventual equality in the profession. It is not surprising that AWM received a mixed welcome from the community; some felt that it was too much an advocacy group. Now, through thirty years of accomplishments, it has gained acceptance from most critics. This acceptance has arisen from work done in concert with women and men from other organizations having related goals and with sisters everywhere: organizing scientific sessions, as well as discussions and other activities at mathematical meetings, publishing the widely read *Newsletter*, and running a plethora of special-purpose projects in partnership with multiple private and governmental funding agencies. AWM has continued to promote, through the work of its volunteer leadership, the interests of women and girls in the study and practice of mathematics.

Vignettes from the work and lives of many mathematical women are woven into part II as it documents cooperative actions. Mathematics is a global endeavor. Mathematicians working in the United States, wherever born or educated, frequently interact with mathematicians in other countries. Early on, AWM had an international membership. Conversations across national boundaries revealed that cultural expectations—the message that doing mathematics is "unfeminine"—remain so strong almost everywhere that too many young women continue to be deterred from pursuing mathematics and other scientific careers. An interesting counterpoint to the usual reporting came from Papua New Guinea: "Females from matrilineal communities of the Island States of the South Pacific outperform males in mathematics."[1]

Part II concludes with some detail about sharing ideas and cooperation on a global scale, with one goal being to improve the lot of beginning mathematicians everywhere. Women who have received international acclaim for their mathematics are applauded for their achievements. Finally, the focus moves back from organization and joint activity to the immediate and personal, with commentary from all continents but Antarctica on women's quest for their rightful place under the mathematical sun.

1. Gurcharn Sing Kaeley, 1992, p. 8, "IOWME Study Group Presentations at ICME-7" by Carole Lacampagne, AWM *Newsletter* 23(1), 1993, 7–8.

Women and Mathematical Ability

RALPH P. BOAS

Based on a letter to the editor in the section "Gender, Mathematics, and Science," AWM *Newsletter* 16(6), 1986, 10. Boas (1912–1992) was then Henry S. Noyes Professor Emeritus of Mathematics, Northwestern University. He served both as vice president of the AMS and as president of the MAA.

I am constantly mystified by the number of women who conduct experiments that purport to show that women are inherently inferior in mathematics. I have myself been conducting a long-running experiment: for roughly fifty years I have been teaching college mathematics. I have yet to see any difference in the mathematical ability of men and women, except that, on the whole, the women are more capable. This slight difference I attribute to Society's constant pressure to keep women out of mathematics unless they are very determined. Indeed, the very best calculus class I have ever taught was an extremely well-prepared class of women. I can only attribute the sociologists' activities in this direction to their having bad cases of "mathematics anxiety," which consists not just of inability to do mathematics, but of worrying about this incapacity.

Now lots of people are incapable in some area or other. I, for example, have neither musical nor athletic talent, but I don't worry about it. I don't have music anxiety or football anxiety. I would like to be able to sing, but it doesn't bother me that I can't. When I meet a professional singer at a party, I don't make silly remarks like "Oh, I never could sing a note." Why are people so defensive about their mathematical deficiencies?

I suspect that the reason is that mathematics is perceived as being important to Society in a sense in which music and athletics are not. Mathematics and English are the two subjects that are taught all through school. There are reasons for this emphasis: English is needed so that we can communicate with each other; mathematics, so that in the first place we can cope with the arithmetical demands of daily life; and second, because without it, Society would have little science and still less of the technology that we love so much.

It is socially acceptable to be incapable in an activity that is not felt to be important in everyday life. People who are incompetent in mathematics feel defensive—that's presumably why they make inane remarks when they meet mathematicians. They go out of their way to show that they can't be blamed because they, as a class, lack talent. If you lack a talent, it is comforting to believe that you belong to a class that is generically incompetent. If you are a woman, the class of women is the most obvious class, especially since there really are fewer female than male mathematicians, for rather obvious social reasons. There are also fewer female than male dentists, but I don't think anyone has tried to show that women as a class lack ability for dentistry.

AWM's First Twenty Years: The Presidents' Perspectives

LENORE BLUM

Based on a talk presented at the Twentieth Anniversary Celebration of the AWM, January 1991, and excerpted from "A Brief History of the Association for Women in Mathematics: The Presidents' Perspectives," which appeared in the AMS *Notices* 38(7), September 1991, 738–754, and was reprinted in the AWM *Newsletter* in two parts, 21(6), 1991, 11–22, and 22(1), 1992, 12–25. Then a research scientist at the International Computer Science Institute, Berkeley, and Letts-Villard Research Professor of Mathematics, Mills College, Blum is now Distinguished Career Professor of Computer Science, Carnegie-Mellon University.

Part 1: How It Was . . .

I would like to begin by re-creating some of the atmosphere twenty years ago. So I must start with a warning: the next few minutes may be a bit depressing, perhaps even somewhat hard to take. But bear with me, it really will get better.

First, for my journey back in time, I checked out the AMS *Notices* for 1971. The Joint Mathematics Meetings (JMM) that year were held in Atlantic City; the program in the January issue was quite revealing. Of the more than fifteen invited hour speakers[2] none were female; of the more than 300 AMS ten-minute talks, about 15 were given by women (5%). I became curious and looked at the Personal Items section, which contains short descriptions of individuals' professional activities and achievements as well as job promotions and appointments. Only 5 of the approximately 145 blurbs seemed to mention women (less than 4%). Of the thirty-one promotions listed, three were female (10%); at the instructorship level, women seemed to do relatively better, getting three of the nine appointments (33%). As I went down the list—as the positions became less prestigious—the percentage of women increased.[3]

---

2. AMS, MAA, and Association for Symbolic Logic combined.
3. As if to confirm this trend even more dramatically, I noticed further on that, of the four deaths reported in that issue, two were women (50%)!

In the February 1971 issue of the *Notices* a letter from Elizabeth Berman pointed out some recent "advice" on how to find employment: "Women find the competitive situation in the government somewhat more advantageous to them, since it is relatively hard to secure a well-qualified mathematician for many higher level government jobs. In many such cases women are welcomed if their qualifications are better than those of the available men." A gloomy picture of the status of women in academia was painted by mathematician Ruth Silverman in a letter that appeared in the June 1971 *Notices*:

> As a result of surveys on many campuses it becomes apparent that there is a pattern of discrimination against women in all fields. . . . Women are predominantly at the bottom of the pyramid, irrespective of qualifications . . . and suffer a substantial salary inequity. Many academic departments have no full-time female faculty at all. . . . In many departments women with Ph.D.'s hold positions below the rank of Assistant Professor. . . . Women tend to be hired on a marginal, temporary, or one-year basis. . . . Often women teaching part-time have the same teaching load as men teaching full-time. There are departments that make it a policy not to appoint women who are married to members of the faculty. . . .

Silverman goes on to recommend that "in the forthcoming annual [AMS] salary survey data be collected . . . comparing salary levels by sex."[4]

Now, if you were a female graduate student at the time, there were certain departments where you probably were not. For example, Princeton did not start admitting women to their graduate program in mathematics until the fall of 1968. Marjorie Stein (Princeton Ph.D., 1972) was the first woman to complete her degree requirements there, although a Japanese woman had been admitted some years earlier by mistake.[5]

But wherever you were, you may very well have heard the following "joke": "There have only been two women mathematicians in the history of mathematics. One wasn't a woman and one wasn't a mathematician." (It may not be so surprising that in those years we were often accused of not having a sense of humor.) Sometimes at least, this mathematical in-joke was told well-meaningly (however misguidedly)—as it were, a friendly gesture to break the ice. Certainly, that's how I had interpreted it several years earlier at a party given by my department chairman when I was a graduate student at MIT. It clearly was a manifestation of the time, of the awkwardness everyone felt with the few women around. (It did not occur to me until some years later that it was also a callous dismissal of two of

---

4. This practice was initiated by the AMS, but some years later.

5. Apparently the admissions committee, unfamiliar with Japanese first names, did not recognize hers as female.

the most important mathematicians in recent history.) The effect was nevertheless to help alienate us from our history, reinforce self-doubts, and keep us mostly unaware of strong women contemporaries who could very well have served as important role models and mentors had we known of their existence early on.

I do not want to give the impression that all professors and thesis advisors were hopeless. Some of us were fortunate to have supportive advisors during those important years. Lipman Bers is a stunning example of a professor who did much to encourage young women in mathematics. As he put it, "It never occurred to me that women can be intellectually inferior to men."[6]

What We Did . . . (In the Beginning)

I think it is fair to say that the AWM had its birth at the JMM in Atlantic City in 1971. As founding member Judy Green remembers (and Chandler Davis, early AWM friend, concurs):

> The formal idea of women getting together and forming a caucus was first made publicly at a Mathematics Action Group (MAG) meeting in 1971 . . . in Atlantic City. Joanne Darken, then an instructor at Temple University, . . . stood up at the meeting and suggested that the women present remain and form a caucus. I have been able to document six women who remained. . . .[7] It's not absolutely clear what happened next, except that I've personally always thought that Mary [Gray] was responsible for getting the whole thing organized. . . .

What I remember hearing about Mary Gray and the Atlantic City meetings was another event, one that would also alter the character of the mathematics community. In those years the AMS was governed by what could only be called an "old boys' network," and Mary challenged that by sitting in on the AMS Council meeting in Atlantic City. When she was told she had to leave, she responded she could find no rules in the bylaws restricting attendance at council meetings. She was then told it was by "gentlemen's agreement." Naturally Mary replied, "Well, obviously I'm no gentleman." After that time, council meetings were open to observers, as the process of democratization of the AMS began.

6. "Lipman Bers," in *More Mathematical People: Contemporary Conversations*, ed. Donald J. Albers, Gerald L. Alexanderson, and Constance Reid (Boston: Harcourt Brace Jovanovich, 1990), 12.

7. They were Judy Green (a graduate student at Maryland at the time), Joanne Darken, Mary Gray (already at American University), Diane Laison (then an instructor at Temple), Gloria Olive (a senior lecturer at the University of Otago, New Zealand, visiting the United States at the time), and Annie Selden. Harriet Lord (then a graduate student at Temple) was at the MAG meeting but unable to stay for the women's caucus.

Local groups of women had meetings, before and after Atlantic City.[8] Through the boundless efforts of Alice Schafer, a member of the Boston area group, an office was established for AWM at Wellesley College.

Berkeley and Me

In the beginning, I was quite ambivalent about the emerging women's movement in mathematics. As I replied to Linda Rothschild, "I was pretty 'unconscious' about such things [in '68]. It didn't hit me until I got to Berkeley."

The spring of 1971 was a particularly bleak time for me professionally. But, then again, I was in Berkeley. It was the era of People's Park, Cambodia, and Vietnam. I found it quite exciting. In the math department, Moe Hirsch, John Rhodes, and Steve Smale had organized a colloquium series, "Social Problems Connected with Mathematics." I agreed to chair a colloquium on "Women in Mathematics" when Steve asked. Since I didn't know much about women in mathematics, I found three women who did: Ravenna Helson, a research psychologist who had done a study on women mathematicians and the creative personality; Sheila Johannsen, a historian knowledgeable about the history of women in mathematics; and Betty Scott, chair of the statistics department at Berkeley, who had just coauthored a report of the Academic Senate on the status of women on the Berkeley campus.

The colloquium panel was a great success. The lecture hall was packed. And the presentation was an eye-opener for me. It had never occurred to me that there might be common personality traits among women mathematicians or that statistics could be a powerful political tool. I found Betty Scott's study a masterpiece, fleshing out cold data with poignant case studies. And then there was data that spoke clearly for itself. For example, she gave these numbers on faculty ladder positions in the Berkeley math department (academic year and percentage of women): 1928–1929, 20%; 1938–1939, 11%; 1948–1949, 7%; 1958–1959, 3%; 1968–1969, 0%.

After that event, I became known as *the* expert on women in mathematics, on the West Coast at least. More important, I started to meet regularly with some of the women math graduate students.[9] This little group became a Berkeley center of AWM.

---

8. In 1969, Alice Schafer, then at Wellesley, and Linda Rothschild, then a graduate student at MIT, organized a Boston area group of women mathematicians and students to meet every few weeks to discuss common problems and goals. Bhama Srinivasan joined when she started teaching at Clark in 1970. The original group also included Bernice Auslander, Kay Whitehead, Caroline Series (then a graduate student at Harvard), Eleanor Palais, and Linda Almgren Kime.

9. Laif Swanson, Joan Plastiras, and Judy Roitman were among them.

The First Decade (1971–1981): Building the Foundations, a Time of Many Firsts

MARY GRAY (1971–1973): THE MOTHER OF US ALL

Without a doubt, Mary Gray is the founder of the AWM and the "mother of us all." I first remember seeing a small announcement for the new organization, the Association of Women in Mathematics, placed by Mary in the *Notices*, February 1971. The first issue of the AWM *Newsletter* (clearly written by Mary) appeared that May, listing Mary Gray as chairman. By the second issue, "of" was changed to "for," and later "chairman" was replaced by "president."

The *Newsletter* has since become the very embodiment of the AWM. From the start, it was our forum for discussing the role of women in mathematics; for exposing discrimination; for exchanging strategies, encouraging political action and affirmative action; for informing, supporting, honoring; and, of course, for listing jobs (beginning with the February 1972 issue). It has been our key linkage with each other, with credit due largely to Mary and subsequent editors, Judy Roitman and Anne Leggett.

Mary set down goals and agenda for the early AWM. In an article ("Uppity Women Unite!") in the January 1972 MAG *Newsletter* she wrote:

> We have some plans to improve the status of women in mathematics. . . . There are two categories of problems, those involving the general female population and those involving professional women mathematicians. We must go back to the elementary schools—rewrite textbooks, use films, etc., and retrain the teachers and counselors. The goal is to show girls and boys that girls *can and should* learn mathematics. . . . As a small first step, careful attention must be given by the mathematical community to the mathematical training of elementary schoolteachers, to see that they learn to like mathematics as well as learning mathematics. . . .

Mary goes on:

> What do women want? Let me be specific as far as women mathematicians are concerned: Equal consideration for admission to graduate school and support while there . . . for faculty appointments at all levels. . . . Equal pay for equal work. Equal consideration in assignment of duties, for promotion and for tenure, . . . for administrative appointments at all levels in universities, industry and government, [and] . . . for government grants, positions on review and advisory panels and positions in professional organizations. Because of past

injustices, special efforts will have to be made for some time. . . .
AWM is ready to help. Now is the time for discrimination to end.

What seems quite amazing now is that these were considered radical demands!

Mary Gray informed us (sometimes in far greater detail than many of us cared to know) of legislation on discrimination and affirmative action and urged us to become involved. She was not afraid to say things straight, to take on the establishment single-handedly. Challenging the system, she successfully ran in 1976 as a petition candidate for vice president of the AMS. As Bhama Srinivasan says, "Mary had the courage, and willingness, to take the initial steps and the initial hostility . . . [charting a course] which eventually wiped out the 'old boys' network.' "

Not all women mathematicians were enthusiastic about the AWM in the beginning. For example, as Cathleen Morawetz puts it, "I did not want the AWM to speak for all women mathematicians. I joined them later, but at that time they were terrible attackers. . . ."

Nevertheless, Cathleen played an important role in changing consciousness about women in mathematics. "I was on a committee for disadvantaged groups in the Math Society, and I thought there should be a separate committee for women. I was terribly afraid when I went before the Board of Trustees—or it may have been the Council. Anyway, when it came my turn to speak, I said 'There's a problem with women. You may not have noticed that there are not many women mathematicians.' "

Cathleen continues, "At that point Saunders Mac Lane said, 'Well, mathematics is a very difficult subject.' I was not up to coping with that, but Iz Singer picked up the ball. The committee was formed and I was made chairman."[10]

In 1973, the Committee on Women published the first *Directory of Women Mathematicians*. Rebekka Struik, University of Colorado mathematician, was instrumental in the production of further such directories.

ALICE SCHAFER (1973–1975): AWM INCORPORATED

In terms of its organizational structure, I picture AWM as an evolving continuum (built with boundless energy and grassroots networking). There is considerable overlap between one presidency and the next. This dynamic was already in evidence in the first transition from Mary Gray to Alice Schafer:

10. "Cathleen S. Morawetz," in *More Mathematical People: Contemporary Conversations*, ed. Donald J. Albers, Gerald L. Alexanderson, and Constance Reid (Boston: Harcourt Brace Jovanovich, 1990), 235.

When I took over the presidency . . . I asked Mary what I could do; she suggested getting AWM incorporated. . . . That was done through a lawyer in Boston, who I had been told would charge very little, so I was amazed when he charged $500, which was really big money for AWM, and so, in the *Newsletter*, I asked for a contribution of a dollar from each member. Some gave and AWM did finally pay the bill. When it came to obtaining tax exemption status from the IRS, the lawyer said he would do it and I said first I would try. He said I could not do it, but I did. . . .

In the early days, money was indeed a problem. Alice continues: "Do you recall that one time the March *Newsletter* was printed in such small print in order to save money that many people could not read it?"

For those of you who were not around during those years, and for those of us who may have forgotten, Alice goes on to paint a colorful, and almost slapstick, picture of what we were up against and how she handled it:

One of the . . . funny things that happened . . . during my presidency is that when the meeting was in San Francisco [January 1974] AWM was still being harassed by the male mathematicians. Lee Lorch, friend of AWM, came to tell me that some of the men were going to attend the AWM meeting, which I was chairing of course, and were going to break it up. He thought I ought to be warned. I was glad of the warning and told him that teaching in high school for three years (before I had enough money to start graduate school) ought to prepare me for that! . . . That meeting was the first time AWM had ever sponsored mathematical talks; before that it had all been consciousness-raising. I had invited Cathleen Morawetz and Louise Hay to give short talks on mathematics . . . and of course their talks were good. The men . . . never said anything. . . .

During this period, in addition to building its own internal structure, the AWM was also beginning to establish itself as a legitimate professional society, to be reckoned with among its peers. By the end of Alice's term AWM was about to be admitted as an affiliate member of the Conference Board of Mathematical Societies (CBMS), the umbrella society of mathematical organizations.[11] When I became president, all I had to do was put on the finishing touches, and there we were, on the same council (and on the same CBMS letterhead) as the AMS, MAA, NCTM, SIAM, and so forth. An amazing feat for an association only four years old!

11. As I recall, shortly after the AWM applied for affiliate membership in the CBMS, a mysterious math society, apparently originating in the Midwest and founded solely to cause difficulties for AWM, decided it also was worthy of CBMS membership. Its application caused something of a commotion, prompting the CBMS to reevaluate its membership criteria. This delayed AWM's entrance for about a year, but in the end, AWM was able to meet the stiffer requirements.

Lenore Blum (1975–1978): Exploring New Territory

In August 1975, I became president of the AWM and served in that capacity for three years. Since Mary had already captured the attention of the mathematics community head-on, and Alice had set up the foundation for a working organization, I was free to explore new territory. It seemed clear that the provincial view of mathematics—including who a mathematician was, and what a mathematician did—was a prime factor in the exclusion of women, as well as others, from the field, and that this view was limiting to the discipline itself. So, to make this statement, as well as to educate myself further, I decided to use the public forum that had proved so successful in Berkeley.

In those years, the academic job market for mathematicians was very tight. Many young people were in a terrible bind, given the prevailing view that the only respectable work for a mathematician was in academia. Since women mathematicians had been finding creative alternatives to academic employment for years, their experiences could be particularly useful, perhaps even change an image. I organized the panel "Women Mathematicians in Business, Industry, and Government" for the 1976 JMM in San Antonio (and a similar one in Seattle, the summer of 1977). Here I met for the first time Marjorie Stein, a mathematician working at the U.S. Postal Service (Statistical Service Requirements Division); Jessie MacWilliams, a coding theorist at Bell Labs; Mary Wheeler of Rice, also a consultant for oil companies, working on numerical solutions to P.D.E.'s; and Marijean Seelbach, a topologist and functional analyst working on optimal control theory at NASA–Ames. These energetic women had clearly found unusual and challenging career paths for themselves using their mathematical training and skills. It was inspiring.

Many of us were eager to explore further our history, so I organized some "History of Women in Mathematics" panels at the JMM with AWM members as speakers. What was so powerful about these sessions, even historic in itself, was that for the first time women mathematicians were talking about women mathematicians (their lives and their work) to women mathematicians. By understanding their work, possibly even identifying with their lives, the speakers were able to convey uniquely meaningful, deeply personal portraits of the women who had come before us. The sessions were charged!

In Toronto (summer 1976), Mary Gray talked about Sophie Germain and her work, Linda Keen about Sonia Kovalevsky,[12] and Martha Smith about Emmy Noether's mathematics and her tremendous influence on her field. As an added treat, Emiliana Noether came to talk about her husband's

---

12. See "From Earlier Times," I.2, for the stories of Germain and Kovalevsky.

aunt (-in-law).[13] In St. Louis, Teri Perl told us about *Lady's Almanac*, a popular women's magazine published in England from 1704 to 1841, devoted in large part to mathematical questions and solutions.[14]

But perhaps one of the most moving occasions was when Sylvia Wiegand spoke of her grandmother, mathematician Grace Chisholm Young. Sylvia read a poignant letter from her grandfather to her grandmother explaining why, at that time, their joint work should bear his name only. [See the full story in I.2.]

A historic panel, "Black Women in Mathematics," organized by Pat Kenschaft and Etta Falconer, was held in Atlanta, January 1978. Of the few black women in the United States holding Ph.D.'s in mathematics at the time (others of course held degrees in mathematics education), six were on the panel.[15]

These AWM sessions at the national meetings were immensely popular. We were clearly identifying and addressing subjects of interest and issues of concern to the mathematics community at large (well before these issues were recognized by the establishment as legitimate, even critical). As a consequence, we began to broaden our constituency, attracting people who had perhaps felt uncomfortable with the more political tone of earlier days.

But political issues were nevertheless still very much on our minds. We provided testimony for congressional investigations, wrote university presidents and newspaper editors and letters (often signed jointly by the three presidents, Mary, Alice, and me) protesting objectionable images of girls and women in textbooks, the media, and advertising. (In school math books, girls were still calculating the perfect recipe, while boys calculated the time to get to the moon. Flyers depicting a naked woman contemplating a calculator were still being distributed at the Math Book Exhibits in 1976.) In 1978, a masterful combination of teamwork and old-girl networking resulted in decisions by the AMS and MAA not to hold national meetings in states that had not ratified the equal rights amendment (ERA). It was a time of heady issues, but also

13. See "From the Twentieth Century," I.1, for Noether's story.

14. See AWM *Newsletters* and Teri Perl's *Math Equals* (Menlo Park, Calif.: Addison-Wesley, 1978). A number of biographies of women mathematicians by women mathematicians have appeared in the *Newsletters* over the years. For example, in the July 1978 issue, Bhama Srinivasan writes about Ruth Moufang (1905–1977), dedicating her article to the many mathematicians who have exclaimed, "You mean Moufang is a woman?"

15. The panelists were Geraldine Darden, Elayne Arrington-Idowu, Eleanor G. Jones, Evelyn Roane, Dolores Spikes, and Etta Falconer. Their stories are published in the September 1978 and May–June 1980 *Newsletters*; see "A Dual Triumph," III.1, for two of them. Also see Pat Kenschaft's article, "Black Women in Mathematics in the United States," *American Mathematical Monthly* 8(1981), 592–604. Lee Lorch plays an important role here. Three women who studied with him at Fisk (during the period 1950–1955) went on to get Ph.D.'s in mathematics: Etta Falconer, Vivienne Malone-Mayes, and Gloria Hewitt.

a time of great excitement and great fun. It was a time of newly found camaraderie, of friendships, of support and respect among women mathematicians.

I was clearly a beneficiary of this "sisterhood" during my presidency. Besides the two former presidents to guide me, Judy Roitman was Newsletter editor and Judy Green, AWM employment officer (a title deemed appropriate as she had taken it upon herself to analyze employment data and monitor the legitimacy of job advertisements). Both Judys were co–vice president. Judy Roitman and I had become great friends during the early Berkeley days, and I welcomed her presidency in the November 1978 issue by saying ". . . our friendship has grown with, indeed, has been intertwined with, our involvement in . . . the AWM." Judy Green often played the role of political advisor and telephone consultant, as well as AWM liaison with MAG and NAM.[16]

Discrimination/Affirmative Action. Before going ahead, I would like to take a few moments to address the twin issues of discrimination and affirmative action that were so central to our lives in those years. AWM gave women mathematicians courage to speak out publicly, even file complaints and charges about their own situation. While many in the mathematics community believed that there was an influx of women faculty as a result of affirmative action, the data in the early years showed quite the contrary. For example, between 1973–1974 and 1974–1975 the percentage of women in regular math faculty positions actually went down, and no significant rise became evident until very much later.[17]

Affirmative action rulings often produced backlash and abuses. In order to satisfy affirmative action guidelines, many math departments resorted to "papering the files." The names have been removed from the following letter from a woman mathematician on the East Coast to a math department on the West Coast, not so much for anonymity, but rather to stress genericity:

> Dear Professor X, This is the third consecutive year that I have been invited to apply for the position of Assistant Professor in the mathematics department of [West Coast University]. I assume it is the third year of [WCU]'s Affirmative Action program. As I mentioned in my last response to such an invitation, I have been Associate Professor since the first year. It is hard to believe that [WCU] is serious about its Affirmative Action program if it makes no attempt to match the experience of

16. She would surely have been AWM president at some point had she not preferred backstage roles.

17. See Judy Green's article in the April 1975 AWM *Newsletter*, mine in the May–June 1976 *CBMS Newsletter*, and Mary's and Alice's in the *Notices*, October 1976.

the candidate considered with the positions available. Would you be interested in a job as Assistant Professor?

Sincerely, Y[18]

JUDY ROITMAN (1978–1981): A SUMMING UP

The early years of the AWM were a time of activism, of speaking out, of politics, of confrontation, of heroes and villains—when most issues seemed clear-cut. Judy Roitman provides her perspective on the decade:

> I can summarize my time in AWM office by saying that I was one of the last—perhaps *the* last—president of an amateur AWM. What do I mean by this?
>
> The AWM grew out of the feminist movement of the 1970s, which was marked by confrontation, attention to, and expression of, personal feelings and individual incidents, and ignorance of history. Having finally read some of this history,[19] I suspect that had we known how closely we were following in the footsteps of earlier feminists, and how little change their tremendous efforts made, we probably never would have bothered. So the early job of the AWM was just to look around us and report the obvious—the situation for women was terrible—and the apparently not-so-obvious—it didn't have to . . . be that way. We spent a lot of time popping up at meetings (departmental, local, national) saying over and over again that women could be perfectly good, even great, mathematicians if given the opportunity . . . and that there were several steps the mathematical community could take to improve things for both women and minorities. It was an easy kind of agitation—you just had to look around you and report what you saw. . . .
>
> But while this style had its successes, it was based on a sort of shooting from the hip. That is why I characterize it as being amateur. . . .[20]

It was also a time of lassoing people in. In addition to AWM national meetings, members were organizing and meeting regionally.[21] In the

18. Y has received numerous prestigious honors and awards.

19. In this excellent book: Margaret W. Rossiter, *Women Scientists in America: Struggles and Strategies to 1940* (Baltimore: Johns Hopkins University Press, 1982).

20. Not completely. It should be noted, for example, that the AWM bylaws were written and passed during Judy Roitman's term. Creatively, they stipulated both formal structure and procedures for AWM's governance, while at the same time leaving room for flexibility.

21. Sue Montgomery and Ruth Afflack in Southern California; Rebekka Struik in the Rocky Mountain region; Jessie Ann Engle, Judith Longyear, and Vera Pless in the Midwest; Pat Kenschaft in New Jersey; Linda Keen in New York; and Lida Barrett and Bettye Anne Case in the South, to mention only a few.

mid-1970s, AWM instituted an Open Council, encouraging the participation of members representing a wide range of self-identified constituencies and areas of interest. By 1981, AWM had grown to over 1000 members (from the United States and fifteen other countries), its influence and political power ranging far beyond these numbers.[22]

The 1970s were certainly a time of increased consciousness about women in mathematics. It was also a time of many firsts, including Barbara Osofsky's AMS Invited Address at the 1973 JMM—the first major address at a national meeting by a woman since Anna Pell Wheeler's Colloquium Lectures in 1927[23]—and Sloan fellowships awarded to Joan Birman and Karen Uhlenbeck in 1974. Julia Robinson's election to the NAS and Dorothy Bernstein's election as president of the MAA both occurred in 1975. Karen Uhlenbeck suggested the concept of an AWM lecture series at the JMM. In 1979 she chaired the selection committee for the inaugural Noether Lecture delivered by Jessie MacWilliams at the San Antonio meeting in January 1980.

It was also a time of solid program development and achievements. During those years, many of us were involved in designing and implementing educational programs to increase the participation of girls and women in mathematics. Other organizations—such as the Math/Science Network, headquartered in the San Francisco Bay Area, and Women and Mathematics (WAM), founded by the MAA—to which many AWM members belonged, were also very much part of this effort. Since the old system was clearly not working for us, we were motivated to explore new paradigms in teaching: developing hands-on activities and materials stressing problem-solving skills, promoting team teaching and cooperative learning, providing role models and information to students (as well as their parents and teachers) about why mathematics was important for their future.[24] Educational programs we developed in the 1970s are now models for educational reform in the 1990s.

Part 2: The Second Decade (1981–1991): A Coming of Age

The second decade in the life of the AWM can be characterized variously as a period of maturing, of increased self-assurance, of establishing and strengthening institutional mechanisms, of gaining acceptance by the mathematics

22. For example, then and over the years, AWM-supported candidates in AMS elections have been likely to win.

23. In the interim, women had been invited sporadically to speak at sectional meetings: Pauline Sperry (1933), Emmy Noether (1934), Olga Taussky Todd (1959), Cathleen Morawetz (1969), Mary Ellen Rudin (1971), and Mary Elizabeth Hamstrom (1972).

24. In 1979, at the summer meeting in Duluth, Judy Roitman organized an AWM panel, "Mathematics Education: A Feminist Perspective," to discuss these new programs and strategies. Speakers included Deborah Hughes Hallett, Diane Resek, and myself.

community. These themes and phrases kept recurring in my conversations and correspondence with the AWM presidents of the 1980s. This decade was a time when AWM grew up.

BHAMA SRINIVASAN (1981–1983): NOETHER SYMPOSIUM, SPEAKERS BUREAU, RESEARCH AND EDUCATION

During Bhama's presidency, AWM sponsored its first major mathematical conference, the Noether Symposium at Bryn Mawr College, in honor of Emmy Noether's 100th birthday. There were nine scientific lectures as well as a panel discussion. The symposium proceedings, *Emmy Noether in Bryn Mawr* (edited by Bhama and Judith Sally), were published by Springer-Verlag in 1983. The event "was not only scientifically successful but a specially moving occasion," Bhama remembers. Three women[25] who had studied with Noether at Bryn Mawr spoke at the symposium. They painted a picture of a mathematically charged, particularly precious time, dominated by Noether and fully integrated with women. One of them, Ruth McKee, recalled how it was to be in Noether's classes: "The strange phenomenon was that from our point of view, she was one of us, almost as if she too were thinking about the theorems for the first time. There was a lot of competition and Miss Noether urged us on, challenging us to get our nails dirty, to really dig into the underlying relationships, to consider problems from all possible angles. It was this way of shifting perspective that finally hit home . . . suddenly the light dawned and Miss Noether's methods were the only way to attack modern algebra. . . ."[26]

Tensions began to surface during the early 1980s: were we an organization of research mathematicians, or did we represent the interests of all women in mathematics, particularly in education? Now that we were not as preoccupied with political issues as in the early years, it seemed we were having an identity crisis! Bhama recalls, "I was concerned about how to balance our various (and sometimes conflicting) constituencies

25. Olga Taussky Todd, Ruth Stauffer McKee, and Grace Shover Quinn.

26. McKee went on to stress how Noether's methods were directly applicable to her own work and living. Her remarks seem particularly relevant today, as the math community seeks words and ways to communicate to policy makers, and the public, the value of mathematics: "Miss Noether's methods of working and thinking became the basis for my analytical work at the research agency of the Pennsylvania State Legislature for almost thirty years. It is probably heresy for me to mention this in front of so many theoretical mathematicians but there is a great need in government for abstract imaginative thinkers to help solve all sorts of problems. For example: What are the basic cost factors in a given government funded program? What is the taxpayer's money really accomplishing? During my career we searched for answers to these questions in such areas as the construction of public school buildings, the operating of State mental hospitals, the faculty workload at various levels of education, highway engineering as directed toward traffic safety. We chewed over the characteristics and searched for the basic independent variables when considered from all possible points of view. Other times the problem was to find the relevant variables to determine an equitable distribution of appropriations. What was the most important factor? population density? financial need? or simple geography? . . ." (See the symposium proceedings.)

and interests. So I set up a number of new committees[27] to address these issues and involve many more women in the workings of the AWM."

Also during this period, the AWM Speakers Bureau—funded initially by grants from Polaroid, then Sloan, and directed by Judy Wason—became fully functional.[28] Intended to improve the visibility of women in mathematics, the Speakers Bureau provides lists of speakers and topics appropriate for high schools and colleges.

LINDA ROTHSCHILD (1983–1985): A PERIOD OF TRANSITION, THE WHITE HOUSE, A MATHEMATICAL MENTOR

Linda Rothschild speaks of her presidency as "a period of transition: AWM was becoming established as a 'serious' and 'respectable' mathematics organization at that time (for better or for worse!). . . . Even the White House recognized AWM as a serious organization by inviting its president to a luncheon for women's professional group leaders in honor of Women's Business Day."

Keeping with AWM tradition, Linda organized a panel, "Mathematics and Computers," for the 1983 JMM in Denver, well before this topic became fashionable in the larger mathematics community.[29] "Of the various panels I put together for the national meetings," Linda writes, "perhaps the most applauded was the one honoring Lipman Bers on his seventieth birthday [at the 1984 JMM] for his contribution to nurturing the success of so many female graduate students." Echoing the sentiments felt by many of us, she adds, "If only there had been ten others like him, think how many more women mathematicians there might be! Professor Bers had had 40 Ph.D. students of whom 16 were women. . . ."[30]

What was it that made Bers such a good advisor of women students? Linda Keen provides some insights: "He gave us all, and probably the

27. These included the Committee on Mathematics Education, chaired first by Evelyn Silvia and then by Sally Lipsey, and the Maternity Committee, chaired by Anita Solow.

28. Special credit for securing funds is due Eleanor Palais, longtime chair of the AWM Fundraising Committee, and to Mary Gray and Alice Schafer.

29. The panelists (Nancy Johnson, Louise Hay, Lucy Garnett, Marci Perlstadt, and myself) talked about personal computing, running a math department with computers, evolution from mathematician to the field of computers, and computers in, and influence on, mathematical research—all quite novel topics at the time.

30. For remarks on previous mathematical mentors of women in the United States, see Judy Green and Jeanne LaDuke, "Women in the American Mathematical Community: The Pre-1940 Ph.D.'s," *Mathematical Intelligencer* 9(1), 1987, 11–23. Of their group of 229 pre-1940 Ph.D.'s in mathematics, more than a third were advised by eight mathematicians: Charlotte Angas Scott and Anna Pell Wheeler (at Bryn Mawr), and six men—Frank Morley (at Johns Hopkins) and A. B. Coble (at Johns Hopkins and Illinois), Aubrey Landry (at Catholic University), Virgil Snyder (at Cornell), and Gilbert Ames Bliss and L. E. Dickson (both at Chicago, where together they advised 30 women Ph.D.'s). It is not hard to surmise that each of these men felt secure in his position in mathematics. Like Lipman Bers, all but one were at one time president of the AMS!

women needed it more, a confidence in our own abilities. He took it for granted that we would expect to have families and that we would continue anyway."

LINDA KEEN (1985–1987): KOVALEVSKY SYMPOSIUM, ROBINSON MEMORIAL

Linda Keen recalls her stint as AWM president: "The first highlight was the Sonia Kovalevsky celebration run by the AWM at Radcliffe together with the Mary Bunting Institute [in October 1985].[31] This was a two-part affair. The first was a program for high school seniors. . . . There were talks about mathematics as well as talks about careers. The students had . . . a chance to talk informally to a number of women mathematicians." This event was to become the model for the many Sonia Kovalevsky High School Days sponsored since by AWM.

"The second part of the affair," Linda continues, "was more 'my baby.' It was a serious mathematical conference on the theme of mathematics that had grown out of Kovalevsky's work. There were about ten speakers, more than half women." Three special sessions[32] in conjunction with the Kovalevsky Symposium were held at the AMS meeting in Amherst two days earlier. *The Legacy of Sonya Kovalevskaya*, a collection of papers presented at both events and edited by Linda, was published in 1987.

"Then there was the Julia Robinson Memorial session sponsored jointly by the AWM, the AMS, and the MAA [New Orleans, January 1986]. It was really a super affair with great talks." Constance Reid, Julia's sister and biographer of mathematicians, spoke about Julia's life.[33] Lisl Gaal gave a brief description of Julia's thesis, and Martin Davis a retrospective of her mathematics. Julia Robinson was a role model for many of us long before we understood what that expression meant. As vice president of the AMS, Julia intervened when the council would not consider a motion to move a meeting from a non-ERA state because the motion was not on the agenda. Julia pointed out this was an emergency situation. The motion passed and the meeting was moved. Linda recalls that when Julia was AMS president "she really made sure women were placed on important committees—and was very supportive to me, both as council member and as president of the AWM."

Gender, Mathematics, and Science. In the mid-1980s, there was a flurry of work by a group of feminist theorists on gender and science. In commentary fairly critical of this work, Ann Hibner Koblitz succinctly

---

31. The celebration was organized by Bernice Auslander and Pamela Coxson with help from the whole Boston group.

32. The sessions were organized by Jane Cronin Scanlon, Lesley Sibner, and Jean Taylor.

33. See "From the Twentieth Century," I.1, for Robinson's story.

summarized the main ideas behind the theory: "Put in its most general guise, the new 'gender theory' says that centuries of male domination of science have affected its content—what questions are asked and what answers are found—and that 'science' and 'objectivity' have become inextricably linked to concepts and ideologies of masculinity." She lists eight criticisms of which I will mention only two, namely, that gender theorists "seem unaware of the increasing numbers of women who have had satisfying lives as scientists" and "employ cartoon-character stereotypes of science, scientists, men, and women."[34]

A letter from Mary Beth Ruskai in the May–June 1986 *Newsletter* expressed concerns felt by many of us "that a few very vocal and visible sociologists are succeeding in promulgating opinions that are detrimental to the advancement of women in science." Ruskai discusses a rash of articles in the popular press where arguments presented by gender theorists invoke a number of stereotypical misconceptions. For example, she points out that "instead of being concerned that women with an aptitude for computing, science, and mathematics were going into other fields," some advocates of the theory see this as a virtue—women are not interested in science because it does not deal with subtleties. Ruskai is critical of the dichotomous distinction between "artistic" and "technical," the cures for math/physics anxiety devoid of proper math preparation, and the recurrent idea "that women are more intuitive than men, where intuition and logic are perceived as opposites."[35]

Ruskai's letter generated more response than any earlier article in the AWM *Newsletter*; a number of responses appeared in the November–December 1986 issue, including the letter to the editor preceding this article. Marianne Nichols expressed an alternate point of view: if we understand "the biases that do exist in math and science today," then we can "see how they limit what we can know and understand. From there one perhaps can begin to expand and enrich these fields."[36]

RHONDA HUGHES (1987–1989): ACCEPTANCE BY THE ESTABLISHMENT, AMS CENTENNIAL

"By the time I became AWM president," writes Rhonda Hughes, "the organization had clearly gained the acceptance of the mathematics establishment (whether we wanted it or not). I could tell, because all sorts of people began to talk to me who had never done so before."

34. For a thoughtful and well-articulated account of this viewpoint by one of its key theorists, see Evelyn Fox Keller, *Reflections on Gender and Science* (New Haven, Conn.: Yale University Press, 1985). In this book, Keller calls for a science "in which difference, rather than division, constitutes the fundamental principle for ordering the world."

35. Mary Beth Ruskai, "Letter on Feminism and Women in Science," AWM *Newsletter* 16(3), 1986, 4–6.

36. AWM *Newsletter* 16(6), 1987, 6–7.

While she was very much a participant in establishment activities, during Rhonda's term AWM was still mindful of its unique perspective and role in the mathematics community. AWM's panel discussion "Response to the David Report" (San Antonio, January 1987) focused on initiatives for women and minorities. Panelist Fern Hunt emphasized the need to increase the diversity of people doing mathematical research, not only from a political or social point of view, but to promote the "diversity of ideas—one of the prerequisites for progress in mathematics." So to the well-known line from *Casablanca*, "Round up the usual suspects!" she would add, "And round up the unusual ones too!"

AWM's presence was very much in evidence at the AMS Centennial Celebration, Providence, 1988.[37] The AWM panel "Centennial Reflections on Women in American Mathematics" focused on a century of contributions and experiences of women mathematicians. Judy Green and Jeanne LaDuke discussed their findings on the substantial presence of female Ph.D.'s in American mathematics before World War II—and the dramatic decline after the war. Mabel Barnes, Olga Taussky-Todd, and Vivienne Malone-Mayes reflected on their own experiences, giving us a rare and personal accounting of this history as well as a unique glimpse into their own lives.[38]

Reflecting on her presidency, Rhonda points to growing pains as well as significant achievements. "In my time, we still seemed plagued by the research-education tension in our membership. This appears to be less of a problem now, with the wide range of programs and activities we've taken on. . . . I am most pleased with the establishment of the Travel Grant program and the Schafer Prize."[39]

JILL MESIROV (1989–1991): LOOKING OUTWARD, THE TWENTIETH ANNIVERSARY

"I think of the past two years as a time when the AWM began to look outward to the rest of the mathematical sciences community," Jill writes in one of our many email conversations. "Our major success in this was the beginning of an ongoing presence at SIAM national meetings. . . . My goal in these efforts was to broaden our representation, influence, and activities beyond the AMS and MAA. And I think this is happening."

37. At the formal ceremonies, Rhonda presented the AMS with congratulatory wishes from the AWM and in appreciation was presented a silver bowl from the AMS (which is now ritually passed down from one AWM president to the next at inauguration time).

38. Their stories appear in "From the Twentieth Century," I.1, and "A Dual Triumph," III.1.

39. For more information, see "Activities and Awards," which follows this article.

"I also think of the last two years as a kind of 'coming of age,' " she continues. "We have really expanded the scope of our activities. . . . I think that our relationship with NSF has become quite strong and vital over the past couple of years. They really view us as giving lots of value for the money they invest in our programs and are very keen these days to fund programs to encourage women and minorities in the sciences. Exxon has also been an important partner for us, giving us yearly grants toward our operating expenses as well as supporting the complete revision of the Resource Center."

As is characteristic of all superachievers, Jill sees projects yet undone, some very close to home. "One area that I wasn't able to make progress on (directly or indirectly)," she points out, "is the two-career family and children issue. One can't do everything, I guess. It's funny because in many ways this is something that is really important to me because it really has an impact on my life everyday."

How Things Are in 1991 . . . (An Assessment)

Has AWM made a difference? Of course. Before AWM, Judy Roitman responds, "It was not uncommon for major women mathematicians to be unemployed; young women were routinely discouraged; the few who persevered were usually treated badly; and role models were few and far between."

The very real involvement of women in the mathematical world today forms a stark contrast to Atlantic City twenty years ago. As Carol Wood, AWM president-elect, puts it, "women are everywhere dense."[40] In the professional organizations and institutions, we are no longer on the outside but rather play key roles within the internal power structure.[41] Witness Deborah Haimo taking over the presidency of the MAA from Lida Barrett as Marcia Sward, MAA executive director, looks on! Women mathematicians have been elected to the NAS, received MacArthur "genius" awards, Sloan Fellowships, Guggenheims, Presidential Young Investigator awards, and, routinely, NSF fellowships and research grants.

During the past twenty years, the percentage of U.S. women receiving Ph.D.'s in mathematics has increased dramatically, from about 6% to over

40. Invited speakers include Christel Rotthaus (AMS/AWM/MAA), Rebecca Herb (AMS/MAA), Maria Klawe (AMS), and Jill Mesirov (MAA). Dusa McDuff is the first recipient of the AMS Ruth Lyttle Satter Prize, established by Joan Birman in memory of her sister to recognize an outstanding contribution to mathematics research by a woman. Dusa's moving and very personal response in the AMS *Notices* 38(1991), 185–187, is highly recommended reading.

41. In the AMS, women are vice presidents, trustees, council members, and chairs of important committees; Julia Robinson was AMS president (1983–1984). Cathleen Morawetz has been director of the Courant Institute; Judith Sunley is director of the Division of Mathematical Sciences at NSF.

20% per year.[42] But, curiously, although there was a significant jump in the number of new female Ph.D.'s in mathematics during the early 1970s, we find that since then the number has stayed amazingly steady, averaging eighty-six per year. What has happened is the number of U.S. male Ph.D.'s in mathematics during this period has dropped by more than half (from 658 in 1974–1975 to 312 in 1989–1990).

How many of these changes are due to AWM and how much to the times in general? "This is a false question," Judy Roitman contends.

> The AWM is the expression in the mathematical community of the broader feminist movement. . . . But without the AWM or some similar group (and I think it was an act of brilliance to form it outside of the existing mathematical organizations rather than a caucus within . . .) the changes for women would have been fuzzier and less specific, driven by affirmative action necessities (which are pretty minimal) and vague changes in public perception, and not directed by our own understanding of what has to be done.

Summing up, Rhonda Hughes voices our shared sentiments: "I think we are entering our twenty-first year more unified and stronger than ever before, with a unique opportunity to have influence on the next generation of mathematicians. What used to be our concerns alone are now the concerns of the entire community, and we can give leadership and vision to the effort to get young people interested in mathematics. After all, we have been thinking about how to do this for a long time."

Activities and Awards

From the 1980s through the present, in partnership with various granting agencies,[43] AWM has sponsored a range of programs to encourage women and girls, from middle school students through senior mathematicians, in their pursuit of mathematics. Prizes and awards are given to honor their achievements. These activities crosscut the periods covered by the two AWM histories and are described briefly here in chronological order by inaugural year of the program.[44]

---

42. See "1990 Annual AMS-MAA Survey," First Report, AMS *Notices* 37(1990), 1217–1230.

43. Financial and in-kind support has been received from granting agencies and others, which have included the Air Force Office of Scientific Research, the Alfred P. Sloan Foundation, the American Mathematical Society, the Department of Energy, the Exxon/Mobil Foundation, the Mathematical Association of America, the Mathematical Sciences Research Institute, Microsoft Corporation, the National Science Foundation, the National Security Agency, the Office of Naval Research, the Society for Industrial and Applied Mathematics, the University of Maryland–College Park, and Wellesley College.

44. See "Across Borders" in this part for descriptions of international activities.

Noether Lecturers

An AWM Noether Lecture has been delivered at each JMM since 1980. These talks are expository gems delivered by distinguished senior women mathematicians. The speakers, along with their titles and affiliations, are given in the list below.

1980 A Survey of Coding Theory
F. Jessie MacWilliams (1917–1990), Bell Laboratories

1981 The Many Aspects of Pythagorean Triangles
Olga Taussky Todd (1906–1995), California Institute of Technology

1982 Functional Equations in Arithmetic
Julia Robinson (1919–1985), University of California, Berkeley

1983 How Do Perturbations of the Wave Equation Work?
Cathleen S. Morawetz, Courant Institute, New York University

1984 Paracompactness
Mary Ellen Rudin, University of Wisconsin, Madison

1985 A Model of a Cardiac Fiber: Problems in Singularly Perturbed
   Systems
Jane Cronin Scanlon, Rutgers University

1986 On Partial Differential Equations of Gauge Theories and General
   Relativity
Yvonne Choquet-Bruhat, Université Pierre et Marie Curie

1987 Studying Links via Braids
Joan S. Birman, Barnard College, Columbia University

1988 Moment Maps in Stable Bundles: Where Analysis, Algebra, and
   Topology Meet
Karen Uhlenbeck, University of Texas at Austin

1989 Large-Scale Modeling of Problems Arising in Flow in Porous Media
Mary F. Wheeler, *then* Rice University, *now* University of Texas at Austin

1990 The Invasion of Geometry into Finite Group Theory (dedicated to
   the memory of Louise Hay)
Bhama Srinivasan, University of Illinois at Chicago

1991 Almost Everywhere Convergence: The Case for the Ergodic
  Viewpoint
Alexandra Bellow, Northwestern University

1992 Oscillators and Networks of Them: Which Differences Make a
  Difference?
Nancy Kopell, Boston University

1993 Hyperbolic Geometry and Spaces of Riemann Surfaces
Linda G. Keen, Lehman College, City University of New York

1994 Analysis in Gauge Theory
Lesley Sibner, Brooklyn Polytechnic University

1995 Measuring Noetherian Rings
Judith D. Sally, Northwestern University

1996 On Some Homogenization Problems for Differential
  Operators
Olga Oleinik (1925–2001), Moscow State University

1997 How Do Real Manifolds Live in Complex Space?
Linda Preiss Rothschild, University of California, San Diego

1998 Symplectic Structures—A New Approach to Geometry
Dusa McDuff, State University of New York at Stony Brook

1999 Aperiodic Dynamical Systems
Krystyna M. Kuperberg, Auburn University

2000 The Mathematics of Optimization
Margaret H. Wright, *then* Lucent Technologies, Bell Labs, *now* Courant
  Institute, New York University

2001 Nonlinear Equations in Conformal Geometry
Sun-Yung Alice Chang, Princeton University

2002 Computing over the Reals: Where Turing Meets Newton
Lenore Blum, Carnegie-Mellon University

2003 Five Little Crystals and How They Grew
Jean E. Taylor, Rutgers University and Courant Institute, New York
  University

2004 Symbolic Dynamics for Geodesic Flows
Svetlana Katok, Pennsylvania State University

Conferences and Symposia

In 1980, the first AWM symposium honored Anna Johnson Pell Wheeler (see AWM *Newsletter* 12(4), 1982, 4–13). The 1982 Noether Symposium and 1985 Kovalevsky Symposium were described in Blum's article above. AWM has also cosponsored national conferences, for example, "Women in Mathematics and Science: Pipeline to the 21st Century" in 1990 with the American Association of University Women (AAUW) and the Society of Women Engineers (SWE). The Symposium on the Future of Women in Mathematics was part of the 1991 Twentieth Anniversary Celebration, where it was exciting to see so many early career to midcareer women giving talks on their work with such confidence and assurance.

The Julia Robinson Celebration of Women in Mathematics was held at MSRI in 1996. Characterized by postdoc and doctoral student workshop participants as "wonderful, energizing, empowering," the conference included talks on Robinson's life and work and on other mathematical subjects, poster sessions, and panel discussions on career issues of importance to the workshoppers. One of them, Sharon Frechette, said, "I was impressed with the message AWM seemed to be sending to young women mathematicians: 'Things might have been difficult for many of us as we were starting out, but it needn't be that way, and we're working to ensure that things continue to improve for women making a career in this field.' "[45] Building on that success, three years later the Olga Taussky Todd Celebration of Careers for Women in Mathematics was organized.[46]

Sonia Kovalevsky High School (SKHS) Mathematics Days

The inaugural SKHS Mathematics Day was held in conjunction with the 1985 Kovalevsky Symposium. In 1987, Alice Schafer ran the first of many SKHS days at Simmons College. In 1993, Mary Gray obtained a grant from the Sloan Foundation and the NSF to fund SKHS days at colleges and universities across the country. Most recently the program has been funded by grants from NSA, Coppin State College, and Elizabeth City State University. At these events, high school women and their teachers are invited to attend a day filled with activities, often including panels and math contests, and always involving hands-on activities.

45. Jean E. Taylor and Sylvia M. Wiegand, "AWM in the 1990s: A Recent History of the Association for Women in Mathematics," AMS *Notices* 46(1999), 32.
46. See "The Taussky Todd Celebration," IV, for a full report.

Travel Grants

The NSF-AWM Travel Grants program was begun in 1988 during the presidency of Rhonda Hughes, while the program was extended by Chuu-Lian Terng to include Mentoring Travel Grants in 1999. This program has helped fund hundreds of women to attend research conferences in their areas or to travel to work with a mentor.

Schafer Prize

Each year at the JMM Joint Prize Session, AWM gives two awards, also proposed by Hughes. The Alice T. Schafer Prize for Excellence in Mathematics by an Undergraduate Woman is supported by an endowment funded by both individuals and scientific societies. The Schafer Prize winners are listed next. See the AWM website, http://www.awm-math.org, for runners-up and honorable mentions.

Eleven of the twelve Schafer Prize winners and co-winners of the 1990s have already earned Ph.D.'s—surprisingly, perhaps, from just five top universities, Duke, Harvard, MIT, Princeton, and Stanford. Ten of these doctoral degrees are in mathematics; Ruth Britto's is in theoretical physics. Britto is now a postdoc at the Institute for Advanced Studies; a recent talk she gave at the University of Texas at Austin was "Field Theories on Non-anticommutative Superspace." Jing-Rebecca Li won the 2002 Householder Prize for the best dissertation in numerical algebra earned 1999–2001. A number of others hold postdocs in strong departments or have tenure-track appointments.

1990 Linda Green, '90 University of Chicago
Ph.D. 1996, Princeton University

1990 Elizabeth Wilmer, '91 Harvard University
Ph.D. 1999, Harvard University

1991 Jeanne Nielsen Clelland, '91 Duke University
Ph.D. 1996, Duke University

1992 Zvezdelina Stankova, '92 Bryn Mawr College
Ph.D. 1997, Harvard University

1993 Catherine H. O'Neil, '94 University of California, Berkeley
Ph.D. 1999, Harvard University

1993 Dana Pascovici, '95 Dartmouth College
Ph.D. 2000, Massachusetts Institute of Technology

1994 Jing-Rebecca Li, '95 University of Michigan
Ph.D. 2000, Massachusetts Institute of Technology

1995 Ruth Britto-Pacumio, '96 Massachusetts Institute of Technology
Ph.D. 2002, Harvard University, Physics

1996 Ioana Dumitriu, '99 Courant Institute, New York University
Ph.D. 2003, Massachusetts Institute of Technology

1998[47] Sharon Ann Lozano, '98 University of Texas at Austin
M.S. 2000, University of Texas at Austin

1998 Jessica Shepherd Purcell, '98 University of Utah
Ph.D. 2004, Stanford University

1999 Caroline J. Klivans, '99 Cornell University
Ph.D. 2003, Massachusetts Institute of Technology

2000 Mariana E. Campbell, '00 University of California, San Diego
Graduate student, University of California, Berkeley

2001 Jaclyn Kohles Anderson, '01 University of Nebraska–Lincoln
Graduate student, University of Wisconsin, Madison

2002 Kay Kirkpatrick, '02 Montana State University
Graduate student, University of California, Berkeley

2002 Melanie Wood, '03 Duke University
Graduate student; 2003–2004, Cambridge University; then Princeton
    University

2003 Kate Gruher, '03 University of Chicago
Graduate student, Stanford University

2004 Kimberly Spears, '04 University of California, Santa Barbara
Graduate student, University of Texas at Austin

Hay Award

The Louise Hay Award for Contributions to Mathematics Education was first
given in 1991. The winners of this annual award have been

47. Earlier prizes were awarded at a summer mathematics meeting; the schedule was changed to
allow inclusion of the winners in the Joint Prize Session. Thus no prize was awarded in 1997.

1991 Shirley Frye, Scottsdale School District, Arizona

1992 Olga Beaver, Williams College

1993 Naomi Fisher, University of Illnois at Chicago

1994 Kaye A. de Ruiz, U.S. Air Force

1995 Etta Zuber Falconer, Spelman College

1996 Glenda T. Lappan, Michigan State University, and Judith Roitman, University of Kansas

1997 Marilyn Burns, Marilyn Burns Education Associates

1998 Deborah Hughes Hallett, Harvard University and the University of Arizona

1999 Martha K. Smith, University of Texas at Austin

2000 Joan Ferrini-Mundy, Michigan State University

2001 Patricia D. Shure, University of Michigan

2002 Annie Selden, Tennessee Technological University

2003 Katherine Puckett Layton, Beverly Hills High School and UCLA Graduate School of Education

2004 Bozenna Pasik-Duncan, University of Kansas

Workshops

The first workshop for graduate students, organized by Jill Mesirov, was held in conjunction with the Twentieth Anniversary Celebration, along with the symposium of recent Ph.D.'s mentioned earlier. The workshops, now funded by NSF and the Office of Naval Research (ONR), are open to all and are held at the JMM and the SIAM Annual Meeting each year. Recent Ph.D.'s speak on their work, and graduate students present posters. Mentoring is an important feature of these workshops, with mathematicians early in their careers matched with midcareer or senior women. Panels and discussion groups are sources of career advice. These workshop features are integrated into AWM conferences such as the Robinson Celebration and the Taussky Todd Celebration.

One motivation for these workshops came from a similar program, "Pathways to the Future," run by Lynne Billard and Nancy Flournoy for young women statisticians since 1988. When Flournoy received the Elizabeth Scott Award, these remarks were made: "Invaluable information on how to prepare and submit a successful grant proposal, how to get ready for promotion and tenure, and how to be an effective teacher and mentor is passed on at the workshops. The Pathways workshops also provide an invaluable opportunity to young women to network and establish connections with other women, young and senior, in the profession."[48]

In 2004 the workshop "After Tenure: Women Mathematicians Taking a Leadership Role" was held. Dedicated to the memory of Ruth Michler, it was designed to prepare women with established careers in the mathematical sciences to become leaders in the profession.

Joint Lecture Series at SIAM Annual Meetings and MAA MathFest

At each MathFest since 1998, there has been a joint AWM-MAA lecture; speakers through 2004 have been Margaret Wright, Chuu-Lian Terng, Audrey Terras, Patricia Shure, Annie Selden, Katherine Puckett Layton, and Bozenna Pasik-Duncan. Beginning in 2005 the AWM-MAA lecture will be named the Etta Zuber Falconer Lecture. The AWM-SIAM Sonia Kovalevsky Prize Lecture series was begun in 2003 with an initial lecture by Linda Petzold; the 2004 lecturer was Joyce McLaughlin.

Essay Contest

In 2001 AWM began sponsoring a contest with prizes for the best essays written by students from middle school through the graduate level on women with careers in the mathematical sciences. Interestingly, a middle schooler won the first grand prize: Alexandra McKinney, Londonderry Middle School, New Hampshire, for "To Infinity and Beyond! A Biographical Essay on Dr. Toni Galvin."

AWM in the 1990s

JEAN E. TAYLOR AND SYLVIA M. WIEGAND

Based on "AWM in the 1990s: A Recent History of the Association for Women in Mathematics" in the AMS *Notices* 46(1), January 1999, 27–38. An expanded version appeared in four parts in the 1999 AWM *Newsletter* 29(1), 8–12; 29(2), 28–31; 29(3),

---

48. Alice Carriquiry, Remarks, Elizabeth Scott Award [online] [accessed 22 June 2003]. Available from the World Wide Web: http://www.stat.missouri.edu/faculty/flournoy/www/Elizabeth%20Scott%20citation.htm.

28–33; and 29(4), 30–37. Taylor was then a professor of mathematics, Rutgers University, where she is now a professor emerita. She is now a visiting member of the Courant Institute for Mathematical Sciences. Wiegand is a professor of mathematics, University of Nebraska–Lincoln. She was a program director for the Division of Mathematical Sciences, NSF. The authors thank Elizabeth Allman, Lenore Blum, Anthony Kapp, Anne Leggett, Roger Wiegand, and the *Notices'* reviewers for their help.

Ever since its founding, AWM has been a passionate organization with a mission: to encourage women to study and to have active careers in the mathematical sciences. The inexhaustible enthusiasm and inspiring example of early AWM volunteers set the standard for significant donations of time and energy by those who followed. As a result the association has become an effective voice and vehicle for the advancement of women in the mathematical sciences.

During the three decades of AWM's existence, the participation of women in the mathematical community has increased considerably. In 1997 the percentage of Ph.D.'s going to women in departments of mathematics and statistics in the United States reached a level higher than ever before: 25% overall (up from 19% in 1990), 29% of those granted to U.S. citizens.[49] Many more women hold entry-level positions, and many more speak at major meetings. In view of these improvements in the status of women in mathematics, is AWM still needed? The answer is yes!

Why AWM Is Still Needed

Problems—sometimes more subtle than in the past—remain for women in mathematics at all levels. For example, the percentages of women receiving Ph.D.'s in the mathematical sciences over the period 1990–1997 varied considerably, making it unclear whether the 25% high point signals a trend. Women's participation at meetings sometimes has measure epsilon.[50] One young woman commented, "I was the most senior woman at the conference I just attended . . . and I was the only woman from the United States. None of the twenty-plus speakers were women." Lenore Blum's prediction at AWM's Twentieth Anniversary Celebration that there would be significant numbers of tenured women in the U.S. top ten departments by 1996 has still not come to pass. While the number of new women mathematicians holding academic positions has increased, women are still scarce among tenured and full professors at most institutions, particularly those in the U.S. top ten. Discouragement, disparities, and lower expectations for women in mathematics are still in evidence.

49. AMS *Notices* 45(1998), 1160–1161.
50. To mathematicians, in this context "measure epsilon" means "close to zero."

Young women and girls still hear the message that "math isn't cool for women." For example, early in the 1990s Mattel created a Barbie doll that said "Math is tough"; fortunately, the doll was eventually recalled, thanks to the outrage expressed by AWM members and others [ND92, JF93].[51] Girls attending a Nebraska high school math camp said they "could not" tell their peers, because math camp was "socially unacceptable." Some high school guidance counselors still steer girls away from mathematics. Sometimes promising women undergraduate mathematics majors are pressured to choose teaching instead of a research career; this happens far more often for women than for men. While teaching is a rewarding and valuable occupation, mathematically talented undergraduate women should be permitted to develop their talents and to pursue the career that suits them best.

As for graduate school, women are more likely to drop out than men. While the percentage of women entering graduate school in mathematics has increased to 50% in some schools, at others it has decreased. Data compiled by Joan Birman show that at the top ten institutions it decreased considerably in 1997–1998 from earlier high points [ND97].

Entry-level job opportunities for women and men now seem to be equal,[52] but women are neither promoted nor rewarded as often as men. In 1993, the median annual salaries of women and men doctorates in the mathematical sciences in the United States were $57,200 for men, but for women only $50,300, or 88% of men's salaries.[53] Generally, women are more numerous in nondoctoral institutions and in positions that are not tenureable. Forty percent of part-time mathematics faculty were women. In 1995 women in the United States earned 45% of the undergraduate degrees in mathematics and 23% of the Ph.D.'s, but only 14% of all tenured mathematics faculty and only 8% of the tenured doctoral faculty in Ph.D.-granting departments were women.[54,55]

Susan Landau, an AWM panelist in January 1995 on the topic "AWM: Why Do We Need It Now?" located sixty-five of the eighty people awarded MIT Ph.D.'s during 1980–1984, including thirteen of the fourteen women. Fourteen of the men were tenured at the top forty institutions, but only one of the women. Twenty-five of the men were tenured at doctorate-granting institutions but only two women. Overall, including departments that do not award the doctorate, thirty-nine of the men had tenure, but only seven of the women [MA95].

---

51. The abbreviations in brackets throughout the article refer to issues of the AWM *Newsletter*.

52. Marie A. Vitulli and Mary E. Flahive, *Notices* 44, No. 3, March 1997.

53. National Science Foundation, *Characteristics of Doctoral Scientists and Engineers in the United States: 1993*, NSF 96-302, Table 38. Arlington, VA, 1996.

54. Donald C. Rung, "A Survey of Four-Year and University Mathematics in Fall 1995: A Hiatus in Both Enrollment and Faculty Increases," *Notices*, September 1997, 923–931.

55. Paul W. Davis, "1996 AMS-IMS-MAA Annual Survey (Second Report)," *Notices*, September 1997, 911–921.

TABLE 1. Women in Mathematics in the Top Ten Departments in 1992

Department	Tenured		Untenured		Tenure-track/Could Lead to Tenure[a]	
	Total	Female	Total	Female	Total	Female
UC-Berkeley	60	2[b]	12	3	2	0
Caltech	13	0	6	0	1	0
Chicago	25	0	24	2	6	0
Columbia	14	1[c]	12	0	0	0
Harvard	17	1	14	3	1	0
MIT	40	0	38	4	12	1
Michigan	49	1	38	6	3	1
Princeton	31	0	28	7	22	5
Stanford	23	0	9	1	2	0
Yale	16	0	11	1	3	0
Total	288	5	192	27	52	7

Note: Reprinted with permission from "Women in Math Update" in "Random Samples" column, Constance Holden, ed., *Science* 257, July 17, 1992, 323. © 1992, AAAS.

[a] These individuals are also included in the "Untenured" columns.

[b] One was a joint appointment with UCLA.

[c] Tenured at Barnard College.

The percentages of women in mathematics departments at elite institutions remain dismal. Some universities, such as Northeastern and Rutgers, have markedly fewer tenured women in mathematics now than twenty years ago. In 1991–1992 there were five tenured women and 283 tenured men in the top ten departments (NAS ranking), while there were 27 untenured women and 165 untenured men.[56] There were no women at Caltech, and no tenured women at the University of Chicago, MIT, Princeton, Stanford, or Yale. Since then there have been some positive changes; for example, Princeton University now has two tenured women, and the University of Michigan has four. MIT has an increased number of untenured women. Still the total number of women is small. (See tables 1 and 2.)

56. "Women in Math Update" in "Random Samples" column, Constance Holden, ed., *Science* 257, July 17, 1992, 323.

TABLE 2. Women in Mathematics in the Top Ten Departments in 1998

Department	Tenured		Untenured		Tenure-track/Could Lead to Tenure[a]	
	Total	Female	Total	Female	Total	Female
UC-Berkeley	60	3	12	3	2	0
Caltech	12	0	3	0	3	0
Chicago	31	0	24	3	8	0
Columbia	17	1[b]	13	2	0	0
Harvard	16	0	14	2	0	0
MIT	36	0	40	10	12	4
Michigan	58	4	44	9	1	0
Princeton	23	2	20	3	14	1
Stanford	22	0	9	1	2	1
Yale	15	0	8	0	1	0
Total	288	10	183	31	42	6

Note: This information was obtained from the departments by the authors. The numbers for Princeton are considerably smaller than those in table 1, presumably because the faculty were categorized differently in 1998 than in 1991.

[a] These individuals are also counted in the "Untenured" columns.

[b] Tenured at Barnard College.

It may seem that outstanding new women Ph.D.'s who obtain jobs at top institutions no longer encounter discrimination, but some of these women notice they are treated differently than comparable men. They report that male students (and even colleagues) accuse them of getting jobs, awards, and attention just because they are women. Some female students, perhaps expecting perfection when they finally see a role model, are also critical of women faculty. Women who are not in top positions feel that their faults are magnified and that they are disparaged far more than comparable men.

In a 1993 letter to the AWM *Newsletter*, Melvin Rothenberg, mathematics professor at the University of Chicago, observed,

> Thirty years ago discrimination against women was rampant and open. More than one distinguished colleague vowed never to accept a woman as a student. Now discrimination is not open, if only out of fear of legal

action. At the same time I wonder how much better it is for women. The top five research departments have literally less than a handful of tenured women. . . . There is no doubt that there exists an environment and attitude at our leading mathematical institutions that many women find hostile and alienating. . . . This environment is deeply discouraging to women graduate students and is a significant factor in limiting their careers. . . . We can and should regard the absence of women in our ranks as a weakness and take appropriate action. [ND93]

For these and other reasons AWM is still needed. Rather than emphasize negatives, however, this article focuses on the accomplishments and the spirit of AWM. Its programs have been enormously helpful to younger women in mathematics. Those interviewed for this article describe their experiences as "exciting and inspiring." As Cheryl Grood, a workshop participant, says, "AWM helped bring me into the mathematical community at each different stage and level in my mathematical career."

AWM's Activities in the 1990s

During the 1990s, AWM continued many of its earlier activities and expanded into new areas. This has resulted in an incredible number of activities conducted by AWM.[57] Unpaid volunteers—Meetings Coordinator Bettye Anne Case, Newsletter Editor Anne Leggett, presidents, treasurers, and numerous committee members—work hard on AWM projects. Case and Leggett, energetic, dedicated women who have served AWM in their posts since 1983[58] and 1977, respectively, have made enormous contributions to the continuation, the memory, the shape, and the dream of AWM.

AWM publishes its bimonthly *Newsletter* under Leggett's direction; women members say each issue "recharges" them and helps them fight feelings of isolation. In 1998 AWM established a website, http://www.awm-math.org, through the efforts of volunteers Tamara Kolda and Barbara Ling. AWM also sponsors AWM-Net, an electronic mail forum for AWM members started in 1994 by Dianne O'Leary. AWM has published *Profiles of Women in Mathematics: The Emmy Noether Lecturers* and the booklet *Careers That Count* (1991).

Among AWM's major events at mathematical meetings are the workshops. Besides dispensing advice to new female mathematicians, the

---

57. Many AWM activities and awards are described elsewhere in this volume and are omitted here.

58. The position of meetings coordinator was added to the AWM bylaws at this time, but Case had been unofficially performing these duties since about 1980 as a member-at-large of the AWM Executive Committee.

workshops highlight their achievements; outstanding graduate students and recent Ph.D.'s receive travel support to attend the meeting, present posters, and give talks. These women value the workshops "for networking with each other, for discussing career difficulties, and for being inspired seeing so many women doing such excellent mathematics." Other AWM activities at meetings include lectures, sessions, prizes, panel discussions, business meetings, and parties—the last are large, joyful occasions with many opportunities for informal networking. Helen Moore, a workshop participant in 1995, remarked about her experience at that meeting: "Every time I talked math with someone, I gained information or insight that advanced my work. . . . [At] every AWM-sponsored event, I gained energy and made plans. . . . And aren't these two areas [our careers and our personal lives] the ones in which AWM strives to make a difference for women?" [MA95]

The association regularly participates in activities of other major organizations of the international mathematical, scientific, and education communities. AWM is a member of CBMS, a consultant and presenter for the Board of Mathematical Sciences, and an affiliate of the AAAS. AWM formed an Affiliated Research Group to assist the national effort led by NCTM to delineate standards for K–12 education in the United States.

Judy Green, AWM treasurer from 1992 to 1996, helped set up the present AWM office after its move to the University of Maryland in 1993.[59] Meetings, membership, and marketing director Dawn Wheeler joined the staff then and immediately conducted a major membership drive. Office expenses to support publications and programs have been partially funded by unrestricted grants from the Exxon Education Foundation. Historical material about AWM is housed at the Wellesley College Library under the supervision of AWM Archivists Alice Schafer and Bettye Anne Case. As former president Chuu-Lian Terng said,

> The list of AWM activities is impressive, but many people probably do not realize that to continue having these programs requires enormous effort by the small AWM staff and by many women mathematicians writing proposals for funding, running the programs, and serving on various committees. During my term, I seemed to be constantly asking people to help AWM, and one of the most rewarding things about my job was that people would say yes and even seem honored to be asked. This says a lot about how our organization is perceived by the mathematical community.

59. Green's efforts in this project were Herculean. She also served AWM earlier as a co–vice president and for many years wrote *Newsletter* reports on the status of women in the profession.

Projects of AWM Presidents in the '90s

Although AWM presidents spend much of their terms (including their years as president-elect and past president) applying for grants, consolidating the initiatives of previous presidents, and responding to various crises, each has managed to put her own distinctive stamp on the organization. Here are the AWM presidents of the 1990s and a few of their projects.

JILL MESIROV, 1989–1991, began the ongoing AWM presence at SIAM National Meetings. She initiated AWM Workshops, the Twentieth Anniversary Celebration, and revision of the AWM Resource Center at Wellesley College. The Schafer Prize (established during Rhonda Hughes's presidency) and Hay Award were first awarded during Mesirov's term.

CAROL WOOD, 1991–1993, increased the influence of AWM in national policy and stabilized the organization during its growing pains. The booklet *Careers That Count* was produced and distributed to schools. At the end of Wood's term, AWM had about 2000 members.

CORA SADOSKY, 1993–1995, organized the move of AWM headquarters to the University of Maryland and the concurrent staff changes. She increased AWM's international connections and involvement in science policy, in particular initiating (in coordination with other organizations) the first Emmy Noether Lecture at an ICM in 1994 and representing AWM at the International Congress of Mathematics Education in 1993.

CHUU-LIAN TERNG, 1995–1997, initiated a fund-raising drive (coordinated by Sylvia Wiegand), emphasized mentoring activities (she and Karen Uhlenbeck began the Institute for Advanced Study/Park City mentoring program for women), and promoted discussion and writing about affirmative action. The Robinson Conference was held during her term.

SYLVIA WIEGAND, 1997–1999, joined with officers of the AMS and other scientific societies to promote government funding for science and mathematics. One of the few presidents from the "heartland" of the United States, she traveled and spoke on behalf of the AWM throughout the United States, at the ICM, and elsewhere.

JEAN E. TAYLOR, 1999–2001, while president-elect, was a midwife to the creation of the AWM website. She worked with others to strengthen the infrastructure of AWM and initiated the Corporate Task Force. The Taussky Todd Conference was held during her term.

Issues of the '90s

Equal Opportunity/Affirmative Action. In 1992, with a bad job market, some universities were rumored to be trying to make up for past inequities by offering no position unless a qualified woman could be hired. AWM President Carol Wood found this awkward for AWM and asked members for advice on what stand to take [JF92]. Then, in January 1994, AWM President Cora Sadosky arranged an AWM panel, "Are Women Getting All the Jobs?" that addressed the fear that the job crisis, in Sadosky's words, "would be much better if it were not for all those women and minorities or all those foreigners who are taking all the jobs" [MA94]. She added, "We strongly believe that this is false and dangerous, that pitting one group of under/unemployed mathematicians against another is just the old tactic of dividing people with similar interests in order to exploit them all." Women were apparently not receiving preferential treatment; 18% of Ph.D.'s from the top-ranked mathematics departments went to women, but only 14% of those getting positions at these institutions were women. Overall, 22% of all mathematics Ph.D.'s were earned by women, and 21% of all entry-level positions at Ph.D.-granting institutions went to women [JF95].

Affirmative action came under attack again in 1995. In response AWM published a series of articles in the *Newsletter*, the Executive Committee passed an official AWM statement in support of affirmative action [JF96], and affirmative action was the topic of the 1997 JMM panel discussion.

Issues for Professional Couples. When AWM began, its focus was women's status in the profession; family-life issues were not often mentioned. In the nineties women with families are more numerous in mathematics; we recognize that successfully juggling families and careers is essential to many women's advancement. Women mathematicians, frequently paired with other mathematicians, often face the difficult "two-body problem" of finding jobs together or living apart. Maternity leave policies at most institutions had been nonexistent or haphazard, so in 1991 AWM drafted a sample maternity leave policy [MJ91]. At the January 1998 AWM panel "Mathematicians and Families," women and men mathematicians gave helpful advice on many family issues [MA98].[60]

Nature versus Nurture. Periodically pseudoscientific "experts" claim that women have inherent mathematical deficiencies. As AWM members and supporters eloquently argued in a series of letters to the newsletter in the nineties, cultural factors surely overwhelm any possible innate differences in mathematical aptitude between males and females [SO90, ND90, MA91].

Sexual Harassment. In [JF92] Marjorie Senechal and Jean Taylor wrote: "Why did women mathematicians wait all these years to say anything

---

60. In this volume, "Pathways in Mathematics," III; "Having a Life," III.4; and "Into a New Century," V, address these complex issues.

about this issue, even to one another? Because, until Anita Hill's testimony, sexual harassment has been a private embarrassment. . . ." Hill had testified at the confirmation hearings of Supreme Court Justice Clarence Thomas that he had sexually harassed her. An AWM Statement on Sexual Harassment was published in the *Newsletter* [ND93, MJ97].

Teaching Evaluations. In an article "Are Student Ratings Unfair to Women?" [SO90], Neal Koblitz analyzed data on student ratings of instructors by sex and concluded that students often rate the same performance differently for women and men. Koblitz's article has been widely circulated by women mathematicians who have found it useful in conversations with chairs, deans, and other administrators, not to mention graduate students and their fellow mathematicians. [The article appears in "Inside the Academy," III.2.]

Denial of Tenure Cases. The case of Jenny Harrison, a University of California at Berkeley mathematics department faculty member who was denied tenure and fought the decision, shook up the academic community and commanded media attention. AWM members were divided about the case but were united in the opinion that AWM takes positions on policy matters, not individual cases [ND92, SO93, ND93].

Lobbying. The 1990s have seen increased activism to encourage adequate funding for mathematics by the U.S. government. In 1997 AWM joined with the AMS, SIAM, MAA, and a hundred other scientific societies in a concerted effort to lobby the U.S. government in support of science (including mathematics) and education. AWM representatives participated in a press conference; spoke to congressional representatives, senators, and aides; and encouraged AWM members to help with this effort. Before this lobbying effort, funding in stable dollars had been decreasing for science and technology. Some legislators adopted science as something positive to promote, something that inspires general approval by the public; as a result, the NSF fared better than expected [JA97].

Milestones

The climate and opportunities for women in mathematics are much brighter than in the past. For example, in 1971 there were no invited addresses by women at the JMM [ND91]. In contrast, each JMM from 1993 to 1998 has featured at least four invited hour addresses by women (including the AWM Noether Lecture). At the summer MathFests in the '90s, two of the Hedrick Lecturers have been women. In fact, the major mathematics organizations have established guidelines that encourage organizers to include women; women often have leadership positions in these organizations or serve on program committees for meetings. Credit is due to individual women mathematicians for their many wonderful accomplishments and to AWM for enlightening the public about the shortage of women in mathematics.

In 1996 Cathleen Morawetz and Margaret Wright, presidents of AMS and SIAM, respectively, were part of a remarkable phenomenon: during that year women presided over eleven major organizations for mathematical scientists and educators in North America plus the umbrella scientific society, the AAAS.

1996: Women Preside*

American Mathematical Association for Two Year Colleges: Wanda Garner; AMS: Cathleen Morawetz; Association of State Supervisors of Mathematics: Mari Muri; American Statistical Association: Lynne Billard; AWM: Chuu-Lian Terng; Canadian Mathematical Society: Katherine Heinrich; Institute of Mathematical Statistics: Nancy Reid; NCTM: Gail Burrill; National Council of Supervisors of Mathematics: Bonnie Walker; SIAM: Margaret Wright; Sociedad Matemática Mexicana: Patricia Saavedra; AAAS: Rita Colwell.

* An earlier version of this list appeared as "Women Presidents" in the AWM *Newsletter* 26(6), 1996, 9, based on an observation of Kenneth Ross, then president of MAA.

In 1998, for the first time in the twenty-four years of U.S. participation in the Olympiad, an international mathematics competition for high school students, the U.S. team included a young woman, Melanie Wood, a silver medalist from Indiana.[61] In another first, the Canadian team included two young women, Mihaela Enachescu and Yin (Jessie) Lei. Altogether among the top twenty countries there were thirty-eight women.

Summing Up

The energy and resources AWM supporters have put into programs improving the representation and climate for women in mathematics have helped to change the face of mathematics. As past president Terng said recently: "About half of the undergraduate degrees in math are now earned by women, and there are many more strong young women researchers.

61. Wood received AWM's Schafer Prize in 2002 as a junior at Duke University.

Many departments are more conscious of the need for putting more effort into nurturing their women students."

AWM hopes to expand its high school programs and to extend its efforts to the elementary grades; it hopes to cooperate more with other organizations to increase the participation of women in mathematics and science. We hope to help persuade more women undergraduates to study mathematics, to expose them to more women in mathematics, and to aid them in the process of learning mathematics so that they can succeed in a wide variety of graduate programs. At the graduate level, AWM will continue to encourage the formation of Noetherian Ring chapters (support groups for graduate students).[62] We hope to offer better mentoring of more recent Ph.D.'s and advanced graduate students. Our entire profession benefits from helping beginning and midcareer mathematicians reach their potential.

As Cora Sadosky observed:

> Our Association really makes an impact on the situation of women in mathematics. And it is a great privilege to work for something that matters. . . . Many gains have been made in the twenty-two years of existence of AWM. Still, women continue to face formidable problems in their development as mathematicians—from elementary school to graduate school to the National Academy and beyond. To successfully confront these problems we need the ideas and the work, the enthusiasm and the commitment of all—students and teachers and researchers and industrial mathematicians—of every woman and every man, who stand for *women's right to mathematics*. [MA93]

Affirmative Action: What Is It and What Should It Be?

CORA SADOSKY

Excerpts from an invited address to the CMS, delivered at the Fiftieth Anniversary CMS Meeting in Toronto; based on "Affirmative Action: What Is It and What Should It Be," AWM *Newsletter* 25(5), 1995, 22–24. Sadosky is a professor of mathematics, Howard University.

People often remark that in the last decade the situation of women in mathematics has dramatically improved, both in Canada and in the United States. And they are right! Katherine Heinrich, who was recently elected

---

62. AWM has recently inaugurated a program of student chapters. See www.awm-math.org.

president of the Canadian Mathematical Society (CMS), is not only a talented mathematician and a great organizer who happens to be a woman, but is an active militant for women's inclusion in mathematics. Another sign is that Ingrid Daubechies, mother of two toddlers, has become the first woman full professor of mathematics at Princeton University, where, for the first time in its history, half a dozen young women at different ranks are teaching (and doing!) math.

So, some people argue, "You see, it was a matter of time, things are improving, why do you persist with your sessions on women in math?"

Well, why indeed? One reason is that what has been achieved, although considerable—against the bleak background of a field where women were invisible only a few years ago—is still quite modest. More important, what has been achieved can be reversed, and a slight change of social climate can provide a pretext for such a reversal. The current powerful drives against affirmative action both in the United States and in Canada are cases in point.[63] Examples show that when there are no reminders about women mathematicians, colleagues tend not to "remember" us.

For example, consider the participation by invitation of women in International Congresses of Mathematicians (ICMs). Singular, yet illustrative of how women in math may be ignored. In more than a century of ICMs there were no women plenary speakers until 1932 (Emmy Noether), and not again until 1990 (Karen Uhlenbeck). Before the '80s only a handful of women among hundreds were invited speakers. After a public protest at ICM-78 on the systematic omission, a few women were invited to speak at ICM-82 (which met in Warsaw in 1983). But no further reminder about women was made there, and none were initially invited to ICM-86, resulting in new protests, some late invitations, and more women at ICM-90. Finally, in 1990, the International Mathematical Union (IMU) passed a resolution "to take into account that many qualified women were available as speakers for ICM-94." This resulted in the unprecedented number of ten women invited speakers (out of 165 lecturers), two of them delivering plenary lectures, at ICM-94. At every ICM not preceded by an explicit reminder to consider women candidates, many outstanding mathematicians were passed over.

Was the 1990 "reminder resolution" of IMU a call to overlook standards? The larger representation of women at ICM-94 in no way diluted its mathematical quality. On the contrary, the plenary addresses of Ingrid

---

63. "On June 23, 2003, the US Supreme Court upheld the use of affirmative action in college admissions in two cases involving the University of Michigan at Ann Arbor, but struck down the mechanics of Michigan's undergraduate admissions policy." Peter Schmidt, "Supreme Court Upholds Affirmative Action in College Admissions," "Today's News," 23 June 2003, *The Chronicle of Higher Education* [online] [accessed 24 June 2003]. Available from the World Wide Web: http://chronicle.com/free/2003/06/2003062305n.htm.

Daubechies and Marina Ratner at Zürich were indisputably among the best. No chauvinistic critic challenged, at least publicly, the excellence of women's contributions.

The discrimination against women was generic, not specific or personal. Yet it had not disappeared, as suggested by the numbers, and there seemed to be an invisible quota system. The eight women nonplenary speakers lectured in eight of the seventeen sessions, *one per session*. It seemed as if the selection panels, although aware enough to consider women candidates, felt that they had fulfilled their duty when the first one accepted. And this is not an isolated occurrence.

A personal anecdote: Years ago, a friendly colleague told me his department was considering hiring a junior person in our field and asked me for a top candidate. After some thought I mentioned one of the best junior researchers in the field. His answer was "But we already have a woman!" and mine to him, "So, would you hire a man for the job? I assume your faculty already has at least one man!"

Do women need special treatment not to be overlooked? For a variety of reasons, not all of them obvious, the answer seems to be *yes*. And it provides the rationale for the prizes, honorary lectures, and mathematical events "for women only." Is that reverse discrimination? I don't think so.

Both AMS and CMS have recently instituted new honors specially designated for women. Consider the example of the Ruth Lyttle Satter Prize of the AMS, dedicated by the well-known mathematician Joan Birman to the memory of her sister, a chemist. Many people frowned at a prize for women mathematicians. Don't we have enough prizes in the Society? Don't they go to deserving people? Yes to both questions! But none of them have gone to women. Is it because there are no deserving women candidates? Maybe so, but when the Satter Prize went first to Dusa McDuff—later elected FRS (Fellow of the Royal Society)—and then to Lai-Sang Young and to Sun-Yung Alice Chang, each a leader in her field, no one perceived the dreaded "lowering of standards." These are top researchers, worthy of the highest honors, who could compete for any prize. But it took the creation of the Satter Prize to highlight their achievements.

These recognitions have an educational effect on the mathematical community. Not only do some individuals get their due, but, more significantly, their merits are publicized. Thus our whole community learns that there are women who have made outstanding contributions, and what those contributions are. This is especially important in the case of younger people, like Lai-Sang Young, who have already influenced their fields, but who may not have been recognized outside them.

Recognition is important if we want to reach people who do not want to discriminate, but who cannot remember the name of a woman mathematician when making nominations for an editorial board or a selection committee.

And let's be fair. How is it with each of us? I can speak from experience. It required conscientious training to "remember" women every time I put together a conference or a program. I knew only men! Well, then, look for women, even if you do not know them personally. Quite likely you do not know them because nobody else "remembers" them, so they go to fewer places where they can meet others.

Do we need to go to extra lengths to find suitable speakers who happen to belong to underrepresented groups? Absolutely *yes*. When the first person who comes to mind is someone well known, who happens to be white and male, we should automatically think about who else (nonwhite and/or female, not as well known) would make us proud to bring to the same event.

The problem is *not* so much die-hard retrogrades who hate women intellectuals on principle and talk about our less-developed brains and our innate inferiority. The problem is honest people who insist they do not discriminate but are against quotas (those favoring women and minorities, mind you) and who may earnestly believe that everything will be okay if only we ignore the issue, using just initials before last names to conceal gender. The problem is the good guys of both genders who are not trained to be good enough. To them we have to provide information and opportunities to get in contact with women mathematicians they can admire, as well as women colleagues they must respect. We want all mathematicians to remember every day that people who are different from them may still love to do what they love doing: mathematics.

A question often asked when discussing affirmative action measures toward women and other underrepresented groups is "Will they feel demeaned by being included through such a measure?" Many times friends said to me, "Certainly you'd be offended to receive an offer because of your gender and not your mathematics!" Sure I would, if the offer was not appropriate to the level of my work. But I would find it much more offensive not to receive an offer for which I was mathematically competent because of my gender!

As I favor affirmative action measures, let me be explicit about my own "golden rule": accept each opportunity you deserve, but reject exaggerated offers making you stick out as a token who is just fulfilling some real or imaginary quota. Of course, you need a high level of self-confidence to follow this rule, and self-confidence is much less abundant among young women mathematicians than among their male counterparts.

This lack of confidence does not come about by chance. Many women have faced a lifetime of suspicion about their talent, their commitment to work, their credentials. So, how would they feel about getting a job in a "female slot"? Badly, unless they are made to feel wanted for other reasons. Departments should be educated to fill this need.

Senior women can make a difference by not frowning at special programs for women, but taking advantage of them and thus making them acceptable, and even desirable, for younger women. Let me give a specific example. In 1982 NSF initiated Visiting Professorships for Women in Science and Technology,[64] allowing a score of women to visit top institutions each year. This program highlights the existence of women scientists in places where there are few or no women faculty. For instance, outstanding women engineers working outside academia have been able to teach engineering students who had *never* encountered a woman engineer before. In the case of mathematics, women researchers have been provided a great opportunity for financial support to work at top institutions.

But at first it took some courage to apply! Who wanted to be singled out as going places as a woman? Fortunately, someone did, the opportunity was put to good use, no stigma was attached, and soon the numbers soared. Now the list of past recipients of this award looks like a "who's who" of women mathematicians in the United States. It became prestigious, and some institutions offer it as a possibility to mathematicians they want to invite who happen to be women.

So, the program turned out to be successful in placing more women at institutions where they will be well received, where they will underscore the existence of women researchers to predominantly male faculties and to graduate students of both genders, and where they themselves will have excellent research opportunities.

Affirmative action programs of this sort are useful, economical, and produce a lot of good, both to individuals and to their communities. It is up to us to see that they do not disappear in the current anti-affirmative-action hysteria sweeping the United States, and that, on the contrary, they are increased at various levels of the educational and research pipelines. The gains of recent decades could be wiped out in one generation, just as the ICM program committees forgot to invite women when they were not "reminded."

We have achieved much. But we are striving for nothing less than the right of all people to do mathematics. For that we have to work together, women and men, so that the mathematical community no longer needs constant reminders of the existence of women in its midst.

---

64. The VPW program was replaced by Professional Opportunities for Women in Research and Education (POWRE) in 1997, which has since been replaced by Increasing the Participation and Advancement of Women in Academic Science and Engineering Careers (ADVANCE).

Across Borders

BETTYE ANNE CASE AND ANNE M. LEGGETT

Because research collaborators are not infrequently from two or more countries, there has been remarkable mathematical communication across national boundaries, even during the Cold War and despite sometimes-keen scientific competition. The students and postdocs of some professors who directed many women are spread around the globe and retain their "mathematical family" contacts. Major professors and postdoc mentors try to introduce beginning mathematicians to the best practitioners in the same research area, regardless of location; they also want to encourage interaction that will allow candidates for promotion and tenure to demonstrate an international reputation. Within mathematical subareas, there may be only a few women, so it is not unusual for friendships and mentoring relationships, as well as long-term collaborations, to develop among them.[65] Due to the international nature both of the mathematics world and of working women's concerns, women mathematicians from outside the United States have joined AWM's ranks since its inception. And wherever there are 2 or 20 or 200 women, lively discussions arise around universal concerns about work and family lives, about pressures and balance.

Mathematical research conferences attracting global attendance have provided a stage where women may join and share; two important "international congresses" held at four-year intervals are the International Congress of Mathematicians (ICM) and the International Congress of Industrial and Applied Mathematicians (ICIAM), with their invited talks by eminent mathematicians.

At the 1974 Vancouver ICM, AWM members organized a discussion comparing the status of women mathematicians in different countries. Concerns were raised there about the small number of women speakers at ICMs. After Emmy Noether's ICM plenary talk in 1932, the next woman to deliver a plenary address was to be Karen Uhlenbeck, in 1990. During those fifty-eight years, only a few women were invited to talk even in the parallel sectional sessions. (The sidebar "Women Invited as Speakers at ICMs" gives information about these over the past thirty years.) At the Helsinki ICM-78, a meeting sparked by AWM members was held to protest the absence of women speakers. Of over 500 women and men ICM participants who attended, only 3 dissented on the passage of a resolution urging the rectification of this situation for the next congress. The first announcement of

---

65. E.g., Caroline Series (UK) and Linda Keen (U.S.) have worked together for many years; women students and postdocs related to Lars Ahlfors, Cliff Earle, Fred Gehring, and Keen's advisor Lipman Bers form an extended multigenerational family.

ICM-86 speakers included no women in traditional research mathematics areas for either parallel or plenary addresses. A list of about 25 qualified women in a number of fields was immediately presented to the Executive Committee of the ICM with a call for reconsideration; subsequently several were invited for parallel sessions. A panel about women's issues was held at the conference; Linda Keen, then AWM president, read a resolution that was endorsed by the 400 ICM participants present.[66]

Women Invited as Speakers at ICMs*

1974   Jacqueline Lelong-Ferrand, Mary Ellen Rudin [18; 27]

1978   No women were invited to speak. [3; 21; 30]

1983   Nancy Kopell, (Ol'ga) Ladyzhenska, Karen Uhlenbeck, Michèle Vergne [3; 12]

1986   Izabella G. Bashmakova (unable to attend, but her paper appears in the *Proceedings*), Sun-Yung Alice Chang, Judith V. Grabiner, Christine Graffigne (joint with Stuart Geman), Caroline Series, Nina Ural'tseva [15]

1990   Karen Uhlenbeck (plenary), Lenore Blum, Shafi Goldwasser, Dusa McDuff, Colette Moeglin, Mary Rees, Eva Tardos [26; 32]

1994   Ingrid Daubechies (plenary), Marina Ratner (plenary), and at least eight women in parallel sessions [9; 25]

1998   Dusa McDuff (plenary) and at least eleven women in parallel sessions [14; 31]

2002   Sun-Yung Alice Chang (plenary), Shafi Goldwasser (plenary), Frances Clare Kirwan (plenary), and at least twelve women in parallel sessions [10; 11]

---

* Names of categories of speakers have changed over the years, and there is no single source for information by gender. Plenary speakers are indicated as such.

At each ICM since 1974, groups of women have organized special activities. In recent years, these have been encouraged by the ICM organizers, listed in the official program, and advertised in the *Daily Bulletin*.

66. The resolution read: "Whereas scientific quality is and should be the foremost consideration in the selection of individual speakers at International Congresses, the full list of invited speakers should take into account the facts that: mathematics is a broad umbrella for a variety of subfields; there are many women with very high mathematical qualifications; good mathematics is done in very many countries of the world. Therefore, be it resolved that the program committees for all future Congresses should represent this mathematical and geographic breadth as well as both sexes and be instructed to keep the above in mind as they perform their work." See [3, 9, and 11].

There is a panel discussion, after which the audience is invited to comment. Generally, the women and men in the audience are friendly and enthusiastic, no matter how determined or concerned. There have been very few audible hecklers, but the handling of one has become the stuff of legend among women attending ICMs: In 1994 in Zürich, a male from the audience embarked on an impassioned tirade against women seeking special privileges. Ingrid Daubechies, a plenary lecturer at that ICM and one of the many who volunteered to reply, defused the situation in a calm and effective manner, offering to speak with him further at a later time. [25, 3]

A milestone of the special activities was the first Emmy Noether Lecture at the 1994 ICM in Zürich. Program approval was given by the organizers for this mathematical talk jointly organized by European Women in Mathematics (EWM), the Committee on Women of the Canadian Mathematical Society, and AWM. Ol'ga Ladyzhenskaya (1922–2004; St. Petersburg University and Steklov Mathematical Institute) spoke on the topic "On Some Evolutionary Fully Nonlinear Equations of Geometrical Nature." At the 1998 Berlin ICM, the Emmy Noether Lecture of Cathleen Synge Morawetz was given a plenary time in the main lecture hall, where she drew a large audience. Morawetz (Courant Institute, NYU) began her talk, "Variations on Conservation Laws for the Wave Equation," by recounting work going back to Emmy Noether—a surprise to some, since Noether is known primarily as a pure mathematician, an algebraist, whereas Morawetz is an applied mathematician. Hesheng Hu, Fudan University, Shanghai, gave the third international Emmy Noether Lecture at the 2002 Beijing ICM, "Two-Dimensional Toda Equations, Laplace Sequences of Surfaces in Projective Spaces, and Harmonic Maps." With the success evidenced by these three lectures, the U.S. delegation to the International Mathematical Union (IMU) negotiated a resolution that will assure Emmy Noether Lectures in 2006 and 2010.[67]

Looking back, it is clear that much of value has been added to the ICMs from actions born of the protests of the '70s and '80s. Qualified women are now more likely to be invited to speak, and the well-attended adjuncts to the meeting program showcase women who do mathematics. A veritable parade of countries is represented by the women who have participated in "ICM Special Activities on Women in Mathematics." They have created additional visibility for the contributions and concerns of women on this largest of the international mathematical stages.

---

67. In August 2002 the following resolution passed unanimously: "The General Assembly recommends continuing the tradition of the 1994, 1998, 2002 ICMs by holding an Emmy Noether lecture at the next two ICMs (2006 and 2010) with selection of the speakers to be made by an IMU appointed committee." See "ICM News," AWM *Newsletter* 32(5), 2002, 18.

ICM Activities on Women in Mathematics*

1974   FRANCE, Michèle Vergne; NORTH VIETNAM, Xuan Hoang; UK, Sheila Brenner; U.S., Lenore Blum, Mary Gray, Judy Green, Judith Roitman, Alice T. Schafer, Bhama Srinivasan, and Rebekka Struik [27]

1978   FEDERAL REPUBLIC OF GERMANY, Hel Braun; FINLAND, Marjatta Näätänen; FRANCE, Yvette Amice; UK, Mary Kearsley; U.S., Lenore Blum, Judy Green, and Bhama Srinivasan; USSR, Helen Kosachevskya [30]

1986   ARGENTINA, Josefina Alvarez; AUSTRALIA, Jennifer Seberry; BRAZIL, Maria Jose Pacifico; DENMARK, Bodil Branner; FEDERAL REPUBLIC OF GERMANY, Gudrun Kalmbach; FRANCE, Marie-Françoise Coste-Roy; IVORY COAST, Josephine Guidy Wandja; NICARAGUA, Consuelo Flores; UK, Caroline Series; U.S., Lenore Blum, Linda Keen, Evelyn Silvia, and Bhama Srinivasan [3]

1990   BRAZIL, Keti Tenenblat; CANADA, Asia Ivić Weiss; CHINA, Hesheng Hu; FRANCE, Marie-Françoise Roy; INDIA, Rajinder Hans-Gill; JAPAN, Aiko Negishi; NEW ZEALAND, Gillian Thornley; SPAIN, Maria T. Lozano; U.S., Carol Wood [13]

1994   CANADA, Verena Huber-Dyson and Asia Ivić Weiss; FINLAND, Marjatta Näätänen; FRANCE, Lucy Mosler-Jauslin; GERMANY, Christine Bessenrodt; INDIA, Raman Parimala; PORTUGAL, Ana Maria Porto da Silva; U.S., Mary Gray, Krystyna Kuperberg, and Cora Sadosky [25]

1998   CHINA, Minping Qian; GERMANY, Christine Bessenrodt and Ljudmila Bordag; MEXICO, Mary Glazman; RUSSIA, Inna Yemelyanova; SWITZERLAND, Claire Baribaud; U.S., Bettye Anne Case, Dusa McDuff, and Bhama Srinivasan. Film, *Women and Mathematics across Cultures*: FINLAND, Marjatta Näätänen, director; DENMARK, Bodil Branner; NORWAY, Kari Hag; and UK, Caroline Series [8; 31]

2002   CHINA, Li Chen and Hesheng Hu; U.S., Sun-Yung Alice Chang, Paosheng Hsu, Paula Kemp, Suzanne Lenhart, and Sylvia Wiegand [17]

* The available records are incomplete; some speakers were added on site.

Another lasting and positive effect arising from the earlier years of woman-organized ICM protests and activities was the birth of European Women in Mathematics (EWM). Participants from Europe who spoke on the panel or attended the associated activities that AWM organized for the 1986 Berkeley ICM founded, later that year, the sister organization EWM. [5; 6; see also http://www.math.helsinki.fi/EWM/, EWM's website.] Serving as a catalyst in the emergence of geographically defined organizations to meet the unique needs of women mathematicians in those areas is the most far-reaching effect of the earlier protests. (See "Voices from Six Continents" at the end of this part for information on the Russian Association of Women in Mathematics and Women in Mathematics in Africa.)

EWM, in addition to its cooperative activities with other groups, holds regular conferences featuring mathematical papers, employment information, and career issue discussions. In 1994 EWM began an ambitious data project, which has been reported at international meetings and is documented on its website.[68] The surprising results show the highest percentage of women mathematicians in Portugal and Macedonia (about 45%) and the lowest (0%) in Iceland, with Denmark and Switzerland each at 2%. Only France, Georgia, Italy, Luxemburg, and Poland show double-digit percentages of women full professors, while there are none in Austria, Estonia, Ireland, Iceland, Slovenia, or Sweden. It does not seem easy to explain why the "cold" countries appear to have fewer women mathematicians than the "warm." These data observations motivated a 1999 AWM presentation at the JMM about the education of girls around the globe, as the cross-fertilization between AWM and EWM continues.[69]

As mathematicians communicate across borders, they also deal with nonmathematical issues. The human rights case of Tatyana Velikanova, a Russian mathematician and computer scientist, was a flash point for both AWM and EWM. She was sentenced to four years in a labor camp and five years of internal exile after being convicted in 1980 of anti-Soviet agitation and propaganda for her activities in the human rights movement in the USSR. Velikanova's children wrote an impassioned letter imploring for help for their mother, and Amnesty International adopted Velikanova as a prisoner of conscience. In the early 1980s, Lenore Blum led an AWM campaign protesting the harsh sentencing [1; 2]. In 1987 EWM requested its members and AWM to appeal on Velikanova's behalf, as her imprisonment had been difficult [19]. In that year, she was pardoned by the Supreme Soviet and released.[70,71]

68. The website is http://www.math.helsinki.fi/EWM, and the data may be found at http://www.math.helsinki.fi/EWM/tilastot.html. Similar statistics for the United States are given in "Pathways in Mathematics," which begins part III of this book.

69. See the following article, "International Views on Education."

70. She refused the pardon, insisting on a repudiation of the conviction, and initially remained in exile in Kazakhstan. Kazhakhstan News Service, 17 December 1987, as cited in "The Cry of the New Martyrs—Tatiana Velikanova and Yelena Sannikova" in "Orthodox America," Nikodemos Orthodox Publication Society, 1998–2000 [online] [accessed 16 January 2003]. Available from the World Wide Web: http://www.roca.org/OA/75/75e.htm.

71. She died in 2002. Sophia Kishkovsky, "Tatyana M. Velikanova, 70, Soviet Human Rights Activist," *New York Times*, 17 October 2002 [online] [accessed 20 June 2003]. Available from the World Wide Web: http://query.nytimes.com/gst/abstract.html?res=F20814FE3F590C748DDDA90994DA404482.

ICIAM is an important stage for applied mathematicians building their careers. [16; 22; 23] Activities at these conferences also enhance communications between women working on related mathematical ideas and provide mentoring opportunities. For the Sydney ICIAM in 2003, two women from each of Australia and the United States who were recent Ph.D.'s spoke in the session "Applied Modeling and Numerical Simulations." In 1999 at the ICIAM in Edinburgh, AWM and EWM joined with SIAM to organize two sessions featuring women speakers. Four senior researchers discussed their employment in industry or work on industrial problems in an academic setting. Another session showcased women within ten years of their doctorates from Australia, Belgium, the Netherlands, and the United States.

Sometimes two or three mathematical organizations based in different countries jointly organize meetings for their members. A memorable example for U.S. and Canadian mathematicians was held at the University of British Columbia in 1993. For that meeting, the AWM, together with the Committee on Women of the Canadian Mathematical Society, organized a panel on affirmative action, with women and men speakers from both countries.

At shorter mathematics conferences, the time is often completely filled with mathematical talks and sessions, with no open periods other than mealtimes. A short program organized for breakfast or lunch is a good way to share concerns common to women mathematicians. Breaking bread together, whether or not there is a formal program, provides the opportunity both to explore issues tangential to the mathematical purposes of the meeting and to network with a supportive community. For example, in 1992 AWM and EWM members discussed strategies for coping with the bleak worldwide job market as they met together for lunch at the joint meeting of the AMS and the London Mathematical Society at Cambridge University. Over lunch at the Zürich ICM, women in leadership roles in the Canadian, European, and U.S. groups made future plans and shared organizational experiences. As a small café in Berlin overflowed with women (and a few men) meeting for lunch after Cathleen Morawetz's plenary talk, the flustered management exclaimed that they "were not expecting so many women mathematicians."

Both in formally scheduled activities and through spontaneous actions based on joint concerns, women continue to work together globally and to build on past successes. For example, they are planning Emmy Noether Lectures and other jointly organized special activities at the ICMs in Madrid for 2006 and in the future. Their continued sharing of views into the situation for women in their own countries, regions, and continents creates a fascinating web of cross-cultural and transnational perspectives.

References

1. AWM, letter to Alexandrov, AWM *Newsletter* 10(6), 1980, n. p.
2. Lenore Blum, "Tatyana Velikanova," AWM *Newsletter* 11(3), 1981, 3–5.
3. ———, "Women in Mathematics: An International Perspective, Eight Years Later," AWM *Newsletter* 16(5), 1986, 9–21.
4. ———, "A Brief History of the Association for Women in Mathematics: The Presidents' Perspectives," AMS *Notices* 38(7), 1991, 738–774.
5. Bodil Branner, "European Women in Mathematics," AWM *Newsletter* 17(5), 1987, 3.
6. ———, "European Women in Mathematics," AWM *Newsletter* 26(6), 1996, 9.
7. Bettye Anne Case, "AWM-EWM Meeting," AWM *Newsletter* 22(5), 1992, 8.
8. Bettye Anne Case and Bhama Srinivasan, "ICM-98: Women in Mathematics," AWM *Newsletter* 29(5), 1999, 18–25.
9. S. D. Chatterji, ed., *Proceedings of the International Congress of Mathematicians: August 3–11, 1994, Zürich, Switzerland*, in 2 vols. (Basel: Birkhäuser Verlag, 1995).
10. Chinese Mathematical Society, "B.2 Plenary Lectures," International Congress of Mathematicians 2002, Aug. 20–28, Beijing. [online, cited 26 June 2003]. Available from the World Wide Web, http://www.icm2002.org.cn/B/Plenary.htm.
11. Chinese Mathematical Society, "Invited Speakers," International Congress of Mathematicians 2002, Aug. 20–28, Beijing. [online, cited 26 June 2003]. Available from the World Wide Web, http://www.icm2002.org.cn/B/Invited_Speakers.htm.
12. Zbigniew Ciesielski and Czeslaw Olech, eds., *Proceedings of the International Congress of Mathematicians, August 16–24, 1983, Warszawa*, in 2 vols. (Warsaw: PWN-Polish Scientific Publishers; New York: North Holland, 1984).
13. Martha Coven and Carol Wood, "The AWM Panel on the Status of Women in Mathematics," AWM *Newsletter* 21(1), 1991, 8–12.
14. Gerd Fischer and Ulf Rehmann, eds., *Proceedings of the International Congress of Mathematicians: August 18–27, 1998, Berlin, Germany*, in 3 vols. (Bielefeld, Germany: Deutscher Mathematiker-Vereinigung, 1998).
15. Andrew M. Gleason, ed., *Proceedings of the International Congress of Mathematicians, 1986: August 3–11, Berkeley*, in 2 vols. (Providence, R.I.: American Mathematical Society, 1987).
16. Carolyn Gordon, "President's Report," AWM *Newsletter* 33(3), 2003, 1–2.
17. Pao-sheng Hsu and Paula Kemp, "AWM at ICM2002," AWM *Newsletter* 32(6), 2002, 6–9.
18. R. D. James, ed., *Proceedings of the International Congress of Mathematicians, Vancouver, 1974*, in 2 vols. (Montreal: Canadian Mathematical Congress, 1975).
19. Gudrun Kalmbach, "EWM Meeting: December 12–13, 1987, Copenhagen," AWM *Newsletter* 18(2), 1988, 5.
20. Olli Lehto, *Mathematics without Borders—A History of the International Mathematical Union* (New York: Springer-Verlag, 1998).

21.———, ed., *Proceedings of the International Congress of Mathematicians, Helsinki, 1978*, in 2 vols. (Helsinki: Academia Scientiarum Fennica, 1980).
22. Joyce McLaughlin, "AWM/EWM/SIAM ICIAM99," AWM *Newsletter* 29(5), 1999, 16–17.
23.———, "AWM at ICIAM 95," AWM *Newsletter* 25(6), 1995, 25.
24. Marjatta Näätänen, "ICM Panel," AWM *Newsletter* 24(6), 1994, 20–21.
25. Cora Sadosky, "President's Report," AWM *Newsletter* 24(6), 1994, 1–6.
26. Ichiro Satake, ed., *Proceedings of the International Congress of Mathematicians, August 21–29, 1990, Kyoto*, in 2 vols., Toyko, New York: Springer-Verlag, 1991.
27. Alice T. Schafer, "Report of the President," AWM *Newsletter* 4(6), 1974, 1–3.
28. Caroline Series, "European Women in Mathematics," AWM *Newsletter* 22(2), 1992, 8–10.
29.———, "EWM Update," AWM *Newsletter* 26(2), 1996, 24–25.
30. Bhama Srinivasan, "AWM Meeting at the ICM," AWM *Newsletter* 8(4), 1978, 3–4.
31. Sylvia Wiegand, "Report on the Berlin ICM," AWM *Newsletter* 28(6), 1998, 3–9.
32. Carol Wood, "ICM-90, Kyoto," AWM *Newsletter* 21(1), 1991, 6–8.

International Views on Education

ELIZABETH S. ALLMAN

Based on "AWM Panel in San Antonio," an article about "The Education of Women in Mathematics: An International Perspective," organized by Bettye Anne Case and Sylvia Wiegand, AWM *Newsletter* 29(3), 1999, 8–9. Allman was then a faculty member at the University of North Carolina, Asheville; now she is an assistant professor of mathematics, University of Southern Maine.

A thought-provoking film produced by EWM was shown at the 1998 Berlin ICM. It began with a color-coded map indicating, by country, the percentages of women among tenured university-level mathematicians in Europe. The striking differences in these percentages were discussed there and had also been a topic of conversation earlier in the '90s among EWM members and others. These data were presented to the audience at a panel at the 1999 JMM by Sylvia Wiegand and Bettye Anne Case. They posed the question for discussion: "How do the educational systems of various countries and cultural mores influence women's interest in learning mathematics and in embarking on a career in that field?" The seven speakers recounted their educational histories and reflected on some of the main influences in their decisions to become mathematicians. While their experiences were different, some common themes emerged.

Chuu-Lian Terng (Northeastern University) spoke of her education in Taiwan, a nation of roughly 21 million people. She learned from friends still living in Taiwan that the same percentages of women in Taiwan are earning higher degrees in mathematics as when she left twenty-eight years ago after earning her bachelor's degree: 30% of the B.A.'s in mathematics are awarded to women; 25% of the master's degrees; and 15% of the Ph.D.'s. She pointed out three main differences between the educational systems in Taiwan and in the United States. In Taiwan, from the ninth grade on, a student's education is completely independent of the family's financial status. All teachers in Taiwan receive the same rigorous training, and there is a fixed curriculum from the first through twelfth grades with the schools centrally administered by the state. From tenth grade on, a tracking system encourages and fosters talent and merit. Also, national entrance exams are graded blindly, so that admission to the best high schools and universities is based on talent; this exam system has allowed talented women to succeed.

Hema Srinivasan (University of Missouri at Columbia) grew up in India. The three main influences on her choice to study mathematics were her schoolteachers and the curriculum, the support of her immediate and extended family, and her classmates and peers. She said that "encouragement at home can withstand a flurry of discouragement from other sources." She described the differences in expectations for boys and girls in India. Young men are expected to do well in school and eventually to succeed in a career. Young women are not necessarily expected to do well in school, as they have other options available to them, including raising a family. As a result, a girl may study math because she likes it and is interested by it, whereas a boy might be more pressed to study simply because he has to. Thus, the gender expectations can actually have a positive influence on a young woman interested in learning.

The third panel speaker, Gail Ratcliff (University of Missouri at St. Louis), was born in Australia and completed her high school and undergraduate education there before coming to the United States for graduate school. Her high school years were a strong influence on her becoming a mathematician. She fell in love with mathematics in high school, and high expectations and the presence of talented female students in her math classes built self-confidence. Her high school was not an elite high school. The state controls the curriculum and assigns teachers to high schools, so that a good, rigorous education is available to all. At the end of high school Ratcliff was well prepared for the statewide exam that determined which university she would attend. Also, in Australia participation in sports is mandatory. Ratcliff believes that participation in sports as a teenager can help young women to become independent, develop a sense of competence, and learn to fit in.

The enrollment of women in Italy is high in the pure sciences, and in fact the percentage of women in mathematics is higher than that in physics or engineering. Anna Guerrieri (University of L'Aquila, Italy) believes that social and historical reasons explain this phenomenon. Traditionally, women have been encouraged to find a job in which they can be in a nurturing position and that will allow time for a family. Becoming a teacher at the elementary, junior high, or high school level fits well with these goals. Also, in Italy it is often possible to teach part-time if desired. In Guerrieri's experience teaching at the undergraduate and master's level, most of her students are women who hope to become teachers. There is little attrition from a mathematics degree in Italy, since students enroll from the beginning in mathematics, and the course of study and exams leads directly to a degree in mathematics. Italian women stay in mathematics, as approximately 50% of the assistant mathematics professors are female. The percentages are lower for associate and full professors, but the situation may change as female assistant professors progress up the ranks. Women are missing from leadership positions such as department head and editors of research journals, but again this may change as the current generation of young women mathematicians matures.

Gloria Hewitt (University of Montana) spoke of the noticeable lack of Native Americans in the United States attracted to the discipline of mathematics. Students internalize the expectations of their teachers, and unspoken biases may either promote or prevent student learning. Thus it is vital that minority students have teachers who believe they can achieve. Role models and mentors are of crucial importance to these students.

Ingrid Daubechies (Princeton University) went to an all-girl public school in Belgium, and her family supported and encouraged her early interest in mathematics and science. She commented that although socially perhaps a single-sex school can be restrictive for a teenager, her schooling did build her confidence.

Claire Baribaud (ETS, Zürich) spoke of the difficult circumstances under which she became a mathematician in Switzerland. Her father, initially upset by her decision, was not encouraging of her studying mathematics. There are very few women professors in Switzerland. Certain subconscious attitudes may discourage women from pursuing academic careers; for example, most Swiss parents expect their daughters to take care of a family while they expect their sons to get good jobs. These attitudes can build self-confidence in boys while discouraging girls from career ambitions. Baribaud also related an anecdote in which a mathematician commented that it was "strange" that Baribaud gave a good talk on her research; this was not expected.

From these histories, it seems clear that one's country of birth can make becoming a female mathematician easier or more difficult. National curricula and entrance examinations may serve to make opportunities available to young women without regard to region or family status. Experiences of the middle and high school years are formative; encouraging young women during that time to pursue an interest in mathematics is essential. Activities outside of mathematics, such as sports and involvement with the other sciences, may also prove helpful. Strong family support and role models sustain young women on the path toward becoming career mathematicians.

Crossing Ocean and Equator

ISABEL SALGADO LABOURIAU

Based on an autobiographical sketch that had its genesis in a discussion held at the 1991 EWM meeting in Marseilles. Labouriau is an associate professor, Center for Applied Mathematics, University of Porto.

I had never thought there was anything special in being a woman mathematician until I arrived at Warwick to do my Ph.D. At a party organized by Caroline Series for women graduates in the Maths Institute I asked why the other women had not come. To my surprise they had all come. That year had been particularly good, with five new female graduate students (there were over twenty men).

As far as I know, there have been exactly three people in my family working in mathematics related fields. Since the other two are my younger brothers, they hardly qualify as role models. On the other hand, my parents are both research biologists, and one grandfather followed an academic career in engineering and geology (not a very important influence, since he died when my father was six in a famous plane crash in Rio). There were plenty of male and female models around for all roles, from housewife to teacher to uneducated people earning more than academics.

I was born in the United States in 1954 while my father did his Ph.D.; my birth interrupted my mother's studies until our return to Brazil. We lived first in Rio, then São Paulo and Brasília, and the four children went on Sundays to the laboratory (plants don't stop growing on weekends), where attractions ranged from microscopes or helping with experiments to climbing trees. We spent the holidays with our grandmothers. With one grandmother I could share my love of books, and I helped the other one make jam. I went to mixed-sex schools where there were very few male teachers and had marks ranging from indifferent to excellent depending on my interest at the moment. Nobody ever said that girls were worse than boys—they usually had better marks in all subjects. Choice of career was not a big problem; I liked all arts and sciences as one and the same thing. During my last year in school a couple of neurophysiologists, friends of my parents, invited me to their lab, where I messed around and helped with experiments. The decision to study maths was made at the last minute. One of the factors in the choice was the quality of the department at Brasília—like most students in Brazil I went to the university closest to where my parents lived. I was lucky to find a very good group of maths students (of both sexes) one or two years ahead of me, so I took advantage of the flexible credit system to do my courses in the reverse of the usual order. We arranged to take the same subjects, knowing that lecturers gave better courses when we were all together.

After my first year in university, my family had to move to Venezuela for political reasons, and I stayed behind. I was in love with mathematics. I tried at the same time to take options in physics and biology but gave up for prosaic reasons such as incompatible timetables. Unfortunately, at that time there was a big row in the maths department in which I got involved as the representative of students. It was no worse than many similar things I've seen later, but I was disgusted with mathematicians and decided to work with applications so as to stay away from that atmosphere. After the first degree I made the mistake of staying in Brasília for the M.Sc. The course was good but a bit of a repetition of my undergraduate days; the only exciting thing was learning that differential equations can be more interesting if you don't try to solve them. I had financial independence and could spend most of my time doing theater and modern dance and hanging around the neurophysiology lab. My original plans for July 1976 were to spend the winter holidays studying for the M.Sc. final exams in August. My supervisor, an algebraist, showed me Zeeman's paper on the heartbeat and nerve impulse, so I went to Rio instead, to a three-week course on catastrophe theory. I managed to convince Djairo Figueiredo to supervise me jointly with Bráulio Magalhães de Castro (one of the neurophysiologists) for an optional M.Sc. dissertation during the next term, where I tried to apply catastrophe theory to mechanoreceptors following Zeeman's paper. In the process of not writing the dissertation I learned a lot about catastrophes, differential geometry, physiology, and the size of my ignorance. The only grant whose deadline was not over was the British Council's, and they turned me down on the grounds that I was not twenty-five years old and did not have two years' professional experience. So I took my exams the following May, applied for all existing grants, got a yes from all of them, but had to wait for a year and a half before going to Warwick. During the wait, money came from a teaching job at a secondary school that turned out to be great fun, and I had the time to take a course in animal physiology.

Living in England was difficult. The so-called temperate climate meant that I lost my body, having been turned into a bundle of clothes with hands and face sticking out of it. Fancy being at a place where most things you value are despised or considered to be impolite, where you cannot figure out what is the acceptable behavior and code of values, and where many people treat your culture as inferior. Adaptation would have been much harder at a traditional university. Warwick was a good choice from this point of view. I never felt discriminated against as a woman there, and had an easier time than most male Latin Americans; I was doing what I wanted and thought the price was worth it. When culture shock was over I had a good time. Ian Stewart agreed to supervise my crazy project of applying singularity theory to nerve impulse. I was a curiosity at the Maths Institute, since I discussed biology and used the computer. Martin Golubitsky, who was a visitor there several times, acted as a second supervisor. I was learning maths, liked what I was doing, and had friends. During my last year there I married a Portuguese colleague, and we started looking for jobs. It was impossible to find one job in Brazil, let alone two (some years later my parents went back to Brazil, and in order to hire them the university needed a special permission signed personally by the president!). The University of Porto was quite happy to take us.

When we came to Porto in 1983 the applied maths department was very small. Our colleagues worked in totally different subjects, but we were all about the same age and had a common project: making the department a good place to work. Most of the time this has meant doing less research and more administration than we wanted to. The situation is improving now: we have a better library and some computer facilities, the undergraduate course has changed a lot, and we started an M.Sc. in 1996. What I feel better about is not having anymore to define everything when I want to tell a colleague what I'm doing.

In the intervening years I have had to learn to work in total mathematical isolation in Portugal. Research could only be discussed by going abroad, and money for traveling was short. Whenever I have managed to travel, I have had help from colleagues in other countries, mostly male (they are more common), and I almost never felt excluded as a woman. Maybe this is only true because I'm not felt to be a competitor by the pure mathematicians. It certainly is made easier because I don't expect discrimination, so I don't invite it. It also seems that "Latin" women have less difficulty in dealing with male colleagues and more training in forming female groups spontaneously. This experience seems to be shared by many female mathematicians in Brazil and in Portugal with whom I have discussed the issue.

My first reaction when asked to write this autobiography was to refuse—I am not a distinguished mathematician, I work in an unknown university in a strange subject. Maybe it is a good idea to know the experiences of unknown mathematicians as well as famous ones. Writing this has been a strange experience, much harder than I expected at first. I was asked to write this after a discussion about role models at the EWM meeting in Marseille. The result is very biased in its emphasis on gender issues, issues that have not been very important in my life, so far.

Voices from Six Continents

Join hands below with women from many lands and climes, as they tell you about their work and lives. The themes of this global potpourri foreshadow those explored later in the book. Reports of talks from mathematics meetings and conferences around the globe and articles authored by international members have appeared in the AWM *Newsletter* since its early days. Selections adapted from them follow, along with parts of the hitherto unpublished transcript of dialogue from an eye-opening video produced by EWM.

Else Høyrup, Denmark, 1975

A very important factor in abstract thinking is *concentration*, but when do girls and women (either staying home or with outside work) get a chance to concentrate—unless as small children or when very old? It is an old expectation that a woman, in contrast to a man, should never just sit still and think, but must always have her hands occupied. (For example, as late as 1962 the rules and customs in an especially intellectual Danish gymnasium school permitted the girls to knit during their classes in the last month before Christmas.) And so girls are often forced to spend a great deal of their time in puttering with small domestic matters and in speculating about and working with their appearance. [5, 3]

Marjatta Näätänen, Finland, 1979

CURRENT POSITION: SENIOR LECTURER, UNIVERSITY OF HELSINKI

There are four Finnish female Ph.D.'s in math, compared with ninety-seven male. The ratio seems to have been an approximately constant 5%, even in the 1970s. The fact that except for the first (in 1951) we all have university careers reflects the attitudes—no repression toward the few women. The way of thinking of the majority of the male faculty members is something like this: Finland was among the very first countries to give her women the vote in 1906; hence women have all rights; it is their own fault if they don't use them. No extra encouragement is justified; neither is repression. But if a reason had to be given to explain the 5% and not 50%

of women among math Ph.D.'s, they would, I'm afraid, mention something biological. [10, 9]

Mary McLeish, Canada, 1982

Within the husband and wife team, trying to keep pursuit of career goals "fair" seems to me to be almost impossible, especially when children are involved. In pure mathematics, if either party slows down for any significant length of time, getting back in the game becomes very difficult. There was a certain amount of pressure throughout my earlier years in mathematics against my role as wife and mother. After changing my area to computer science, I found a very different attitude. My sex did not seem to be noticed, and questions of whether I had children and what I did with them never seemed to arise. [9, 4]

Souad Barnouti, Iraq, 1983

Barnouti pointed out that women were not active in management or administration. Employment is well studied, but we still do not have knowledge of the mechanisms that cause conditions. Technology tends to push women out of work. Whenever a field becomes modernized, then men take over. What is good in an organization for men is not necessarily good for women. Organization tends to be "rational" and asexual, with a goal of maximal profits. (Reported by Lee Alder and Fran Rosamond in [1, 16].)

Maria Jose Pacifico, Brazil, 1986 ICM

CURRENT POSITION: RESEARCHER IN MATHEMATICS,
FEDERAL UNIVERSITY OF RIO DE JANEIRO

Mathematics in Brazil is young, and women participation's is even more so. Indeed, significant participation of women in research in mathematics dates from the last fifteen years or even less. Women account for 7 to 20% of researchers in the main institutions. No woman is yet a full member of the mathematical sciences section of the Brazilian Academy of Sciences, in which there are about fifteen men.[72] Women held few of the top positions in the universities.

What could be the causes of such a situation? I believe that there is no direct discrimination in the opportunities for learning, for research training, and in hiring, and in fact, in our main institutions the atmosphere is quite good in this respect. Still, most of us continue to be more responsible

72. Keti Tenenblat became a full member in 1991. See http://www.abc.org.br/english/orgn/acaen. asp?codigo=keti [accessed 24 April 2004].

for the running of our houses than our companions, and only recently have our own families and the society in general stimulated young women to have professional careers in science. [2, 19]

Caroline Series, England, 1986 ICM

CURRENT POSITION: PROFESSOR OF MATHEMATICS,
UNIVERSITY OF WARWICK

This spring I circulated a letter to women members of the London Mathematical Society asking for their ideas on the subject of the panel. I sent out about seventy letters, roughly half to institutional addresses. Of the replies I received, the general impression was of little change, with many problems stemming from the primary school level.

Lady Jeffreys, the distinguished applied mathematician and former Mistress of Girton College, Cambridge, who is now eighty-three writes: "It is sixty-five years since I began my studies, and it is disappointing that it is still considered rather odd for a woman to be mathematical. Something has to be done in the home ('Your mother couldn't do it either, dear') and in the primary school, giving the girls confidence, which the little boys have. At all stages confidence is important."

I also believe that confidence is very important, and to this end we can be enormously helped by contacts and support from other women mathematicians. We all know examples where the right word at the right moment has changed the course of a career; certainly in my own case several times very small events have been tremendously important. We need to establish more of an "old girls' network" and especially to encourage those starting out on their careers.

Over the last few years I have had the opportunity to work closely with several women and have found that for me this is a very productive and exciting way to do research. One of the things that make life interesting is the difference in relations between men and men, men and women, and women and women. We should understand that collaborating closely with anyone requires development of a personal as well as a mathematical relationship, and in my own case I have found this vastly easier to do with women than with men. [2, 20]

Josephine Guidy Wandja, Ivory Coast, 1986 ICM

CURRENT POSITION: PROFESSOR OF MATHEMATICS,
NATIONAL UNIVERSITY, ABIDJAN

At ICM-83, during the International Commission on Mathematical Instruction session, the general secretary of the African Mathematical Union

(AMU) was invited to talk on mathematics in Africa. He said, among other things, "There are five hundred male mathematicians who have the Ph.D. and one woman who has a Ph.D. in pure mathematics," and he pointed his finger at me. I was so surprised.

When I returned to my country, I wrote to many African universities to find out if there were women teaching mathematics there, but I received no answers. I went to Joss (Nigeria) in March 1986 to attend the AMU Congress. There, I met some women, Dr. Osibodu Bukunola and Prof. Dr. Grace Alele-Williams, who are in math education. I initiated an informal meeting with women who were attending this congress. One result of this meeting was that AMU gave its promise to create a commission on Women in Mathematics in Africa (WIMA); we hope it will help us.

In my country there are no differences in the courses or training received by women or men. But the difficulty is greater for a woman than for a man, to arrive at the Ph.D. Many African universities don't prepare Ph.D.'s, so two or three years in Europe or the United States are needed to complete our graduate work. A man may be able to leave his wife and children behind to do this, but it is very difficult for a woman to leave her children to her husband and go. [2, 20–21]

Bodil Branner, Denmark, 1986 ICM

CURRENT POSITION: ASSOCIATE PROFESSOR OF MATHEMATICS, TECHNICAL UNIVERSITY OF DENMARK

There is now discussion in our society encouraging girls to choose "untraditionally," because a traditional choice of education more often leads to unemployment and because the society needs more people in the "untraditional" jobs. The lack of engineers and computer scientists is mentioned very often. I hope the society is finally forced to change the expectation for how well girls will do in—among other things—science.

I don't see open discrimination. I have certainly always felt welcome in the science community. But I see a lack of expectation for girls in science in the society as a whole, and it is hard to tell how serious an effect that causes. I also think that if we build our system of education too rigidly, then it will have a discriminating effect on women in general. What I mean is that if you—in order to have a career—have to do a lot of things at a particular age (which will often coincide with the time when many form a family), then we see a lot of women dropping out, more or less without a chance to get back in. In Denmark one rigidity of the system is that we have no bachelor's degree. So unless you finish with a master's degree, you have no degree at all. [2, 4–5]

Aiko Negishi, Japan, 1990 ICM

Negishi described the institutionalized bias against women in mathematics in the pre-1945 Japanese educational system. Both boys and girls attended six years of primary school, after which the boys continued with five years in junior high, three years in senior high, and three years at the university. Girls, on the other hand, spent four to five years at a girls' high school, followed by three years in junior college. Moreover, in the years following primary school, girls were instructed in math for only two to three hours per week, whereas boys spent five hours per week in math class. Different textbooks were used for the girls.

Although young Japanese women in mathematics are doing well now, many parents and teachers still think that girls are weak at math, said Negishi, and that they shouldn't study it. There is a need for university role models and for important posts in decision making about teaching and education to be given to women. Nonetheless, Negishi concluded that "things are gradually changing in Japan." (Reported by Martha Coven and Carol Wood in [4, 8–9].)

Asia Ivić Weiss, Canada, 1990 ICM

CURRENT POSITION: PROFESSOR OF MATHEMATICS
AND GRADUATE PROGRAM DIRECTOR, DEPARTMENT OF
MATHEMATICS AND STATISTICS, YORK UNIVERSITY

Asia Weiss described the situation in her country as "rather grim." In 1987, of the degrees given in mathematics, 36% of bachelors' degrees, 27% of master's degrees, and 13% of Ph.D.'s went to women. Of all Ph.D.'s given, 27% went to women; mathematics is therefore particularly bad, Weiss concluded. Women account for 15% of all university faculty positions, but only 5% of math faculty. The number of women in part-time teaching, however, approaches 50%.

One of the reasons why so few Canadian women pursue degrees in mathematics, said Weiss, is that many women fail to take higher-level math courses in high school and graduate ill-prepared for college-level mathematics. There have been a few efforts to remedy this situation, particularly in urban areas, she added. Some blame can be given to professors, who "somehow lose their female math students."

Two Canadian universities have established scholarships for women in the form of first-year and summer research awards. Affirmative action hiring for faculty is in place at York University, where they are "actively looking for female job candidates. If there is an equal choice between male and female candidates, they must choose the woman." The Engineering Society has up to ten fellowships for women only. Child rearing is also now a legitimate reason for asking for research grants. On the downside, however, Weiss noted that the ongoing effort to get women on faculty committees, because there are so few women, means that each has to work an extra five to six hours a week. (Reported by Coven and Wood in [4, 9].)

Rajinder Hans-Gill, India, 1990 ICM

CURRENT POSITION: PROFESSOR, CENTRE OF ADVANCED STUDY IN MATHEMATICS, PANJAB UNIVERSITY

Hans-Gill, who received her Ph.D. from Ohio State University, is on the faculty of Panjab University. When she first came to the school, she was the only woman on the math faculty of twenty; now there are six women on a faculty of twenty-nine. One-third of the Indian delegates at the ICM-90 were women.

Hans-Gill described the situation in India as "full of contradictions." Women do quite well in mathematics and receive half of math degrees awarded up through master's degrees. There is a big drop, however, in the percentage of women who go on to get their Ph.D.'s. Most women who receive their master's then work in the computer industry, teach, or work in prestigious administrative jobs. Many feel the pressure to raise families, and because "there is no tradition of leaving and then returning to mathematics in India," these women are not encouraged to go back and get their Ph.D.'s when their children are older. Master's degrees are typically received by people aged twenty to twenty-three, Ph.D.'s by people around twenty-eight. Women need encouragement, said Hans-Gill, because "they will play a great role in mathematics in the future."

Hans-Gill concluded by noting that many Indians are not given the opportunity to study mathematics seriously. Villages rarely have good schools, particularly for women, and the math teachers in religious schools are "not good." (Reported by Coven and Wood in [4, 9].)

Maria Theresa Lozano, Spain, 1990 ICM

Lozano began by commenting that although 60% of the students at her university are women, only 10% of the math faculty are women. The main reason for this, she explained, is that most women with degrees in mathematics teach in high schools, where the hours are good for women with families. Also, of the few graduate students in mathematics, a very small percentage are women. Most women work only part-time after they are married, and "it's hard to get a Ph.D. working part-time." (Reported by Coven and Wood in [4, 9].)

Hesheng Hu, China, 1990 ICM

CURRENT POSITION: PROFESSOR OF MATHEMATICS, FUDAN UNIVERSITY

Of the 147 members of her Institute, only 19 are women. Hu is the only full professor. Forty-eight of the 215 students in mathematics are women. Very few Chinese women mathematicians are known in the world, said Hu. Of the seventy-nine Chinese delegates to ICM-90, only three were women. There are historical and social reasons for the low numbers of women in mathematics, including the traditional lack of encouragement for women to study the sciences, particularly the "more difficult sciences." One can also look to women themselves for explanations, said Hu, who explained that she is very busy and has "little time for housework." (Reported by Coven and Wood in [4, 10].)

Gillian Thornley, New Zealand, 1990 ICM

CURRENT POSITION: SENIOR LECTURER IN MATHEMATICS,
INSTITUTE OF FUNDAMENTAL SCIENCES, MASSEY UNIVERSITY

The New Zealand government recently put through equal opportunity legislation, which "has caused a few things to move at universities." A special fund set up by the vice chancellor provides two research awards for women faculty, which amounts to teaching relief so that these women may work on their Ph.D.'s. Other grants exist, one of which allowed Thornley to miss an important week of school in order to attend ICM-90.

There has never been, said Thornley, a woman in an associate or full professorship (the top grades among tenured faculty) in a New Zealand university. Some women go overseas to study and stay there or move into other kinds of work. Thirty percent of the women who graduate with degrees in mathematics go into teaching. This situation forces Thornley to ask herself, "Is it right to encourage women to get Ph.D.'s in math in New Zealand if there aren't any jobs for them?" To counter this, various groups involved with mathematics education have been raising awareness of the gender problem in mathematics. Things are getting better, but there are still many problems, she concluded. (Reported by Coven and Wood in [4, 10].)

Doreen Thomas, South Africa, Australia, 1994

CURRENT POSITION: ASSOCIATE PROFESSOR AND HEAD,
DEPARTMENT OF ELECTRICAL AND ELECTRONIC ENGINEERING,
UNIVERSITY OF MELBOURNE

Encouragement plays an important part not only in making a choice but also in succeeding in that decision. I have always had a great love of mathematics. At an all-girls school in South Africa, I was taught by two excellent women math teachers and encouraged to do an additional math course for year 12.

I chose to do a B.Sc. at university. When the time came to decide my subjects, the professor remarked that I might find math rather difficult. That was the challenge, and I made the decision to study mathematics. In my honors year at the University of Johannesburg, I was one of eight women and only two men students—rather unusual. There I was encouraged to excel. . . .

As one of only a handful of women students among over a hundred postgraduates studying mathematics in Oxford, I was in a definite minority. My supervisor, also a woman, took me on as her first Ph.D. student. I had no encouragement from her whatsoever. We met weekly for an hour, and I dreaded the encounter. . . . [6, 18]

In 1987 I saw an advertisement in the University of Melbourne newspaper about the first Women's Re-entry Fellowship. Awarded to women whose career after completing the Ph.D. has been interrupted by having children, it is designed to encourage women to reenter research. I was among the first women to receive an award. This remains an excellent initiative of the University of Melbourne; this year four awards are being offered. It was a turning point in my career and a stepping-stone to a lectureship at the university.

I became a research fellow in the Department of Mathematics. Our project has been very successful, and an active research group has grown working in minimal networks. I was offered a lectureship and enjoyed having contact with the students again. I now choose to juggle a very busy family life with a career as a lecturer. . . .

It has always taken great determination for a female to overcome the difficulties to achieve in a field that is considered to be a male province. In mathematics the wonder is not that so few have attained proficiency in the field, but that so many have overcome the obstacles to doing so. The tragedy is that, even today, we can find remnants of the elitist (or sexist) tradition that has so often surrounded mathematics in the past. [7, 14–15]

Laura Fainsilber, France, 1996

Current Position: Associate Professor of Mathematics,
Chalmers University of Technology
and Göteborg University, Sweden

I have traveled a lot, I have been in different departments with very different atmospheres and at first I did not feel that being a woman in mathematics was at all an issue, or that I should be singled out or anything. I knew I was in a minority, but I did not want to be treated separately. Then in Berkeley I started thinking it was a problem when I saw that most of my friends who were women were flunking the exams and were dropping out of the graduate programs.

When I came back to France, I felt the atmosphere was very different because the way people interact was very different and my situation was also different. I was among students who knew less than I did and I was receiving more attention. That made a difference, I think, in the way I felt about doing math. I felt much more confident because I was getting a lot more contact with professors than I had otherwise, so I decided I would rather stay in France than go back to the States.

I think the difference I felt between Besançon and Geneva is similar to the difference between southern countries like Spain and Italy where there are a lot of women (they do not feel isolated, but there are other problems that come up) and women in northern countries where they are very few (their problems are not the same, but the isolation is tremendous). These two extremes make it difficult but interesting for a society like EWM to bring these people together. The people from the North need to see groups of women who work together, it is impressive.

The structure of academic careers has an impact. In countries that give permanent positions rather early, women have a better chance than in countries where you go from postdoc to postdoc and have to wait a long time while you are having children and while raising them at a time that is really crucial for women—this is difficult for men too.

I found this year just after the Ph.D. very difficult, partly because I was isolated in Geneva, partly finding my bearings. Getting to start to work on my own after the Ph.D. was hard, and a lot of women do get lost at that stage, in addition to the ones that got lost before. I think that in areas where they already have a permanent position at that stage, even if they fumble around for a few years, it does not matter as much, because they have a permanent position. In France, I know women who do excellent work and are quite impressive, but maybe they had a few years where things did not always fall together.

The type of work we do in math is different from other fields in that it is very individual. There are no labs, no experiments going on. The teaching is continuous and clearly visible from the outside, but the research is less well defined. It depends very much on the concentration we can devote to it, the quiet atmosphere, and the self-confidence of the individual.

In France mathematics is essentially a masculine field, so you can position yourself in different ways—go into it because it is a masculine field and you want to, or you can try to ignore it. But when you meet people they say: "You are a mathematician?! You don't look like a mathematician," and you say: "Really?" They are very surprised, and I do not think they have the same reaction toward male mathematicians. You go around the departments, except Besançon and a few other places, and you go to conferences: there are only one or two women in the room, and that makes the atmosphere different from other fields. [11]

Isabel Salgado Labouriau, Brazil, Portugal, 1996

CURRENT POSITION: ASSOCIATE PROFESSOR, CENTER FOR APPLIED MATHEMATICS, UNIVERSITY OF PORTO

You have more women doing math in all the Latin countries. I have no idea why that is; it is a fact that I have no explanation for. When I went to England, it was very funny. I was going to this first world country from the third world, everything was going to be much better—and then I realized that it was not, it was the opposite, in social terms it was much worse in England than it was in Brazil. In Brazil if you want to work, you work and it is natural. After England, Portugal was like going home in many respects, and one of them was working in a department with more women than men, which felt like getting back to normal. [11] [See also the preceding article, "Crossing Ocean and Equator."]

Claire Baribaud, Switzerland, 1998 ICM

CURRENT POSITION: LECTURER, UNIVERSITY OF APPLIED SCIENCE OF WESTERN SWITZERLAND, GENEVA

In Switzerland there are very few women who have a steady job as a mathematician in a university. Very few tenure-track jobs open up at all in universities, and when they do, women are not hired.

There have been affirmative action regulations in Switzerland for several years. Moreover, some scholarships have been created in order to encourage women to go back to the academic world after a hiatus for having children. However, in Switzerland the social system is not well developed, and it is extremely difficult to both be a mother and work at the same time.

In the past few years, nevertheless, more and more industries, banks and financial institutions, and insurance companies have begun to hire women. There have been comments about the preferability of hiring women rather than men for basically these two reasons: women will accept lower salaries; women are more compatible in the workplace than men because they like to discuss matters and are productive in a team effort. [3, 18–19]

Ljudmila A. Bordag, Germany, 1998 ICM

CURRENT POSITION: PROFESSOR, HALMSTAD UNIVERSITY, SWEDEN

I will give you my personal impressions about the deep changes in education that took place after the German unification, especially at Leipzig University. I have a somewhat unusual career, so I can in some sense compare three systems, Russia, former East Germany, and unified Germany. I was born in Russia and studied at St. Petersburg University, and I married a physicist, a citizen of the German Democratic Republic (GDR, known as the former East Germany).

After obtaining my Ph.D., the question arose as to where to look for a job. At that time it was quite difficult for my husband to get work commensurate with his qualifications in Russia. In 1978 we decided to go to Leipzig University, the second largest university in GDR. In Leningrad there were many more women in professorial positions, and the general atmosphere was friendlier to women. At Leipzig University I was in fact not really discriminated against as a woman; however, some time later it became clear to me that I did not have any chance for a career, because I was from Russia. There appeared to be some secret instructions not to give leading positions to citizens from abroad. So I remained in the position that I started with for all those years: a scientific assistant, which was, however, a permanent appointment.

During this time we had two children. We had no problems with our superiors, since having children was considered natural and usual. Child care was very good, and perhaps this was the best feature of life in the GDR. (Child care in East Germany exists more or less as before. Although it is much more expensive, it is still possible.)

After German unification we hoped that we could take part in policy decisions at the university. But the reality turned out to be quite different. Our state, Sachsen (Saxony), was not the richest one, and restrictions and cuts were introduced. This policy was very severe at Leipzig University, more severe than at all other East German universities. Before the unification we had about 9000 employees, including scientists, technical and other staff. A number of institutes were closed, while others were reorganized. For a period of about two years we had a very difficult time. We had to fill out a large number of forms about our former political activities and possible surveillance by GDR security. Many positions were eliminated as the structure of our university was changed to that typical in West Germany. A number of laws protecting employees were suspended; we had to apply anew for our positions. Some of us received temporary positions, while most of the staff lost their positions completely. Now our university has a staff of about 2500 persons.

What is the situation for women after the unification? Now it is easy to know all the women in the scientific staff of the Mathematics Institute. That includes me, a second woman who became an equal opportunity officer, and a third woman who has now regained a permanent position after a long succession of lawsuits. But there are also encouraging events for me after the unification. I now hold held the Dorothea von Erxleben Guest Professorship at the University of Magdeburg, a one-year professorship organized by women at Magdeburg, including Professor Christine Bessenrodt, one of the organizers of this panel. This professorship is given to women in fields of science where they are underrepresented. [3, 19–20]

Mary Glazman (1945–2000), Mexico, 1998 ICM

The big economic crisis in Mexico started around 1982 and has changed Mexico in profound ways. At that time, salaries for academics were low, as usual, but when the crisis began, things really worsened significantly. People with young families became seriously worried. They did not see any future in the academic world, and many people left their jobs. Considering the investment that is made in a country in training its people, especially by public institutions, and the time it takes to establish a stable group, there was great concern among academic authorities about what was happening. Losing its best heads was a luxury that Mexico could not afford.

Around 1983 or 1984, when the crisis in the academic world was reaching its peak, an effort made by an important group of researchers, particularly scientists, brought about a response: the establishment of a national system of research (in Spanish: Sistema Nacional de Investigadores) better known as SNI . . . . There has never been a woman referee on the SNI Math Committee. Women's vision has not been taken into account. It is more difficult for a woman to confront this type of committee and fight for her rights, at least in Mexico, undoubtedly due to the long tradition of cultural conventions in this regard. In the long run, this will change as more women are involved in mathematical work. [3, 21–23]

Minping Qian, China, 1998 ICM

CURRENT POSITION: PROFESSOR, SCHOOL OF MATHEMATICS, PEKING UNIVERSITY

Ten to fifteen years ago, women in cities in China had relative equality with men. This relationship did not extend to the large population in the countryside, where the tradition of thousands of years, together with continuing poverty, meant many girls did not have even a complete elementary school education. However, at that time, girls who did graduate with a high school–level education had equal opportunity with boys to go to universities and colleges; women who became mathematicians were equally treated in terms of getting jobs and promotions. The husbands of such women usually shared responsibilities for housework and children. Many women in China were proud of this favorable situation regarding equality with men, even when compared with women in developed countries. . . .

In the past ten years, however, along with the economic reform, this status concerning proportions and equality has been changing gradually. For the first time there have appeared large numbers of job discrimination cases involving women. I have had some experience with this through some of my women students who were rejected for jobs only because they were female. These women were better, academically, than some men students who got the jobs. Theoretically, such action is against the constitution of the country. When I asked how could they do this, I was told that in those institutions there were already "too many young women" and that the women were rejected not because they were considered low in ability but "because they [women] were practically hard to manage." [3, 23–24]

Inna Yemelyanova, Russia, 1998 ICM

CURRENT POSITION: CHAIR AND PROFESSOR OF APPLIED MATHEMATICS, NIZHNY NOVGOROD STATE UNIVERSITY

A major problem for all faculty is the economic situation and cuts in education spending. Due to the funding crisis there have already been drastic decreases in allocations to fundamental science in Russia, and future severe cuts are projected (e.g., in 1999–2001 the funding is projected to be cut again by 20%); the employment situation is not clear at all. . . .

One bright note for Russian women mathematicians was the recent fifth anniversary of the founding of the Russian Association of Women Mathematicians (RAWM). The association was created to provide information, consulting, and social support for women who had chosen mathematics as a sphere of their scientific activity. Four conferences followed, with more than 600 women mathematician participants from Russia and many other countries. There was some conference support from UNESCO, Mathematica, and ISF. The sixth conference, "Mathematics. Education. Economics," was held May 1998 in Cheboksary in the Chuvash region. Because the population there is not ethnic Russian, there was discussion on the development of Chuvash science and culture along with the discussions on science, methods of teaching, and women's adaptation to the market economy. [3, 24–25]

Audience Participation, China, 2002 ICM

In Beijing at the discussion "Connections and Opportunities for Women in Mathematics," panelists and audience alike participated in a lively session. Zhang Lei, a journalist for the Beijing *Guangming Daily*, wrote an article about the experience; Pao-sheng Hsu, one of the organizers, translated it for the AWM *Newsletter*. From the article [8, 8]:

> Amidst a light-hearted and happy atmosphere, more than fifty women from the United States, Canada, France, China, the Philippines, India, Indonesia, and Iran, etc., who work in the field of Mathematics gathered in one hall and exchanged ideas without barriers. Except for a few enthusiastic men, the participants were almost all female. Even though they were not known or paid attention to by too many people, they made a brilliant line in the scenes of this International Congress of Mathematicians. At the panel, participants gathered their own personal experiences and perceptions, discussed and expressed their opinions enthusiastically on topics such as "living and working conditions of a woman mathematician in your country," "the obstacles of pursuing a career in mathematics," "changes in the past few decades in status and working conditions for women mathematicians," etc.

The exuberant headline of the article says it all: *"Women in the Kingdom of Mathematics Will Not Be Isolated Again."*

References

1. Lee S. Alder and Fran Rosamond, "International Conference on Teaching and Research Related to Women," AWM *Newsletter* 13(2), 1983, 14–18.
2. Lenore Blum, "AWM/ICM-86 Report: Women in Mathematics: An International Perspective, Eight Years Later," AWM *Newsletter* 16(5), 1986, 9–21; 16(6), 1986, 4–5.
3. Bettye Anne Case and Bhama Srinivasan, "ICM-98: Women in Mathematics," AWM *Newsletter* 29(5), 1999, 18–25.
4. Martha Coven and Carol Wood, "The AWM Panel on the Status of Women in Mathematics," AWM *Newsletter* 21(1), 1991, 8–12.
5. Else Høyrup, "Women—and Mathematics, Physics and Technology? Women—and Research?" Parts V and VI, translated by John Lamperti. AWM *Newsletter* 5(3), 1975, 3–4.
6. Susanne Irvine, Kerry Landman, Christine Mangelsdorf, and Doreen Thomas, a public lecture given at the Department of Mathematics, University of Melbourne, Parkville, Victoria, Australia, "Women in Mathematics: part II of III," AWM *Newsletter* 26(6), 1996, 18–20.
7. ———, "Women in Mathematics: part III of III," AWM *Newsletter* 27(4), 1997, 13–16.

8. Zhang Lei (translated by Pao-sheng Hsu), "Women in the Kingdom of Mathematics Will Not Be Isolated Again," p. 8 in "AWM at ICM2002," by Pao-sheng Hsu and Paula Kemp, AWM *Newsletter* 32(6), 2002, 6–9. Originally appeared in *Guangming Daily*, 26 August 2002.

9. Mary McLeish, "Discussion of the Problems of Husband-Wife Teams in the Mathematical Sciences," in "Women Mathematicians in Canada," AWM *Newsletter* 13(1), 1983, 3–4.

10. Marjatta Näätänen, "Women and Mathematics in Finland," AWM *Newsletter* 9(2), 1979, 8–9.

11. Marjatta Näätänen, director, in collaboration with Bodil Branner, Kari Hag, and Caroline Series, "The Madrid Interviews: Women and Mathematics across Cultures," transcript of a film, EWM, 1996.

Complexities Photo Album

Women in the Kingdom of Mathematics Will Not Be Isolated Again. This jubilant headline from a Beijing newspaper during ICM 2000 (quoted in "Voices from Six Continents") is personified on the overleaf: in the commons area of the Mathematical Sciences Research Institute, a lively poster session displays the mathematical work of women graduate students and recent Ph.D.'s. Such sessions are features of many workshops and conferences, including those organized by AWM and EWM.

Career information, mentoring, and presentation of individual research were integral to the Julia Robinson Celebration of Women in Mathematics and the Olga Taussky Todd Celebration of Careers for Women in the Mathematical Sciences (see I.1 and IV). These conferences brought junior and senior mathematicians together in 1996 and 1999.

1 (overleaf) Robinson conference poster session
2 1996 conference namesake Julia Bowman Robinson (1918–1985) in 1975

3 1999 conference namesake Olga Taussky Todd (1906–1995) in 1976

4 Seven of the plenary speakers and the after-dinner speaker with John Todd, widower of Taussky Todd: Richard Varga, Cathleen Morawetz, Helene Shapiro, Christa Binder, Linda Petzold, Fern Hunt, Lani Wu, Lisa Goldberg, and Todd [not pictured: plenary speakers Evelyn Boyd Granville, Dianne Lambert, and Margaret Wright]

Family. One theme of this book, explored in part I and elsewhere, is that of family—both mathematical and by kinship. At these two conferences, the participants enjoyed meeting family members of the conference namesakes. Jack Todd attended the conference bearing his wife's name. Constance Reid attended both of the conferences; her after-dinner comments are part of "Being Julia Robinson's Sister" in I.1. Women enlarge their families of friends as they meet with others at varied life and career stages, enjoying both the mathematics and the camaraderie. Young women sometimes bring their children along; infants—such as the youngest to attend the Robinson conference, pictured on the facing page in the arms of her mother—are a special delight.

5 Conversation at the Taussky Todd Celebration: Isabel Beichl, Evelyn Boyd Granville, and Tamara Kolda

6 Conference organizers: Jean Taylor, Bettye Anne Case, Sylvia Wiegand, Gail Ratcliff, and Dianne O'Leary [not pictured: Susan Geller and Carolyn Gordon]

7 Rebecca Myers with her mother, Perla Myers, at the Robinson Celebration

Senior mathematicians. Noether Lecturers are well-recognized senior women mathematicians. In the case of many of the earlier Noether Lecturers, the extent of their accomplishments had not previously received appropriate recognition in the mathematics community. Both Julia Robinson and Olga Taussky Todd spoke in this lecture series, sponsored by AWM at the JMM, as listed in "AWM Activities" in part II. Linda Keen gave the lecture in 1993 a few months before a reception held at the University of Maryland, at which mathematician Richard Herman, dean of the College of Computer, Mathematical, and Physical Sciences, welcomed AWM's national office to its new headquarters. Nancy Kopell's 1992 talk on neural networks governing motor activity in animals presaged her Gibbs Lecture at the 1999 JMM (see part V for her story). Linda Rothschild received the 2003 Stefan Bergman Prize from the AMS jointly with Salah Baouendi for their collaborative and individual mathematical research. Krystyna Kuperberg had the pleasure of being introduced by her son Greg Kuperberg, also a mathematician, when she delivered her Noether Lecture in 1999.

8 (below) Mary Gray, Linda Keen, Richard Herman, Cora Sadosky

9 (top facing) Nancy Kopell in 2000

10 (middle facing) Chuu-Lian Terng with Linda Preiss Rothschild after Rothschild's 1997 lecture

11 (bottom facing) Krystyna Kuperberg and her son Greg after her lecture

Beginning mathematicians. Women at all career stages enjoy sharing their mathematics; here women are shown early in their careers. Kate Okikiolu was invited to give a research talk in 1996 on the occasion of the twenty-fifth anniversary of AWM, as a representative of the generation of women who would be making strong contributions in the years to follow. One hobby of Okikiolu is dancing, beautifully exploited on a poster depicting ideas related to her mathematical work. (Pamela Davis Kivelson, a California artist associated with the American Institute of Mathematics, has made a series of posters celebrating the achievements of women in the sciences. Dusa McDuff, Katherine Socha, and Karen Uhlenbeck are among women mathematicians involved in attempting to bring these projects to a broad audience.) At AWM workshops at the JMM, Heather Johnston, as a recent Ph.D. in 2000, gave a talk; Helen Moore, then a graduate student, gave a poster presentation in 1995.

12 Kate Okikiolu

13 Anne Leggett with Helen Moore at Moore's poster

14 Heather Johnston

The future. Girls and college-age women are encouraged in various ways to continue their study of mathematics. AWM's Alice T. Schafer Prize is awarded to undergraduate women for excellence in mathematics. Sonia Kovalevsky High School (SKHS) Mathematics Days bring local high school girls and their teachers to a college or university campus for a day of interesting and fun activities, including hands-on workshops, where their eyes are opened to the multiplicity of careers available in the mathematical sciences (see "Activities and Awards," II; also "Gung and Hu Award Conferred on Schafer," III.2).

15　(below) The first Schafer Prize: winner Linda Green, AWM president Jill Mesirov, prize namesake Alice T. Schafer, winner Elizabeth Wilmer

16　(top facing) Carolyn Mahoney

17　(bottom facing) Problem-solving workshop, SKHS Day, St. Joseph's University, 1999

On the cusp of the new century, the composition of the younger population cohorts—the young children now and generations yet to be born—is discussed in "Demographic Trends and Challenges for Mathematics" (see part V). Carolyn Mahoney is shown at "Mathematical Challenges of the 21st Century" keynoting this session on the ethnic and gender composition of the population from which future mathematicians will emerge.

SIAM and ICIAM. At annual JMM and SIAM meetings, and at some international meetings including ICIAM, workshops or mini-symposia are held. These include mathematical talks or posters, along with information dissemination and mentoring activities. Distinguished senior women in the mathematical community help younger women who are establishing careers. For the 1997 AWM-SIAM workshop, Suzanne Lenhart, Marsha Berger, Margaret Cheney, and Rosemary Chang led an informal discussion on various career issues. At ICIAM99, senior women spoke on "Industrial Research Successes," sponsored jointly by AWM, EWM, and SIAM (see "Across Borders," II).

18 Toronto 1996: Graciela Cerezo and Joyce McLaughlin

19 Suzanne Lenhart, Xin Chen, Marsha Berger, Margaret Cheney, Rosemary Chang

20 Philadelphia 1993: Shubbe Rajopadhyye, Barbara Keyfitz, Catherine Roberts, and Cheryl Hile

21 Stanford 1997: Deborah Lockhart

22 Edinburgh 1999: Rosemary Chang (U.S.), Margaret Wright (U.S.), Barbera van de Fliert (Netherlands), and Kerry Landman (Australia)

Mathematics education. The MAA MathFest is held each summer. WAM (the Women and Mathematics Network, an activity of the MAA Committee for the Participation of Women) organized the panel "Expanding the Vision: Increasing the Participation of Women" for the 2001 MathFest held at the University of Wisconsin, Madison. Women involved in a number of projects spoke about them; for example, Florence Fasanelli has been involved with several MAA projects, including the Tensor Program and SUMMA. Earlier that year, an interesting panel was held at the JMM on the topic "AWM and K-8 Education: What Should We Do?"

The first Louise Hay Award for Contributions to Mathematics Education was presented to Shirley Frye in 1991 by AWM, shortly before her retirement from teaching for more than forty years in the public schools of Arizona and Pennsylvania. More recently, Frye received the 2002 Mathematics Education Trust Lifetime Achievement Award, which annually recognizes NCTM members who distinguish themselves by their many years of service and dedication to the mathematics education profession (see "Activities and Awards," II).

23 back: Viji Sundar, Virginia Kasten, Genevieve Knight, and Suzanne Lenhart; front: Elizabeth Yanik, Kathleen Sullivan, and Florence Fasanelli

24 Rhonda Hughes and Shirley Frye

25 Virginia Warfield, Judith Roitman, Shirley Malcom, Jean Taylor, and Erica Voolich

Friends. One of the pleasures of mathematical conferences is meeting friends, old and new. Opportunities abound both for professional interactions and for getting to know other women (of special importance to those who have few local female colleagues). AWM sponsors receptions in conjunction with many meetings. After the Gibbs lecture at the JMM, the lively AWM party includes music when possible, sometimes dancing.

26 Twentieth Anniversary Celebration, 1991: Jill Mesirov line-dancing with mathematician husband, Benedict Gross

27 1993 JMM workshop: Judy Green, Cora Sadosky, Carol Wood, and Lenore Blum

Lee Lorch and Chandler Davis, charter members of AWM, have been honored with certificates of appreciation. Organizations cannot survive without creative, dependable staff such as Hope Daly and Dawn Wheeler. They are pictured here after Daly's award from AWM on her retirement in 2000 from the AMS meetings department, where she gave years of helpful attention to AWM's needs at national meetings. Wheeler keeps AWM's Maryland office humming.

28 Lee Lorch and Chandler Davis

29 Hope Daly and Dawn Wheeler

Women in the last quarter century have achieved considerable visibility on the national and international mathematical stage. Margaret Wright, pictured earlier on a page of women's activities at SIAM, has been that organization's president; four women have been president of the MAA.

At the AMS. Women have served as invited speakers at all levels, including the prestigious Colloquium Lecture series, as officers, and on committees. Julia Robinson and Cathleen Morawetz have both been president. Jean Taylor and Karen Uhlenbeck (see V) were invited lecturers at the 2000 meeting "Mathematical Challenges of the 21st Century." Short Courses at the JMM bring mathematicians working in related areas up to speed on new specialized work. In 1993, Ingrid Daubechies (see V) ran a short course on wavelets, the research area in which she has been highly influential.

At the 1988 AMS Centennial Celebration, the AMS presented other mathematics societies with gifts. AWM received a silver bowl inscribed "Association for Women in Mathematics, American Mathematical Society, 17 Years of Cooperation, 1971–1988." In 2003, outgoing AWM president Suzanne Lenhart continued a tradition by presenting the bowl to new president Carolyn Gordon.

30 (top facing) Karen Uhlenbeck

31 (bottom facing) Ingrid Daubechies

32 (below) 2003 ceremony: Carolyn Gordon and Suzanne Lenhart

International activities. ICM, ICIAM, ICME, and other conferences allow women from around the world to share their experiences, both informally and through special activities. Women in applied mathematics speaking at ICIAM99 are shown on an earlier page. Below, the ICM 1998 panelists and organizers from six countries pose after their well-received session. Mathematical talks called the Emmy Noether Lectures have been given for the last three ICMs; women from many countries and organizations cooperated to choose them. Cathleen Morawetz delivered this lecture in Berlin in 1998. She is pictured with Martin Groetschel, the president of ICM '98, and Irene Gamba, one of her former doctoral students. The 2002 lecturer was Hesheng Hu, here pictured with Paosheng Hsu, one of the organizers of special women's activities at the Beijing meeting, and distinguished mathematician S. S. Chern. Both Morawetz and Chern have received National Medals of Science. For more information on the international scene, see the last four articles in part II.

33 back: Bhama Srinivasan (U.S.), Ljudmila Bordag (Germany), Minping Qian (China), Inna Yemelyanova (Russia), Christine Bessenrodt (Germany), Mary Glazman (Mexico); front: Dusa McDuff (U.S.), Claire Baribaud (Switzerland), Bettye Anne Case (U.S.)

34 Martin Groetschel (Germany), Cathleen Morawetz (U.S.), and
Irene Gamba (U.S.)

35 Pao-hseng Shu (U.S.), Hesheng Hu (China), S. S. Chern (U.S.)

From the beauty and strength of woven knots to

36 Knots: Mathematicians—Jennifer Chayes, Mathematical Research Institute at Microsoft; Joan Birman, Columbia University; Vaughn Jones, University of California, Berkeley; William Thurston, University of California, Davis.

From the beauty and strength of woven knots to the dance of life—knots are everywhere, not just for our pleasure but as part of the intrinsic structure of nature, in the twisting of DNA strands and the braided forms of stringy matter.

Mathematicians tease out the different patterns of knots and discover that these determine the possible shapes of three-dimensional space.

III. CHOICES AND CHALLENGES

This part begins with questions about the status of women in the profession, the choices available to them, and the challenges they face. A summary of recent U.S. data includes comparative information about women from other sciences, discusses some of the social and political factors that impact women's careers, and introduces the relevant literature. The data in turn suggest further questions; these echo throughout the career accounts in this part's first three chapters. The fourth chapter returns explicitly to women's attempts to find fulfillment in both their careers and their personal lives. From a feminist perspective, these authors explore and analyze the reciprocal effects of geography, relationships, and parenthood on career advancement.

The women appearing in this part, with their diverse personalities and interests, illustrate the individuality of the process of career development. The first chapter involves both the choices made by and the special challenges faced by some of the earlier African-American women to pursue mathematics Ph.D.'s. Chapters 2 and 3 form a careers casebook. Chapter 2 illustrates scholarly and professional concerns and the varied responsibilities in academe. Chapter 3 is devoted to career-building in government and industry. The stories in the latter chapter took a surprising turn: most of the women profiled now have different employers than they did when these stories were told. Some have gone back to academe, others to a different sector of government or industry (one is a co-inventor of a crawfish-processing machine). As these writers discuss the societal, political, and personal factors that shape careers, they often show how constraints may be transformed into opportunities.

Pathways in Mathematics

BETTYE ANNE CASE AND ANNE M. LEGGETT

This paper explores the career pathways traveled by doctoral women mathematicians in the United States today, along with the educational pathways leading to the doctorate. Discussion points include the alternatives available to women, what obstacles might stand in their way as they pursue their options, and the impact of personal life. It is much easier to enter the professions today, but both life issues and the culture of the workplace impact women strongly. These issues are discussed in the first three sections of this paper: "Choices," "Trajectories," and "Balance."

Next, the paper turns to questions that are more quantitative in nature. Are entry-level positions, given the attainment of the doctorate, as available to women as to men? Is there evidence of a "glass ceiling" in the world of mathematics, as it is alleged to be for so many professional women? What sort of pictures do the available statistics paint? A tangled web of data is analyzed in these sections: "From the Data" and "At the Top."

The paper concludes with summary comments and an autobiographical excursion in "Reflections" and "Coda."

Choices

Women today have more career options than in the not-so-distant past, when choices were more circumscribed. Sociocultural expectations, clearly of major impact on the career choices available to and made by women and men, have changed in positive ways. In the first half of the twentieth century, few women obtained their doctorates in mathematics [16]. Women seeking work in mathematics operated under many constraints, as did women in other sciences. Many of these women were remarkably productive, given the restrictions imposed by their environment. Zuckerman, following Rossiter [35], says of women scientists in general: "The historical record shows that many accomplished women were either ignored or actively discouraged. Those honors which came to them at all came very late" [41, 27]. A good illustration of this point appears in the dramatic story of Julia Robinson's appointment as a professor only after she became a member of the National Academy of Sciences (NAS) (see I.1).

During World War II, many conventions were suspended under pressure of necessity, as branded into the American psyche by Rosie the Riveter saying "We Can Do It" on the well-known poster. The contributions at that time of women trained in mathematics at all levels were less visible but no less vital. At war's outbreak, Ruth W. Stokes, then a professor at

Winthrop College, a southern school for women, wrote to R. G. D. Richardson, secretary of the AMS:[1]

> Aberdeen, Model Basin, and . . . the Department of Interior in Washington and Signal Corps Office in Arlington . . . are simply begging for our graduates. They will employ all I can send. . . . Each . . . would have gladly given me a position at once [at a salary higher than my teaching salary] . . . but [one] said I could be of greater service training large numbers here. . . . Resignations are coming in here daily. Two, no three, of the young [women] have gone into the Navy [3].

In contrast, after World War II, career choices for women were much curtailed. Women were, with peacetime, usually expected to vacate their often much enjoyed jobs in favor of the returning men. So in the 1950s, woman's role as "housewife," whatever her educational and professional attainments, was again common and encouraged. In the latter half of the twentieth century, many women worked hard to improve their collective destinies. The housewife of the early '50s, who generally did not have outside employment, gave way in the vernacular to the homemaker, used more inclusively, as in "my mom is a librarian and a homemaker." In today's semantics we have, regardless of employment, "soccer moms."[2] Among the "stay-at-home moms" who do not work outside the home, some quit fast-track careers, part of an "opt-out revolution" [4]. The terminology has come full circle, but with many of the former constraints removed or relaxed.

At the beginning of the twenty-first century, women are receiving about 30% of the Ph.D.'s in the mathematical sciences. Today's U.S. woman Ph.D. need no longer fear losing her job if, for example, she marries or has a child. But laying some fundamental concerns to rest has not leveled the playing field. The choices and career characteristics of women have been investigated by historians and sociologists of science (e.g., [26; 37; 43]). Sonnert [37, 1–3] asserts that "women scientists, as a group . . . have less successful careers than men scientists" and cites two theories advanced by scholars—the deficit model (women are treated differently in science) and the difference model (women act differently in science). Data analyzed in an NAS study show that the "numbers of women grew in all fields, sectors of employment and faculty ranks," but "women, although they have made great progress . . . in the past 25 years, are still more likely than their male counterparts to be in positions of lower status and lower pay" [26, 8].

1. Stokes was a 1931 Ph.D., Duke University. Richardson, while a dean at Brown, served as an unofficial job broker during his tenure as secretary (see Bernstein's story in III.2 for a related anecdote). He arranged the interviews that Stokes mentions in this letter of September 4, 1942.

2. More than one mathematician profiled in this book answers to this designation.

Although the beginning woman mathematician is usually most familiar with academic employment, a rich spectrum of possibilities is available. The precedent among mathematical societies for a broad view of
appropriate employment goes back at least to the 1888 founding in New
York of a precursor of the AMS, when it was agreed that "it was desirable
to bring together mathematicians resident in New York and the neighborhood, teachers of mathematics at institutions of collegiate rank, actuaries
of local insurance companies, practicing engineers, and others genuinely
interested" [1, 4]. Well over a century later, women brought together by
AWM to speak about their mathematical careers mirror a surprisingly similar range of opportunities as they speak to eager audiences of women and
men, of recent Ph.D.'s and those who mentor them. [See III.2 and III.3]

The desire to disseminate job information has at times been heightened
by dire unemployment statistics for new doctorates, for example, in the
mid '70s and the early '90s, both with double-digit unemployment percentages [22, 804]. Interesting and nonlinear career patterns emerged as
women shared their career autobiographies with the mathematical community. In light of these career patterns, a 1995 panel raised the question
"Do Women and Men Have Different Career Trajectories?" [15].

Trajectories

The question "Different Trajectories?" includes within it a variety of sub-
questions. Are women and men more likely to be productive at different
career stages? Is it true that geographic constraints imposed by family life
have more impact on women's careers? Does childbearing affect women's
careers in ways without counterpart in those of men? Despite the fact that
mathematicians have long worked for a wide range of nonacademic employers, beginning graduate students often expect to follow in their professors'
footsteps. How viable a prospect today is the standard linear model—
from Ph.D. to junior position to tenure to a march up the professorial
ladder—that has been the expectation for men for many years? Murray,
writing about women receiving their Ph.D.'s in the 1940s and 1950s,
called this progression "The Myth of the Mathematical Life Course." She
says that these women were "caught between two cultures," responding
in different ways, but that "most often . . . their mathematical lives were
an improvisational blending of the old and the new" [29, 15–19]. More
women than men of those mathematicians born before 1950 exemplify
such an approach to career.

Will multiple career changes become the typical pattern for the future
mathematician? Will moving back and forth between academe and other
sectors become more common? Do the extraordinarily successful women
of this book with nonlinear career trajectories presage the future for

mathematicians of both genders? Along with many of the other senior women profiled in this book, Olga Taussky Todd (I.1 and IV), was a notable trailblazer in this respect.

Will the profession adjust its preparation of all doctoral students to encourage breadth of study in mathematics, so that they will be equipped to meet the needs of future employers? The necessity of the latter is already recognized by policy and funding agencies, among them the NSF through an alphabet soup of focused initiatives[3] and the NAS through conferences and publications such as *From Scarcity to Visibility* [26]. The impact of globalization, as well as successes and failures in international relations, will be felt in the employment market. In the academy in the near future, it is difficult to predict whether the number of retirements anticipated (a plus for the job candidate) or economic volatility, along with changes in the structure of the academy (both negatives), will have greater influence on the shapes of careers.

Another parameter affecting career trajectories is reflected in the debate on age versus productivity, as Claudia Henrion reported at the "Different Trajectories" panel. The traditional view is heard as Felix Klein, on his fiftieth birthday, says to his student Grace Chisholm, "Ah, I envy you. You are in the happy age of productivity. When everyone begins to speak well of you, you are on the downward road."[4] An early study by Stephen Cole used paper and citation counts to define productivity and work quality in the sciences. He found no correlation between age and number of published papers; scientists over the age of fifty were found to be slightly less likely to publish highly cited work [11]. In "Age and Achievement," Stern examined the record for mathematicians, who were not included in Cole's study, and concluded that "age explains very little, if anything, about productivity"[5] [38]. She found no clear relationship between age and quality of work as indicated by citation count. Neither Cole nor Stern provided gender breakdowns of their data.

Henrion reported that the prominent women mathematicians she interviewed in the 1990s felt that their best work had been done later in life, "typically in [their] forties, fifties, even sixties." Some have speculated that women peak mathematically about ten years later than their male counterparts, she says. Henrion continues, "These interviews suggest that

---

3. Such programs include the Focused Research Group (FRG), Grant Opportunities for Academic Liaison with Industry (GOALI), Interdisciplinary Grants in the Mathematical Sciences (IGMS), Vertical Integration of Research and Education in the Mathematical Sciences (VIGRE) programs, with its successor Enhancing the Mathematical Sciences Workforce in the Twenty-first Century (EMSW21), and Advancement of Women in Academic Science and Engineering Careers (ADVANCE).

4. See the lead article of I.2.

5. Stern examined mean numbers of papers published by age groups of mathematicians over the five-year period 1970–1974 and found: under age 35, mean of 5.12 papers; 35–39, 7.33; 40–44, 6.24; 45–49, 3.49; 50–59, 5.22; 60+, 6.11.

for women, the relationship between age and productivity may in fact be inversely proportional. . . . Those with children found that their research focus was much improved as their children aged and demanded less of their time. . . . While all of these comments are anecdotal, what is striking is the uniformity of the response" [18]. The adage "mathematics is a young man's game" is still prevalent in the culture (and encouraged by the age-forty upper limit for the Fields Medal).[6] As Henrion points out, strong departments want to hire "potential hot stars; [youth being] considered a sign of precocity and brightness," and these factors mitigate against women [18, 14]. Will the strong research contributions by both women and men at ages past those often considered the most productive years for mathematicians open eyes—and doors?

Are men mathematically more productive than women? In a study of science publication rates by gender, Cole and Zuckerman report that a common explanation for differential performance is that women far more than men bear the burdens of marriage and child care, and further, that "whether or not this is true, the belief that it is so affects women's career opportunities, their decisions, and the way they are actually treated" [10, 157]. If men in academia assume a balanced share of family responsibilities, will it also become "expected" that fathers will have lower productivity?

Balance

Cyclic fluctuations in the job market exacerbate the tension created for workers needing to balance family and lifestyle issues with career demands. The life partners of many women mathematicians are academics or professionals in other fields with a national job market, heightening the difficulty of finding employment in the same geographic area. Some couples have found creative solutions to this problem. Joan Hutchinson, an author in "Inside the Academy," has for many years shared a tenured faculty position with her husband, Stan Wagon. Other couples have lived apart when necessary or had commuting marriages (e.g., Green and Pour-El in "Having a Life").

Another issue that is of vital, often paramount, interest to many women mathematicians as they try to establish careers is whether or not to have children, and if so how to time their births. The responsibility of caring for an ailing partner or aging parents is more likely to occur after the tenure process, but may have a chilling effect on career advancement for mid-career and peak-career women. While this book was in preparation, many

---

6. The Canadian mathematician J. C. Fields donated funds to establish these medals, up to four of which are awarded at each International Congress of Mathematicians to eminent researchers not over forty years of age, to recognize both existing work and the promise of future achievement.

contributors mentioned family issues in the direct and immediate—"I'll get back to you as soon as my new baby sleeps through the night," "I have to arrange a caregiver for my dad," "I'll proof it this afternoon during my son's soccer game"—or the abstract—"All the women my age are involved with their grandchildren. . . ."

The federal Family and Medical Leave Act (FMLA) of 1993 may allow a worker unpaid leave for a serious health condition or a family obligation (birth or adoption, health condition of a spouse, child, or parent).[7,8] The problems with implementation of the FMLA in academe are legion, in part due to the semester or quarter plan of the academic year and the fact that professors are not interchangeable. The American Association of University Professors (AAUP) urges institutional policies

> that enable the healthy integration of work responsibilities with family life . . . including family-care leaves and institutional support for child and elder care and . . . policies, such as stopping the tenure clock, that specifically relate to pretenure faculty members who are primary or coequal caregivers for newborn or newly adopted children, responding to the special and age-related difficulty of becoming a parent during the pretenure years. . . . A more responsive climate for integrating work and family responsibilities is essential for women professors to participate on an equal basis with their male colleagues in higher education. . . . [2]

Many universities have formulated such policies, attempting to assure meeting both the faculty member's needs and the institutional mission.[9]

A recent national survey of chemistry and English faculty by researchers at Penn State explored issues at "the intersections of work, family, and community" [8; 13]. The focus was on bias avoidance behaviors, where individuals "deny themselves the opportunity to take on family commitments . . . or try to hide the performance of family tasks from co-workers and employers, all for the purpose of being perceived as committed and thereby securing career advancement." Results indicate that women respondents have fewer children than their male counterparts and perceive less support for dual commitments to work and family. Surprisingly, "although the evidence is somewhat mixed, it does not appear the research institutions are less conducive to family formation relative to teaching institutions . . . Rates

7. The AAUP has recently published a guidebook, *The Family and Medical Leave Act: Questions and Answers for Faculty*. For ordering information, see www.aaup.org/catalogue/02FLMA.htm.

8. The Supreme Court ruled May 27, 2003, that state workers have a right to sue their employers in federal court for alleged violations of the FMLA; see "Supreme Court Backs State Workers' Family Leave Rights," in "News for Working Families" [online]. AFL-CIO, 2003 [accessed 28 June 2003]. Available from the World Wide Web: http://www.aflcio.org/issuespolitics/worknfamily/ns05292003.cfm.

9. For example, the University of Colorado Faculty Handbook includes a section on compensation and leave, supplemented in 1995 by an eleven-page memorandum from the vice president of administration. Vanderbilt University has a comprehensive parental leave policy that includes gay parents.

of parenting among women fall as we move to the more teaching oriented colleges and universities offering only bachelor's degrees" [13].
These last conclusions are corroborated for scientists in general in
Scarcity: "Although the gender difference is narrowing, men are more
likely than women to be married and to have children at the time they
receive their Ph.D. . . . Female Ph.D.'s have children later in their career
than do men. . . . Marriage and children are associated with increased
rates of full-time employment for men, but declining rates for women"
[26, 3–4].

One way to aid faculty, sometimes institutionalized, is a practice called
"banking hours." Essentially, a faculty member may shift part or all of her
teaching obligations for one or more school terms, in return for teaching
extra courses or for taking on additional departmental duties, either before
or after, with no loss in salary. A drawback is that stopping the tenure
clock under this arrangement is more problematic than with an FMLA
unpaid leave; the faculty member's research and service obligations would
typically, at least on the record, continue during the period of no teaching.
One department reports using banking to enhance stated policies as an
aid in recruiting prospective faculty members who desire to start families.[10] A midlevel university department, through an innovative blend of
banking hours and special assignments, enabled a woman assistant professor to avoid taking unpaid leave. Twice in five years, arrangements were
made, in the first case including the delay of the tenure clock. (The department chair advocated strongly with the academic dean for approval.)
Seven years after her initial appointment, she was tenured, had children of
ages four months and five years, and had good research productivity,
although likely slowed by her banking and special assignment debts. A
level playing field? No. But a good faith effort.

The academy and individual institutions and employers are slowly
learning how to deal appropriately with the myriad of life circumstances
complicating individual academic progression. Enlightened administrators are needed in the effort. One such was Mina Rees, appointed dean of
the graduate school of City University of New York in 1961 (see III.3). In
1965 she noted, upon receiving an American Association of University
Women Achievement Award:

> It may be because the Graduate Dean is a woman, or it may be for com-
> pletely objective reasons, that ours is proving an ideal university to draw
> into advanced graduate work the most obvious source of unused talent
> in a society that desperately needs additional numbers of persons with
> training through the doctorate, namely, women. . . . we have welcomed

10. Conversations with Rebekka Struik and the University of Colorado implementation memorandum for FMLA.

qualified women who have applied even to the extent of considering the need for a baby-sitter a proper reason for providing financial assistance. [17]

Attention is now being given by professional societies and funding agencies to providing information and encouragement for women to move up administrative ladders. Two conferences in academic year 2003–2004 had that theme: "Women of Applied Mathematics: Research and Leadership" and "After Tenure: Women Mathematicians Taking a Leadership Role." Ideally, following the example of Mina Rees, sympathetic women administrators would aid not only in institutionalizing family-friendly policies but also in creating university-wide awareness that no written policy can anticipate every circumstance, given variations in individual needs and among departments. By finding inventive ad hoc solutions when necessary and being open to appeals from those whose departments prove insufficiently flexible, these administrators could further improve the lot of women and men, both faculty and students, who face career and life balance issues.

From the Data

What is the current status of doctoral women mathematicians, according to the best available data? Are there discernible trends in the opportunities available to them? The information given here provides some partial answers to these questions; it is based on the statistical analyses disseminated by professional societies and federal agencies. First some information is given about the diverse classification schemes cited in this report.

Departmental Groupings within the Mathematical Sciences. When not otherwise attributed, data are from the First, Second, and Third Reports that together comprise the "Annual Survey of the Mathematical Sciences" (Annual Survey). This joint effort of the AMS, the American Statistical Association, the Institute of Mathematical Statistics, and the MAA employs a stratification of mathematical sciences departments that has been used in the community for over thirty years (e.g., [22, 935]). The department groups for the Annual Survey are based first on the highest degree offered in the mathematical sciences. When it is the baccalaureate, the department is in Group B; for Group M departments the master's degree is the highest level. There are a large number of M and B departments, some small, and the Annual Survey questionnaire is sent to a stratified random sample; standard errors are given in the reporting [22, 926].

The Annual Survey questionnaire is sent to all doctoral mathematical sciences departments, and the high return rate yields a near census. The reporting divides them into groups, two of which are subareas in the mathematical sciences. In 2003, there were 286 such departments, with Group IV including 86 departments limited to statistics, biostatistics, and biometrics; and Group Va, 23 in applied mathematics. The remaining doctoral departments are named either as mathematics or by a joint nomenclature (e.g., mathematical sciences) and are then separated into groups based on numerical rankings produced by the National Research Council (NRC).[11] The top group is further divided into departments from public and private universities. The subgroups are I (Public), 25 highest-ranked departments at public universities; I (Private), 23 highest-ranked at private universities; II, 56 middle-rank departments; and III, 73 lower-ranked. Lower classifications in NRC rank may result when a department awards very few doctoral degrees; 4 of the Group II departments and 21 of the Group III departments produced no Ph.D.'s during the academic year 2002–2003. For fuller explanation of the groupings, see [21, 218 and 233; 25, 238 and 253]. These groups for the Annual Survey are a much finer sieve than the Carnegie classifications, where categories are based on university rather than departmental characteristics; they use the same NRC database [7]. The top Carnegie Research I classification includes most Group I departments and about 30 from Group II.

In the following sections, the mathematical sciences departments classified in Groups I, II, III, Va, M, and B will be referred to collectively as "math" departments, with the subdivision abbreviations "doctoral math" (I, II, III, Va), "master's" (M), and "bachelor's" (B).

Another source of data specifically about mathematical sciences departments, which will be cited below for comparisons and confirmation, is published by the Conference Board on the Mathematical Sciences (CBMS). This latter survey provides useful information about faculty and students in U.S. departments, including some gender data, and adds information about two-year colleges. It separates reports of data appropriately for statistics departments and mathematics sciences departments, but as a report focused on undergraduate programs in mathematics it groups all mathematical sciences doctoral departments. The report is issued at five-year intervals; the Year 2000 report gives some comparisons to 1995 [28].

The Annual Survey does not in general aggregate data across the math and statistics departments, and the tables below follow that practice (see [25, 246–247]). Statistics departments (IV) consistently have a

---

11. The NRC collects data and prepares rankings of departments about every ten years; these are then published by the National Academy Press. The Annual Surveys cited use groupings based on [14].

larger percentage than do doctoral math departments of women students (both U.S. citizen and overall) and of graduates who go to industry. They have a higher percentage of U.S. citizen doctoral recipients, although their numbers of undergraduates are very small relative to their graduate student enrollments.

The Pipeline. The ambitious Third Report includes undergraduate and graduate enrollment data, and undergraduate degree information; the First and Second Reports describe the cohort of new doctorates by field of thesis, citizenship, gender and first jobs. The percentages of females at increasingly more advanced educational stages are given in table 3. At the earliest reported levels, junior/senior majors, the pipeline is wide. Bachelor's and master's departments report majors and undergraduate degrees around 50%, as do doctoral III departments. The doctoral II departments

TABLE 3. Females as a Percentage among Students in Fall 2002 and among 2002–2003 New Doctorates

	Junior and Senior Majors	Under-graduate Degrees	First-Year Graduate Students	All Graduate Students	New Doctorates
I (Pu)	37	41	28	26	22
I (Pr)	28	25	29	22	19
II	42	44	38	36	29
III	48	50	40	38	45
Va	33	50	36	34	21
Doctoral math	41	43	35	31	26
M	48	50	41	46	—
B	49	47	—	—	—
Math	46	46	—	—	—
IV	40	33	56	54	41

Notes: The first and second columns are computed from table 5A in [22, 933]; the third and fourth, from table 6A in [22, 934]; and the last, from table 4B in [21, 221].

I (Pu) is highest-ranked public doctoral math departments; I (Pr), highest-ranked private doctoral math; II, mid-ranked doctoral math; III, lower-ranked doctoral math; IV, doctoral statistics; Va, doctoral applied math; M, master's level; B: bachelor's only. In columns 1 and 2, undergraduate majors and degree awards for computer science majors within mathematical sciences departments are not included.

and the public universities of doctoral I have 40% female majors, and more than 40% of the undergraduate degrees go to women. The pipeline is narrower from the beginning at the private universities of Group I, where women earn only one in four mathematics bachelor's degrees.

These relationships between departmental groups and female doctoral production percentages are consistent both with the lower representation reported in *Scarcity* for women mathematics undergraduates in Carnegie Research I institutions [26, 220] and with the 1988–1995 data analyzed by Radke Sharpe and Sonnert [33]. Another corroborative report is based on the CBMS survey made every five years. That survey groups departments by the highest degree offered, and the aggregate totals include four majors in math departments (degrees awarded vary from department to department, but overall may be grouped as mathematics, mathematics education, statistics, and computer science).

Looking at the data since 1990, do any trends emerge? The CBMS report says, "If one compares CBMS 2000 data to the findings of the CBMS survey in 1990, one sees essentially no change in the percentage of bachelors degrees awarded to women." CBMS reports 42% for 1995 and 43% for 2000 [28, 70]. Table 3 reports 46% women over the three levels of mathematics departments, but this does not include computer science (CS) majors within math departments. If CS majors in math departments are included with the math majors to match the CBMS database, then the Annual Survey yields 42%, strongly consistent with the CBMS data. These reports establish that the percentages of female majors over all departments have remained relatively constant, hovering in the neighborhood of 43% during the years 1992 to 1999 and dropping slightly since [22, 933; 24, 936].

In numerical terms, the doctoral math departments report 6800 total female undergraduate majors; master's departments were estimated to have 6000, and bachelor's departments, 10,100. This gives a picture of where women mathematics majors are receiving their instruction. Of the math undergraduate degrees awarded to women, nearly half, 47%, were from bachelor's departments, with 21% from master's and 32% from doctoral departments. Although the 86 doctoral statistics departments report very few undergraduate majors and degrees, women are 40% of 1000 junior/senior majors and earn 33% of 300 undergraduate degrees [22, table 5A, 933].

A snapshot of the entire graduate student cohort for 2002–2003 is given in the last three columns of table 3. Columns three and four give percentages of women students enrolled that fall and the last column, percentages of the doctoral awards to them. What signs for the future may be seen? Overall, the female percentages in the younger cohorts allow for the possibility that, even after the greater attrition that is statistically expected for women, new doctorate percentages of women will improve [21, table 8, 225; 25, table 8, 245; 32, Appendix B, table 4-11].

New Doctorates and First Jobs. The data below give partial answers to such seemingly simple questions as "What percentage of mathematics doctoral awards go to women?" (From which doctoral area groups? U.S. citizens only or all awards?) What is the historical record? (Neither the composition within departments, nor of departments within groupings, permit direct comparison; both data collection and analysis methods have varied.) The bullet points below present relevant summary statements which can be made based on the Annual Survey data. However one counts, there was significant improvement in the percentages of new doctorates awarded to females in the last quarter of the twentieth century. This provides hope that, as more new doctoral women become available, women will also establish a larger presence wherever math and statistics Ph.D.'s are employed.

- Comparing the 1975–1976 and 2002–2003 percentages of women in math and statistics among U.S.-citizen new doctorates shows an increase from 12% to 32% [21, table 8, 225; 25, table 8, 245].[12] The all-time high was in 1998–1999, with 187 degrees awarded to women, yielding 34% [25, 239].
- For combined math and statistics departments, women receive 30% of new doctorates; the 26% females from math departments is pulled upward in the aggregate by the statistics departments at 41% [21, 219].
- There has been seven-year stability in the neighborhood of 30% for U.S. women earning doctorates in math and statistics [21, 225].
- After four years of downturn in the numbers of U.S. doctoral awards to both men and women, a small upturn in numbers is noted [21, 219]. Considering time-to-degree experience for graduate students and recent first-year graduate student numbers, it appears that this gain "may be the first indication of a moderate trend upward over the next few years" [21, 220].
- The percentage of female new doctoral recipients by fields of thesis ranged from 14% in optimization/control to 41% in statistics and 70% in mathematics education [21, 224].
- Overall unemployment rose slightly from the previous year's comparable report, to 5% for men and 6% for women. A dramatic decline continues in the numbers of new doctoral recipients with jobs in business and industry with 97, a 50% decline over four years from 206 in 2000 [21, table 5B, 222].

12. It is impossible to determine what the exact percentage in 1975 would have been, using the same criteria now in use. The earlier numbers include two groups of departments that are heavily male: Group Vb, operations research and management science, was included before 1998–1999, and computer science departments were included before 1982–1983. Thus the 12% reported for 1975 most likely represents an undercount [25, table 8, 245].

TABLE 4. Females as a Percentage of New Doctoral Recipients
Produced and Hired by Doctoral-Granting Departments,
Combined Data for 2000–2001 and 2001–2002

Group	I (Pu)	I (Pr)	II	III	IV	Va	Total
Produced	24	23	29	32	42	23	30
Hired	23	23	30	32	35	14	27

Notes: The percentages are computed from table 2B in [22, 802] and table 2B in [24, 805].

I (Pu) is highest-ranked public doctoral math departments; I (Pr), highest-ranked private doctoral math; II, mid-ranked doctoral math; III, lower-ranked doctoral math; IV, doctoral statistics; Va, doctoral applied math.

Table 4 gives on row 1 the percentages of women receiving degrees by department group for the final reports of 2001 and 2002 combined. Row 2 gives the comparable percentages reported as hires in the fall term of the academic year following those awards, in whatever capacity. This two-year average and use of Second Report data give a more reliable picture than the separate one-year averages since annual percentages fluctuate considerably when there are small numbers involved, of both departments and individuals. Hiring in particular is dependent on so many factors that both hiring numbers and percentages of women hired bob up and down. Examples of this fluctuation from the smallest of the groups are: doctoral I (Private) averaged 23% hiring (table 4, line 2) for the two years together, whereas the single-year reports were 17% and 29%; applied mathematics departments Va, 14% in table 4, reported by-year percentages of 18% and 10%. [22, table 2B, 802, and table 3D, 805; 24, table 2B, 805, and table 3D, 807].

An NSF initiative expected to significantly improve numbers of U.S. citizen doctorates may also affect the distribution of doctoral and post-doctoral awards among departmental groups. The Vertical Integration of Graduate Research and Education in the Mathematical Sciences (VIGRE) program and its recent successor, Enhancing the Mathematical Sciences Workforce in the 21st Century (EMSW21), make awards primarily to doctoral I departments, which historically have low percentages of female doctoral students (see tables 3 and 4). It is too early to tell whether these extremely attractive financial aid offers will cause the desired Ph.D. production increase or a shift in production among department groups or both. Also, effects on gender distributions of new doctorates should be monitored.

Do these women with new doctorates find jobs, and where? A glance at table 4 shows that the ranked doctoral math groups produce and hire

TABLE 5. Percentage of Females in Faculty Cohorts Fall 2002: Tenured,
Tenure-Track, and Non-Tenure-Track Faculty

	Tenured	Tenured + Tenure-Track	Tenure-Track	Non-Tenure-Track
I (Pu)	7	9	17	18
I (Pr)	5	6	12	20
II	7	9	21	31
III	11	15	27	26
Va	10	10	12	18
Doctoral math	8	10	21	23
M	19	22	30	39
B	21	23	29	36
Math	15	18	27	28
IV	15	21	36	41

Notes: The data in this table are computed from tables 3A and 3B in [22, 929].

I (Pu) is highest-ranked public doctoral math departments; I (Pr), highest-ranked private doctoral math; II, mid-ranked doctoral math; III, lower-ranked doctoral math; IV, doctoral statistics; Va, doctoral applied math; M, master's level; B: bachelor's only.

females at about the same rates. Since the majority employer is the academy, and many graduate students expect to find employment there, it is a cause for concern when the percentage of women faculty new hires is much smaller than the percentage of women doctorates produced. This is the chronic situation for the specialized statistics[13] and applied mathematics groups, IV and Va, as seen in those columns of table 4.

A bright note, which it is reasonable to hope is a foreshadowing of more prestigious jobs for women in the future, may be seen in the percentages of women hired in Group I and II departments as shown in table 4; most of these are as multiyear postdocs. These positions are necessary for those who desire eventual tenure in doctoral departments, where job candidates are expected to have completed, preferably in a highly ranked department, a postdoc during which both teaching and research programs have been strengthened and validated. Improvement in the hiring of women in doctoral department tenure-track positions can be achieved only as more

13. Although statistics departments have a higher percentage of women faculty overall than math departments, they hire as faculty a significantly smaller percentage of women than their new doctoral production percentages.

women have the opportunity provided by these plum first jobs. The number of postdoc positions available has increased steadily, as has the proportion of women obtaining them. For contrast, note that in 1973 the *Survey of Earned Doctorates* found 79 postdocs, 4 awarded to women [31, Appendix table 55, 181–182]. The Annual Survey in fall 2002 reported 101 postdocs, with 25 women—a dramatic percentage and numerical increase for women from 1973 [22, 809].[14] With the availability of additional postdoc appointments, it appears that master's and bachelor's departments may be hiring more individuals completing a postdoc. Within the last five years, there has been a 27% decrease in the number of new doctoral recipients hired by master's and bachelor's departments; part of the decline may result from restructuring by administrators seeking to use more temporary instructors at lower cost, regrettable but beyond the scope of this analysis. Faculty members at liberal arts colleges suggested in an AWM *Newsletter* forum that research maturity, often best evidenced by a postdoc, is important before joining their faculties [36]. Lisa Traynor (Bryn Mawr College) said: "It would have been very difficult to continue as I have with my research if I had not spent a few years [as a postdoc]" [40]. (See also III.2.)

On a percentage basis, master's (33% women) and bachelor's (34%) departments are strong employers of women new doctorates, but the total numbers are smaller than the doctoral departments. There were 148 new doctoral women hired (some, perhaps most, as postdocs) over the two-year period by doctoral departments, 35 by master's departments, and 75 by bachelor's departments. Somewhat surprisingly, two-year colleges, research institutes, and other nonprofits hired 92 women in that two-year period, at a percentage of 37% of their new doctoral hiring [22, table 2B, 803; 24, table 2B, 805].

What other workplace opportunities for new doctorates are there? The likelihood of a man's going to business and industry is somewhat higher than a woman's; government is a smaller employer, but the proportion reverses. For fall 2003, the numbers of new doctorates employed in doctoral departments, government, and business and industry were down, balanced by increases in the other academic departments and research institutes [21, 219]. This shift is too recent to be reflected in the NRC reporting that since 1975, there have been shifts in the sciences in general from academia to industry. Less industrial employment in general is problematic for women because of the promise of the "narrowing in gender differences in this sector of employment" [26, 5].

A look at the actual numbers of women awarded doctorates puts the human face on the data. In the spring of 2003, about six months after the

---

14. It should be noted that these data are self-reported, and that the two surveys use different data-collection techniques.

award of their doctorates, the women reported their employment. U.S. academic positions included 80 in doctoral departments, 54 in master's and bachelor's departments together, 3 at two-year colleges, and 47 in research institutes or other academic departments. Of the others in the cohort of 296, 53 were in business, industry, and government; 25 were employed outside the United States; and 10 were unemployed or not seeking employment. Remarkably, only 24 of the cohort were not located by this survey [22, 803].

The Annual Survey Second Report compiles salary information from individual new doctorates in academic and other positions, with historical data back to 1965. The current distributions that plot medians, quartiles, and outliers are useful both to employers seeking to offer competitive salaries and to applicants. They also provide comparative beginning positions in the job market for women and men with six different types of appointments (academic teaching or teaching and research, 9–10 months; also 11–12 months; academic postdoctorate; academic research; government; business and industry). Median salaries are the same for women and men postdocs, and differences are within 5% of total salary except in government, where women's median is $7100 higher than men's, and business and industry, where the advantage reverses, with a difference of $2900 [22, 808–811].

The Workforce. The news above is encouraging about females in the mathematics pipeline, from undergraduates through postdocs. Women new doctorates find first employment in numbers and at salaries not greatly divergent from those of men showing similar potential, a major improvement relative to the not-so-distant past as experienced by today's senior mathematicians. Information on a large sample of all employed doctoral scientists, weighted more heavily toward those at early stages of their careers, is available in *Characteristics of Doctoral Scientists and Engineers in the United States (Characteristics)*. In 2001, women are reported as about 15% of the mathematical scientists in that cohort. The subcategories by type of employment report the percentages of women among employed doctoral mathematical scientists as universities and four-year colleges, 15%; other educational institutions, 29%; private for profit, 15%; self-employed, 17%; private not for profit, 15%; federal government, 10%; state and local government, 14%; and other, numbers too small to report [30, tables 13 and 14]. (Except for self-reported information from new doctorates about their jobs and salaries, the Annual Surveys say little on employment outside academia.)

Tables 5 and 6, based on the Annual Surveys, look at the employment of women in academia [22, 929–931]. Two complementary questions about academic employment of women mathematicians are: By type of

TABLE 6. Percentage of the Tenured and Tenure-Track Females Employed by Departments in Various Groups

Group	I (Pu)	I (Pr)	II	III	Va	M	B	IV
Tenured	4.9	1.6	6.5	8.4	1.0	25.7	44.9	7.1
Tenured and tenure-track	4.1	1.4	6.0	8.3	0.8	25.0	46.0	8.4

Notes: The data in this table are computed from table 3B in [22, 929].

I (Pu) is highest-ranked public doctoral math departments; I (Pr), highest-ranked private doctoral math; II, mid-ranked doctoral math; III, lower-ranked doctoral math; IV, doctoral statistics; Va, doctoral applied math; M, master's level; B: bachelor's only.

department and permanence of position, what are the percentages of women? (See table 5.) What percentage of women academic mathematicians is found in each type of department? (See table 6.) To examine the employment of women among the doctoral full-time faculty at colleges and universities, tenure—or stability of employment expectations—is paramount to consider. The qualifications required to earn tenure and the job situation once tenured differ from institution to institution, but in general dismissal requires proven just cause following contractual procedures. The climb up the professional "ladder" usually begins with a position that is said to be on the tenure track. After a probationary period, usually of six years, the faculty member is eligible for tenure; if tenure is not granted, the faculty member's contract is terminated. Some faculty positions, for example, most lectureships and postdoctoral appointments, are not on the tenure track. (Although it is possible to move up to a higher-ranked department after the initial tenure-track appointment, it is unusual.)

Table 5 reveals that only 8% of tenured mathematics doctoral faculty are women (for statistics it is 15%), with master's and bachelor's departments estimated at 20%. On the brighter side, due to social and Ph.D. supply factors, the academic cohort of women, especially in the sciences, is relatively young [26, 3]. Data in the Annual Surveys and *Characteristics* support the possibility of slow increases in both numbers and percentages of tenured women. *Characteristics* estimates that 22% of tenured faculty within ten years of the Ph.D. are women (math and statistics combined), compared to only 9.5% for those tenured who have held the Ph.D. for at least ten years [30, table 22]. Looking at the 1996 and 2002 Annual Surveys, the percentage of women among the tenured mathematicians is up from 1% to 5% in every department group and up overall from 11.5% to 15%; the female doctoral statistics percentage has doubled to 15%. The untenured tenure-track math cohort was 27% women in both surveys, with the 2002 survey reporting 36% for statistics. Untenured tenure-track percentages of women continue to lead those of the tenured by double digits, so continuation of the 1996–2002 patterns of awarding tenure will allow percentages to inch up slowly (tenure-track women are not yet numerous enough to pull up the overall ladder position percentage very much (column 2 of table 5)). [12, tables 3E and 3F, 917; 22, tables 3A and 3B, 929]

Serious consideration needs to be given in the academy to the implications of the high female percentages—20% to 40% women—in non-ladder positions in every type of department, as shown in the last column of table 5. Monitoring the percentages of new doctorate women taking temporary positions is difficult because the postdocs must be parsed out, and a "temporary" position is only negative when a "permanent" one was desired but not available. But by any measure disproportionate numbers of women hold these tenuous jobs, without expectation of tenure, despite their considerable contributions to the profession. This pattern is consistent with that for the sciences in general [26, 6]. Some non-ladder faculty are midcareer to late-career women who had no better jobs available when they were applying for positions, or for various family and personal reasons were precluded from obtaining the type of first job that could lead to a professorial career. These Ph.D. mathematicians often perform essential departmental functions such as the orientation and training of graduate teaching assistants or the development of computer-assisted instruction, appropriately utilizing their education and experience, although typically undercompensated.

There are 2808 Ph.D. women on ladder appointments in U.S. math and statistics departments. Of these, 813, up from 616 in 1996, are in doctoral departments [12; 22]. Table 6 [22, table 3B, 929] shows that 7 of 10 of these women are employed in master's and bachelor's departments. This is consistent with much of the Radke Sharpe and Sonnert analysis of years before 1995 that states: "Women's representation continues to be inversely correlated with the prestige of the institutional groups. Women had the greatest representation in the least prestigious institutional settings, and the lowest representation in the most prestigious mathematics departments" [34, 13]. Tables 5 and 6 also track with *Scarcity*'s report that in 1995 only 11% of the mathematics Research I (roughly Groups I and II) workforce (all types of positions, including non-ladder and postdoc) was female [26, 1–6]. *Scarcity* reports that women are more likely than men to be in off-track positions and quotes Long and Fox: "the less prestigious the type of institution, the more likely the employment of women and minorities" [26, 144–145; 27].

Salaries provide another measure of professional achievement. The Faculty Salary Survey in the First Report each February gives then-current faculty salary data as ranges within department groups by faculty rank. These detailed charts and tables have no breakdown by gender [21; 24; 25]. Because annual raises have not kept pace with the salaries offered to recent doctorates, no extrapolations may be made to more senior faculty. Again, some information is available in *Characteristics*. For full-time doctoral mathematical scientists employed in universities and four-year colleges, 2001 median annual salaries for women and men at fewer than ten years since the doctorate were the same; for those ten or more years past the Ph.D., women's median salary was 17% lower than men's. Over all types of employment, the median salaries of women were 18% lower than those of men [30, tables 53 and 61]. A careful attempt to analyze salary gender differences in *Scarcity* found that for full professors in the social and behavior sciences at Research I universities, fields in which many women work, men were earning nearly 15% more than women in 1979, with the gap reduced only to 10% by 1995. They noted that discrimination comes not only in the form of differential salaries for equal work, but also through differential access to higher-paying jobs. Further, "The male/female earnings gap in science is not fully explained by the individual or contextual factors that have so far been measured. Even analyses that methodically control . . . leave a large residual of wage gap unexplained . . . . Clearly more research is necessary . . . to assess the effects of discrimination against women at each step in the stratification process" [26, 215–217].

From overall science data, it is expected that women will fall off the ladder for personal reasons at rates higher than men. Nevertheless, the aggregated doctoral tenure-track percentages, together with the increasing numbers of women who will join the recruiting pool after their postdocs, give reason for optimism that the slow movement toward larger percentages of tenured females will continue. But what of the most prestigious departments? Is there room at the top?

At the Top

Data published in *Science* about women in the "top ten" mathematics departments showed that in 1991–1992, there were only five tenured women. Adding the seven women on tenure track gave twelve women out of 340 ladder faculty—3.5% women [19]. A follow-up in 1998–1999 showed that there were still fewer than 5% women ladder faculty at these departments [20; 39]. (See tables 1 and 2 in "AWM in the 1990s," II.) In the forty-five doctoral I (Private) and applied math departments, there are but forty-four tenured women. In these department groups as well as in the superelite "top ten," despite the good efforts of many, senior women have not yet moved "From Scarcity to Visibility." The NAS publication of that name warns that as the numbers of women increased overall in the sciences "traditional measures of status, such as becoming a full professor in a research university, did not expand proportionally" [26, 1–2]. There are few cracks in the "glass ceiling" in these top ten departments. Glass ceiling issues outside the academy have frequently been discussed in the popular press. Catalyst, a women's advocacy group, takes a census of the Fortune 500 at regular intervals. In 2002 it found that 60% of those companies had women holding a quarter or more of the corporate officer titles, while only six of them had women CEOs and seventy-one still had no women corporate officers [6].

Moving from the "top 10" to the "top 100," do the percentages of women in the tenured faculty improve? Sadly, not enough. There is, at 8%, serious underrepresentation of senior women for the doctoral math groups overall. (See table 5.) Are there enough senior women colleagues to serve as role models and mentors for younger women faculty and graduate students? Hardly. The average over all the doctoral departments is fewer than two tenured women per department. The numbers are better at the public university doctoral I departments, where although the percentage of senior women is still low at 7%, the departments are large enough that they average about 3.5 senior women per department. In general, in the doctoral departments there may be only one or two women to emulate. The junior woman, at a career and life stage where she may very much want a female mentor, may have difficulty finding one. She may well find herself bereft of the aid and comfort that another woman can provide in the mentoring relationship, both in terms of confiding personal information affecting her work and of seeking guidance.

What of the future? The experience to date in the sciences has been that "women are somewhat less likely than men to be promoted to tenure and move much more slowly through the senior faculty ranks" [5, 182]. The latest Annual Survey finds 21% of the doctoral math tenure-track faculty to be women, which is encouraging. However, the top private and applied departments lag behind, with only 12% women on the tenure track. The number of women on doctoral I and II tenure tracks is a healthier 18%, although the numbers of these junior women are so small that when combined with the 7% of tenured women faculty, the result is under 9% women on ladder in these departments. Over time, both the percentages and numbers of women in doctoral math departments seem likely to improve, but unless the overall number of tenure/tenure-track positions available increases,[15] the changes will continue to be slow.

15. "It may be more difficult for women to attain senior faculty positions than it was for men in the 1970s and 1980s when the number of academic positions was growing" [26, 5].

Reflections

The "principle of the triple penalty," as enunciated in 1975, asserts that women scientists must overcome barriers to their entering science, psychological reactions to perceived discrimination, and the consequences of actual discrimination [42]. A more recent analysis [9, 303] lists a number of factors—"the early large negative reaction of not accepting an optimal first job because of personal and family constraints, . . . mild discouragement from a grant rejection which could be related to the job placement, . . . [and] the birth of a child before the tenure decision"—that may have career-dampening results, including tenure delay, lower salary, and difficulty with promotion to full professor. Although many claim that the failure of women to rise in the professions at the same rates as men may be explained by differential performance, *Scarcity* asserts that "differences in the positions held by women are likely the cause of lesser productivity, rather than the other way around" [26, 6]. The data above clearly show that in mathematics the percentage of tenured faculty who are women is very low at the research departments that are considered to be prestigious; those women who are tenured mathematics professors are mostly not in these top departments. (See tables 5 and 6.)

More females than males are lost from the pipelines of mathematics and other sciences at every stage of study, as the decreasing percentages of females in successive cohorts at school and at work indicate. Mathematics is among the slowest of the science and engineering fields to move toward fuller representation of women. The transition from the Ph.D. to the full-time labor force is a critical juncture at which relatively more women than men are lost. What can be learned on the means for integrating women into the field from the sciences that have been more successful in doing so? What patterns can be discerned for women doctorates across the sciences during the last quarter of the twentieth century in nonacademic jobs, and for tenure, promotion, recognition, and reward?

Each female mathematics student or woman mathematician moves during her life journey over her own pathway from Undergraduate Education to Graduate Education, to First Job on to Midcareer and to Senior Mathematician. At each stage of her journey, what measure will she take of the profession's inclusiveness? The status of women in the mathematics profession has increased as their "scale of presence" at these locations along the pathway has improved from absence to scarcity (e.g., tenured women in top math departments), then on to visibility (as for tenured women in other departments) to parity (or nearly so for undergraduate degrees) and even to abundance (in some undergraduate programs and statistics doctoral programs). Ample measure, in numbers of women peers, is important as she moves along, affecting not only her comfort but also the support available to her. Will measures improve enough that fewer capable women leave the mathematical pathway? Perhaps only time will tell.

Coda

Reflecting on the poignant memoirs of part V and the information, anecdotes, and statistics here, the question "why us?"—"why are we mathematicians?"—arises. Environment, culture, adaptation, and availability—geography and happenstance—influence the venue of endeavor, as illustrated throughout this book. But they do not explain who becomes a mathematician. Are there childhood indicators of this statistically unlikely event? Surely the stories of candles in the night to hide the study of mathematics from family disapproval, common to the biographies of Mary Somerville and Sophie Germain, serve as dramatic childhood precursors of later interests.

Among the cohort of today's active women mathematicians, first acquaintance with plane geometry is a common early influence on the development of interest in mathematics, as for each of the authors of this paper. Musing on similarities of detail of childhood experience with authors in part V led Leggett (first of ten siblings) to reminisce; she shared a draft of an article along these lines for the AWM *Newsletter* with Case (an only child). Case notes that she, too, liked to play with Erector sets given to boys, made forts with mixed groups of boys and girls, and was fascinated by George Gamow's *One Two Three . . . Infinity*. From Leggett's newsletter article [23]:

> I loved arithmetic from day one; I remember happily drawing circles around groups of sticks on a worksheet the first day of first grade, first following dotted lines on the page and then figuring out where to put lines (by counting things out) all by myself! It was a lot of fun to do the 100 addition or multiplication facts very very fast; I realized quickly that if they were presented in order, you didn't even need to think, just let the pencil fly! However, until I took geometry, I believed that teaching people how to read would be life's highest calling. After geometry, I began to think that teaching math would be a great job.
>
> Although I loved my dolls, I was very interested in "scientific" things. One Christmas my brother Bill and I asked for a chemistry set together. We had an enormous amount of fun with it, singly and together. We collected seeds from weeds/grasses by running the stems backward through the holes of a plastic shaker top—I was convinced this would have made wonderful bread for the pioneers. I theorized about making cloth from pussy willow fuzz; it seemed obvious it would make a lovely soft sweater.

It has been written elsewhere that learning basic carpentry skills helps the development of visualization skills. I well remember a box that my brother Bill and I made from leftover two-by-fours from some household project; we planned and measured carefully as we built one of the heaviest boxes with less than a cubic foot of interior capacity on the face of the earth. We were so proud of our plans when that heavy lid fit so snugly into place.

I bought a Visible Woman™ with Christmas money one year and tried to use my oil paints to give the organs their proper colors, which didn't really work out. Another year I asked for a microscope, but it was never quite the same after I tried to clean what I thought were spots from the lens, not realizing they were floaters in my eyes. A "dissecting kit" came with the microscope, although I soon broke my dissecting needle on too tough a subject. And then there was my globe—I took the stand apart to "see how it worked," but the metal rod representing the earth's axis was slightly bent, and I could never get it properly inserted again. Despite the failure of many of these experiments, I had a wonderful time.

My mother told me all about finding areas of oval tabletops by using lots of little triangles and showed me a poster on the Pythagorean theorem she had made during her high school days. She told us about infinity to stop the game "who can name the biggest number" (clearly we were driving her crazy); no one on the playground would believe us. We learned about negative numbers, although we didn't call them that, by playing card games. My dad knew I liked numbers, so he put a receipt book and some other things like that in my stocking one year; I thought this was so totally cool. Later, Dad brought me a book on genetics from a business visit to an educational publisher. I had a great time making charts for the Mendelian crosses and so on.

Although at times I wanted desperately to fit in better, I liked being "smart" and never considered abandoning the study of math. Having decided that teaching math was the best career imaginable, eventually I learned that despite all the gender issues, being a math professor at a university was exactly where I would "fit in."

References

1. Raymond C. Archibald, *A Semicentennial History of the American Mathematical Society, 1888–1938* (New York: AMS, 1938).
2. American Association of University Professors (AAUP), "AAUP Statement of Principles on Family Responsibilities and Academic Work as Association Policy"

[online]. Washington, DC, 2001 [accessed 1 August 2002]. Available from the World Wide Web: http://www.aaup.org/statements/REPORTS/re01fam.htm.

3. American Mathematical Society, Secretary, "Correspondence of Roland George Dwight Richardson, 1921–1942," History of Science Collection, Special Collections, The John Hay Library, Brown University, Providence, RI.

4. Lisa Belkin, "The Opt-Out Revolution." *New York Times Magazine,* cover article, October 26, 2003.

5. William T. Bielby, "Sex Differences in Careers" in [43], 171–187.

6. Catalyst, "Catalyst Census Marks Gains in Numbers of Women Corporate Officers in America's Largest 500 Companies" [online]. New York: Catalyst, November 19, 2002 [accessed 3 January 2004]. Available from the World Wide Web: http://www.catalystwomen.org/press_room/press_releases/2002_cote.htm.

7. Carnegie Commission on Higher Education, *A Classification of Institutions of Higher Education: A Technical Report* [online]. Berkeley, CA: The Carnegie Commission [1973, 1976, 1987, 1994] 2000 [accessed 27 November 2003]. Available from the World Wide Web: http://www.carnegiefoundation.org/Classification.

8. The Center for Work and Family Research, "CWFR Mission" [online]. Social Science Research Institute, Penn State University [last update 25 October 2002; cited 6 June 2003]. Available from the World Wide Web: http://www.ssri.psu.edu/cwfr/ mission.htm.

9. Jonathan R. Cole and Burton Singer, "A Theory of Limited Differences: Explaining the Productivity Puzzle in Science" in [43], 277–310.

10. Jonathan R. Cole and Harriet Zuckerman, "Marriage, Motherhood and Research Performance in Science" in [43], 157–170.

11. Stephen Cole, "Age and Scientific Performance," *American Journal of Sociology* 84 (1979), 958–977, cited prior to publication in [38].

12. Paul W. Davis, "1996 AMS-IMS-MAA Annual Survey (Second Report)," AMS *Notices* 44 (1997), 911–921.

13. Robert Drago and Carol Colbeck, "Preliminary Results from the National Survey of Faculty" [online]. The Mapping Project, Penn State University [revised May 17, 2002; cited 6 June 2003]. Available from the World Wide Web: http://lsir.la.psu.edu/workfam/prelimresults.htm.

14. Marvin L. Goldberger, Brendan A. Maher, and Pamela Ebert Flattau, eds., *Research-Doctorate Programs in the United States: Continuity and Change* (Washington, D.C.: National Academy Press, 1995).

15. Carolyn Gordon, Claudia Henrion, and Joyce McLaughlin, "Burlington Panel: Do Women and Men Have Different Career Trajectories?" AWM *Newsletter* 25(6), 1995, 12–18.

16. Judy Green and Jeanne LaDuke, "Women in the American Mathematical Community: The Pre-1940 Ph.D.'s," *Mathematical Intelligencer* 9(1), 1987, 11–23.

17. ———, "Mina Rees: 1902–1997," AWM *Newsletter* 28(1), 1998, 10–12.

18. Claudia Henrion, "Burlington Panel" in [15], 12–16; see also *Women in Mathematics: The Addition of Difference* (Bloomington: Indiana University Press, 1997).

19. Constance Holden, ed., "Women in Math Update" in "Random Samples" column, *Science* 257, July 17, 1992, 322–323.

20. Cathy Kessel, "Education Column," AWM *Newsletter* 33(3), 2003, 11–12.

21. Ellen E. Kirkman, James W. Maxwell, and Colleen A. Rose, "2003 Annual Survey of the Mathematical Sciences," *Notices of the AMS* 51(2), 2004, First Report, 218–233.

22. Ellen E. Kirkman, James W. Maxwell, and Kinda Remick Priestley, "2002 Annual Survey of the Mathematical Sciences," *Notices of the AMS* 50(7), 2003, Second Report, 801–813, and Third Report, 925–935.

23. Anne Leggett, "A Mathematical Memoir," AWM *Newsletter* 24(1), 2004, 15–17.

24. Don O. Loftsgaarden, James W. Maxwell, and Kinda Remick Priestley, "2001 Annual Survey of the Mathematical Sciences," *Notices of the AMS* 49, 2002. First Report, 49(2), 217–231; Second Report 49(7), 803–816; Third Report 49(8), 928–938.

25. ———, "2002 Annual Survey of the Mathematical Sciences," *Notices of the AMS* 50(2), 2003, First Report, 238–253.

26. J. Scott Long, ed. *From Scarcity to Visibility: Gender Differences in the Careers of Doctoral Scientists and Engineers* [online]. National Academy Press, 2001 [accessed 25 May 2003]. Available from the World Wide Web: http://www.nap.edu/books/0309055806/html.

27. J. Scott Long and Mary Frank Fox, "Scientific Careers: Universalism and Particularism," *Annual Review of Sociology* 21(1995), 45–71.

28. David Lutzer, James W. Maxwell, and Stephen B. Rodi, *Statistical Abstract of Undergraduate Programs in the Mathematical Sciences in the United States.* Conference Board on the Mathematical Sciences. AMS 2002. Also available from the World Wide Web: www.ams.org/cbms/cbmssurvey.html [accessed 2 December 2003].

29. Margaret A. M. Murray, *Women Becoming Mathematicians: Creating a Professional Identity in Post–World War II America* (Cambridge, Mass.: MIT Press, 2000).

30. National Science Foundation, Division of Science Resources Studies, *Characteristics of Doctoral Scientists and Engineers in the United States: 2001*, NSF 03-310, Project Officer, Kelly H. Kang. Arlington, VA 2003. Also available from the World Wide Web: http://www.nsf.gov/sbe/srs/nsf03310/start.htm [accessed 11 June 2003].

31. National Science Foundation, *Women and Minorities in Science and Engineering* (Washington, D.C.: U.S. Government Printing Office, 1986).

32. National Science Foundation, *Women, Minorities, and Persons with Disabilities in Science and Engineering: 2000.* Arlington, VA, 2000 (NSF 00-327). Also available from the World Wide Web: http://www.nsf.gov/sbe/srs/nsf00327/start.htm [accessed 6 June 2003].

33. Norean Radke Sharpe and Gerhard Sonnert, "Proportions of Women Faculty and Students in the Mathematical Sciences: A Trend Analysis by Institutional Group," *Journal of Women and Minorities in Science and Engineering* 5(1), 1999, 17–27, Begell House, Inc. Reprinted in AWM *Newsletter* 30(4), 2000, 17–24.

34. ———, "Women Mathematics Faculty: Recent Trends in Academic Rank and Institutional Representation," *Journal of Women and Minorities in Science*

and Engineering 5(3), 1999, 207–217. Begell House, Inc. Reprinted in AWM *Newsletter* 31(1), 2001, 13–20.

35. Margaret W. Rossiter, *Women Scientists in America: Struggles and Strategies to 1940* (Baltimore: Johns Hopkins University Press, 1982).

36. Stephanie Frank Singer, "Notes on Working at a Liberal Arts College," in "Research and Teaching at Liberal Arts Colleges," AWM *Newsletter* 26(6), 1996, 6–7.

37. Gerhard Sonnert, *Who Succeeds in Science? The Gender Dimension* (New Brunswick, N.J.: Rutgers University Press, 1995).

38. Nancy Stern, "Age and Achievement in Mathematics: A Case-Study in the Sociology of Science," *Social Studies of Science* 8(1978), 127–140. Reprinted in AWM *Newsletter* 18(2), 1988, 12–20.

39. Jean Taylor and Sylvia Wiegand, "AWM in the 1990s: A Recent History of the Association for Women in Mathematics," AMS *Notices* 46(1), January 1999, 27–38.

40. Lisa Traynor, "The Compatibility of a Liberal Arts College and Research," in "Research and Teaching at Liberal Arts Colleges," AWM *Newsletter* 26(6), 1996, 7–9.

41. Harriet Zuckerman, "The Careers of Men and Women Scientists" in [43], 27–56.

42. Harriet Zuckerman and Jonathan R. Cole, "Women in American Science," *Minerva* 13(1), 1975, 82–102, as cited in [26].

43. Harriet Zuckerman, Jonathan R. Cole, and John T. Bruer, eds., *The Outer Circle: Women in the Scientific Community* (New Haven: Yale University Press, 1992; originally published by W. W. Norton and Co., New York, 1991).

1. A Dual Triumph

It is instructive to consider the extraordinary determination to succeed shown by some African-American women mathematicians born before 1950 who faced the additional challenges posed by racism. "Pathways in Mathematics" explored the status of women in the profession, following the previous part's focus on collective efforts and activism in response to societal and cultural challenges to the rights and privileges of women. In this chapter, Vivienne Malone-Mayes points out that the African-American community did not place as many restrictions on women as did the general culture; she said: "Girls were conditioned from my earliest recollection to prepare to *work* . . . [with the hope] that through education [they] could escape the extremely low paying jobs designated for Black women." The separate and joint effects of sexism and racism on life and career form a recurring theme. Noted politician Shirley Chisholm, the first African-American woman to be elected to Congress and the first woman candidate for president of the United States, had this to say on the subject: "Of my two 'handicaps' being female put many more obstacles in my path than being black. Sometimes I have trouble, myself, believing that I made it this far against the odds."[16]

Two of the three first African-American women known to have received Ph.D.'s in mathematics appear in I.1. Marjorie Lee Browne's 1950 Ph.D. from the University of Michigan led to a teaching career during which she influenced many future mathematicians, including one whose story appears later in this chapter. The earliest known award of the mathematics Ph.D. to an African-American woman was to Euphemia Lofton Haynes in 1943 by the Catholic University of America. The surviving member of the three, Evelyn Boyd Granville, was a plenary speaker at the 1999 Taussky Todd Conference. In that talk (see part IV), she recounts the earning of her 1949 Yale Ph.D. and continues with a description of her multifaceted career, in which she is still active after several "retirements."

By 1970 there were still fewer than twenty African-American women known to have earned the mathematics Ph.D. Of those, Gloria Conyers Hewitt was among the earliest members of AWM, and Malone-Mayes was the first African-American to serve on its Executive Committee. In 1978, several of these women spoke at the AWM discussion "Black Women in Mathematics." Contributions from that session of Geraldine Darden and Elayne

---

16. Shirley Chisholm, *Unbought and Unbossed* (Boston: Houghton Mifflin, 1970), xii.

Arrington-Idowu are included here, along with an earlier article by Malone-Mayes. Also appearing here is a memorial tribute to Malone-Mayes, which states, in words that might apply equally well to many of these pioneers: "With skill, integrity, steadfastness, and love, she fought racism and sexism her entire life, never yielding to the pressures or problems which beset her path."

Black and Female

VIVIENNE MALONE MAYES

Based on "Black and Female," a talk at the panel "Noether to Now—The Woman Mathematician," moderated by Alice T. Schafer, 1975 Summer JMM, as reported in the AWM *Newsletter* 5(6), 1975, 4–6. Malone Mayes (1932–1995) was then a professor of mathematics, Baylor University.

When you are both Black and Female, it is difficult to distinguish which of these traits may account for the way you are received by others. I shall briefly review my career as a student and as a professor in an attempt to use hindsight as a tool in determining the influence these traits may have had on my professional growth. In many instances, it will be quite difficult to conclude whether these events happened because I am Black or because I am a woman or because I am both Black and female.

My precollege education took place in Waco, Texas, in strictly separate and strictly unequal schools. The law dictated that separate but equal educational opportunities be provided for both races. The separation provision of the law was rigidly enforced, but the "equal" provision was conveniently ignored. I shall not detail the inherent educational disadvantages of a separated school system. These details are available in the text of every lawsuit filed by the NAACP in its fight to abolish a dual school system. Read these texts and, where appropriate, insert my name as the plaintiff, for each of us was victimized educationally by the segregated schools.

Despite the intellectual setback Blacks suffered in the segregated schools, within these schools certain strong, positive impressions were made. Although the faculty was predominantly female, there was variety in the makeup of these women: some were beautiful and others were plain, some were mothers and some were not, some dressed well and some didn't, some were rigid disciplinarians and dogmatic in their views while some were tolerant. In every Black school I've attended there's always been at least one Black woman teacher or professor with whom

I could identify and who was a model I'd like to emulate. Black girls were expected to excel in their studies. No difference was made between boys and girls. The moral lectures given by teachers and designed to stimulate students to aspire to high and lofty careers were directed equally to boys and girls. Girls were conditioned from my earliest recollection to prepare to *work*. Every girl expected to work. Her hope was that through education she could escape the extremely low-paying jobs designated for Black women. Boys expected girls to work. Within our homes were working mothers. My father's most affectionate name for my mother was "partner."

Today Black men select wives who can help increase the family income. There is no loss of status because a man has a working wife. A few of these men have middle-class incomes, but their heritage has approved and accepted the concept of the working woman. Historically, the acceptance of the working wife was imposed on Black men by the discrimination they suffered on the job market. But acceptance of this arrangement has become entrenched and is no longer related to the historical reason.

Perhaps I was personally influenced most by a Black professor I had at Fisk University, Evelyn Boyd Granville, a graduate of Smith and a Ph.D. from Yale University. Her dissertation advisor was Einar Hille. She was no more than six years older than the students she taught. But she set high standards and demanded a quality performance from her students. She was a "lady" in the traditional sense, and also had superb credentials. I believe that it was her presence and influence that account for my pursuit of advanced degrees in mathematics. I am also deeply indebted to Lee Lorch, whose superior teaching skills made my pursuit of a Ph.D. a successful one.

Girls held the majority in my upper-level math classes at Fisk. I recall only two men graduates in mathematics my senior year out of a total of eighteen students. After Fisk, I taught at two Black colleges. In both instances, girls outnumbered boys in every class. Today, this trend continues in Black colleges.

Lest you believe there is complete acceptance of women in all areas by Black people, I'd like to point out one occupation in which the Black woman is *still not* accepted by the community. This occupation is that of preacher. There are a few exceptions, but by and large, no denomination has encouraged or supported giving women an equal opportunity to serve as pastors in Black churches.

I have given details of some experiences I have had in an all Black situation so that you may get some feeling for the cultural shock I received in my transition from an all-Black to a predominantly white institution.

My first recollection of my tenure as a graduate student at the University of Texas was a summer class that met at 7:00 A.M. in which I was the only Black and the only woman. For nine weeks, thirty or forty white men ignored me completely. I never initiated any conversations as there was no encouragement to do so. It seemed to me that conversations before class on mathematics between classmates quickly terminated if it appeared that I was listening. My rapport with students in the other two classes was not much better. This was my first experience attending school in a vacuum. My mathematical isolation was complete; I was not acquainted with any Blacks who had interest in these subjects. My teachers were fair and did not give me any preferential treatment.

For years I felt this initial response to my presence was because I was a Black attending a predominantly white school in the south. However, other Black women have told me they received similar treatment in northern schools. Was my vacuum created because I was black or female or both?

Friendships eventually developed after several semesters had passed, in which time a group of us had taken the *same* courses. Most of my classes had two women in them. She and I became very good friends.

In spite of the growing acceptance of me by classmates, certain privileges or opportunities that would have accelerated my mathematical maturity were withheld:

1. I could not become a teaching assistant. Why? Black.
2. I could not join my advisor and other classmates to discuss mathematics over coffee at Hilsberg's Café. Why? Because Hilsberg would not serve Blacks. Occasionally, I could get snatches of their conversation as they crossed our picket line outside the café. (It was only after Hilsberg was required by law to serve Blacks that I noticed that women were seldom included in these informal over-coffee problem-solving sessions.)
3. I could not enroll in one professor's class. He didn't teach Blacks. And he believed that the education of women was a waste of the taxpayer's money.

After graduation from graduate school, I joined the faculty of a predominantly white university. By this time I expected to have classes with a male majority. The students have never shown any prejudice toward me as a Black or as a female. As evidence I submit the splendid cooperation I have received in their response as students and of those who have worked as my assistants.

I never had any complaints about salary or promotions. I have received financial support from the administration for innovative and experimental projects under my direction. An additional safeguard of my welfare has been yearly visits by representatives of the government. They have checked salaries and promotions to determine if I was being subjected to any discrimination. These reports have always been encouraging to me.

One area in which I have been most frustrated is in counseling of gifted women students. Fortunately, the situation is changing. But until recently, their response to my enthusiastic encouragement of what they could accomplish was met with apathy. Their response was a complete puzzle to me. One girl revealed quite frankly that she didn't want to have any special plans for her life so she'd have no difficulty in accepting the plans her husband would have for *their* lives. Perhaps the greatest service the women's movement will give is to awaken women that they should participate in the decisions that shape their destiny.

In Remembrance

ETTA ZUBER FALCONER AND LEE LORCH

Based on "Vivienne Malone-Mayes: In Memoriam," AWM *Newsletter* 25(6), 1995, 8–10. Falconer (1933–2002) was then a professor emerita of mathematics, Spelman College, while Lorch is a professor emeritus of mathematics, York University.

Vivienne Malone-Mayes was born February 10, 1932, in Waco, Texas, and died there on June 9, 1995. An excellent student all her life, Vivienne graduated from the segregated A. J. Moore High School in Waco in 1948 at only sixteen years of age. Starting at Fisk University in Nashville, Tennessee, immediately, she earned the B.A. in 1952 and the M.A. in 1954. Her friendships with, among others, Charles Costley, Joyce Gould, Gloria Hewitt, and ourselves had their beginnings at Fisk.

At Fisk, she had courses from Evelyn Boyd Granville, the second African-American woman to receive the Ph.D. in mathematics (from Yale in 1949). Of Granville, an inspiring and exacting teacher, she wrote: "I believe that it was her presence and influence which account for my pursuit of advanced degrees in mathematics." [See IV.]

It was a hard decision to make, harder to implement. It could not be done all at once. First, she returned to Waco to serve (1954–1961) as chair of the mathematics department at Paul Quinn College, operated by the African Methodist Episcopal Church. Seeking as always to expand her knowledge, she applied to take some courses at Waco's Baylor University, only to be rejected explicitly on grounds of race (1961).

The University of Texas, already required by federal law to desegregate, had to admit her. It was a lonely and stressful time for her. Writing in 1988 in the AWM *Newsletter*, she observed that "it took a faith in scholarship almost beyond measure to endure the stress of earning a Ph.D. degree as a Black, female graduate student." But earn it she did, drawing on her vast reserves of courage and determination as well as on her undoubted abilities.

In 1966 she became the thirteenth African-American woman known to have received the Ph.D. in mathematics. Her thesis, supervised by Don Edmondson, was entitled "A Structure Problem in Asymptotic Analysis."[17] Later her research interests shifted to summability theory, where she published jointly with B. E. Rhoades.[18]

In graduate school she was very much alone. In her first class, she was the only Black, the only woman. Her classmates ignored her completely, even terminating conversations if she came within earshot. She was denied a teaching assistantship, although she was an experienced and excellent teacher. [Further details are given in her talk "Black and Female," reprinted above.] As she has commented, "Opportunities which would have accelerated my mathematical maturity were withheld."

Overlooking all this, one of her professors, complaining against the civil rights demonstrations, said to her: "If all those out there were like you, hardworking and studious, we wouldn't have any problems." Her reply: "If it hadn't been for those hell-raisers out there, you wouldn't even know me."

Mathematical talent was not enough for success even though Vivienne had this in abundance. It took enormous courage and determination as well. It took all these attributes together for her to become the second Black person and the first Black woman to get a mathematics Ph.D. from the University of Texas.

Her invited talks sponsored by AWM and published in the AWM *Newsletter* describe not only her own journey but also that of the collective of Black women, indeed of all women and all Blacks. Written many years ago, they cast a penetrating light on the past and the present. They are must reading still today . . . and for everybody.

Surviving all this with her customary strength, good humor, and stability, she became in 1966 the first Black faculty member at Baylor University, the institution which had rejected her as a student only five years previously. There she spent the rest of her teaching career, retiring because of ill health in 1994.

For some years her career path was smooth. In 1975 she described her experiences in positive terms in her talk "Black and Female." She was less satisfied beginning in the 1980s, especially within the department. There were no longer any visits by federal inspectors. The Reagan-Bush years

17. Part of this work was published in "Some Steady State Properties of $\int_0^x f(t)dt/f(x)$," *Proceedings of the American Mathematical Society* 22(1969), 672–677.

18. Vivienne Malone-Mayes and Billy E. Rhoades, "Some Properties of the Leininger Generalized Hausdorf Matrix," *Houston J. Math.* 6(1980), 287–299.

saw the budgets of the civil rights agencies cut drastically, and inspection visits fell victim. Vivienne felt that this weakened her position and cited several specific complaints.

Throughout the years, good and bad, she maintained a high level of activity in mathematical, community, and religious organizations. She was the first Black elected to the Executive Committee of AWM and served on the Board of Directors of the National Association of Mathematicians (oriented toward the Black community in the mathematical world). She was a member of the AMS, the NCTM, and the MAA, where she was elected director-at-large for the Texas Section. In addition, she served as director of the High School Lecture Program for the Texas MAA.

Her dedication to the community at large was just as great. We have already mentioned her antiracist picketing; her articles situate her academic struggles within the broader antiracist movement. She served on advisory boards for civic and charitable organizations and was both choir director and organist for her church.

She enjoyed her friends and kept in frequent telephone contact with many, however far-flung. Her poor health did not keep her away from the January JMM. Her last winter meeting was in 1993 in San Antonio. This was a particularly joyous reunion. It brought Vivienne together with Gloria Hewitt, the two of us, and, above all, Evelyn Boyd Granville, whom she had not seen in many years but whose inspiration she had never forgotten. There had been frequent occasions for two or three of us to be together, along with other friends, but this made for a very special occasion that Vivienne enjoyed enormously, as did we all.

This was to be our last face-to-face contact with Vivienne, whose poor health, perhaps weakened by the accumulated racist- and sexist-induced stress of the years, soon worsened. But the telephone calls continued until just a few days before a heart attack claimed her life.

She had made of it a good life. She could well have said with Terence, the Roman playwright, "Nothing human is alien to me." All her life, to its very end, she was part of the struggle to make the path smoother for those who followed. She made her presence in the national mathematics community felt and respected. In the organizations embracing the entire mathematics community she was to be found and heard. In the organizations specifically devoted to the problems of minorities and women, there she was too. With skill, integrity, steadfastness, and love, she fought racism and sexism her entire life, never yielding to the pressures or problems that beset her path. She leaves a lasting influence. From her life, the world has gained much. In her premature death we have all lost. Inspired by her life, we are bereaved at the loss of a loving and beloved friend.

References

Carswell, Catherine. "BU Math Professor's Life Filled with Firsts," *Waco Times-Herald*, February 26, 1986; reprinted, AWM *Newsletter* 16(4), 1986, 8–9.

Houston, Johnny L. "Spotlight on a Mathematician," NAM *Newsletter*, Fall 1995.

Kent, Nita Sue. "Blacks at Baylor: After 20 Years," *Baylor Line*, November 1982, 9–13.

How I Decided to Pursue a Ph.D. in Mathematics

GERALDINE DARDEN

Based on a talk at the panel "Black Women in Mathematics," organized by Patricia Kenschaft and Etta Falconer, 1978 JMM, as reported in the AWM *Newsletter* 8(3), September 1978, 9–11. Darden was then a professor of mathematics, Hampton Institute and later served as chair. In 1981 she moved to Bell Laboratories, from which she retired in 2001.

There have always been black women in mathematics, but the first two Ph.D.'s I knew of were Evelyn Boyd Granville (Yale, 1949) and Marjorie Lee Browne (Michigan, 1950). These two are representative of other black women who are struggling to help youngsters learn the beauty of mathematics in this space age.

I grew up in the South, went to all-black elementary and secondary schools, and attended a private black college in Virginia, Hampton Institute.[19] I am not bragging, but I was a good student and was valedictorian of both my elementary and secondary graduating classes. I was the top student in my class in the Department of Mathematics at Hampton Institute. Even so, nobody ever suggested that I go to graduate school in mathematics. My professional role models were the usual ones that black children had during that time (this was during the 1950s)—teachers, doctors and nurses, and lawyers. Most of the professionals I knew personally were teachers, so the natural vocation for me to choose was teaching. I wanted to be a secondary teacher, so I earned a B.S. in mathematics education from Hampton Institute in 1957. I accepted a job as a mathematics teacher in a junior high school, but within six months I began to wonder if I had made a mistake in my choice of profession.

---

19. The name has since been changed to Hampton University.

In 1957 the Russians sent *Sputnik* up, and Washington became very generous with funds to study the sciences. One result was that in 1958 I was awarded a grant to study mathematics in a summer institute at North Carolina Central College for six weeks. There I met two people who had a great influence on my life. One was Walter Talbot, and the other was Marjorie Browne. When I reached Dr. Browne's office upon my arrival, I said to her, "Good morning, I'm Geraldine Darden."

She was sitting at her desk, and she looked up and said, "Why aren't you in graduate school?"

I looked around to see who else was in the room, because we had never met. She did not know me. Obviously she had seen my undergraduate record (we had to submit one to be considered for the institute). Finally I said, "Are you talking to me?"

She said, "Yes."

I said, "What did you say?"

"Why aren't you in graduate school?"

I was still a little flustered, so I thought a moment and answered, "Well, I had to go to work when I finished college."

"So you could earn that big car you are driving out there?"

"No, I come from a large family and I needed a car to go to and from work."

"Okay."

During that summer I participated in one of the most intense studies in linear algebra that anyone has ever had in any institution, and she encouraged me from that point on. She was an excellent teacher—she knew her algebra and was very demanding. As a result I was accepted into an NSF Academic Year Institute at the Department of Mathematics, University of Illinois at Champaign-Urbana. By the end of the year I had earned a master's degree in mathematics education.

The Institute at Illinois was very good. Although I did very well, I was still not sure that I could join a regular graduate mathematics program and compete. So I accepted a job at Hampton Institute and found that I really did enjoy teaching; it was my junior high teaching experience that had made me doubt that I wanted to teach. Now I knew that college teaching was right for me.

But to continue teaching at the college level, I would have to return to graduate school, and there was that graduate degree in mathematics again. I asked the president of Hampton for a leave of absence to earn a master's in mathematics—I thought I would start out gradually. He called me into his office and said, "No, I won't give you a leave to study for a master's degree. I will give you a leave if you go back to study for a Ph.D. Do you still want the leave?"

I said "Yes." I was admitted to the Ph.D. programs at both the University of Illinois and Syracuse University and for family reasons decided to go to Syracuse.

Because some credits from Illinois were transferred to Syracuse, I did not have to take the first-year algebra class but was advised to take the advanced algebra class. I put in many hours, and at the end of the year when I went to pick up my final exam results, I asked my professor, Arthur Sagle, how I had done.

He said, "You did pretty well. You came in third."

I said, "Are you sure you know who I am?" (I was the only black female in the class.)

He laughed and said, "No, you did well!" There were some people in the class who had excellent reputations in the department, and having come in third, I knew I had come in ahead of some of them. He continued, "What are you going to do? What are you going for?"

I said, "A master's."

He said, "You're not going to get a Ph.D.?"

"I don't know if I can do that," I replied.

"You can do it!"

It was at that moment I really believed I could do it. From then on, even though there were many times when I doubted I *would* do it, I believed that I *could* do it. I received my master's (1965) and my Ph.D. (1967) in mathematics from Syracuse University.

I took time to relate this to you because I want to urge those of you who are teachers—on any level—please do not make prejudgments. We don't see many minority students in science or mathematics; if you see a good student, please encourage that student. If you do for them as Browne, Talbot, and Sagle did for me, I think you will see many more black women in mathematics.

A Double Dose of Discrimination

ELAYNE ARRINGTON-IDOWU

Based on a talk at the panel "Black Women in Mathematics," organized by Patricia Kenschaft and Etta Falconer, 1978 JMM, as reported in the AWM *Newsletter* 10(3), 1980, 5–6. Arrington was then an assistant professor of mathematics, University of Pittsburgh, where she is now a senior lecturer.

I recently heard someone state that Blacks just don't make good mathematicians. Even more recently, I heard someone say that he had never met a good woman mathematician. These statements caused me to reflect on the two most prevalent forms of discrimination in our society—racial and sexual—and on the fact that as black women most of us have had to take a double dose. What bearing has that double dose of discrimination had on our ambitions and abilities to become practicing, contributing mathematicians?

I can only tell you of my personal experiences. I was born and raised in the North, near Pittsburgh, Pennsylvania, and I always attended integrated schools—in general, about 80% white and 20% black. Those of us who were black knew that we could never be cheerleaders or majorettes, and certainly not angels in the school Christmas play. Racial discrimination was there, and we had to accept it. I even had to accept it when there was no valedictorian the year I graduated first in the class. But all in all, I don't think I felt any lasting bad effects from my school days because, in general, the teachers seemed to encourage everyone to do her best academically.

When I graduated from high school, I went to the University of Pittsburgh School of Engineering. About that time I received a lot of letters and a big dose of sexual discrimination. I first received a letter telling me that I had been awarded an Honor Scholarship by the university. I later received a letter informing me that I had been chosen by the university to receive the Mesta Machine Company Scholarship, and because that provided more benefits than the Honor Scholarship, the latter would be rescinded. The choice had been made on the basis of high school grades and college board scores. Several days later, I received yet another letter. I was to be given back the Honor Scholarship because Mesta Machine Company refused to give their scholarship to a female. Mesta hoped that the recipient of their scholarship would spend many years of employment with them as an engineer, and a female just wasn't a suitable candidate.

This experience had several adverse effects that illustrate what discrimination can do. First of all, I felt resentful—I resented the white male who received "my scholarship," as I thought of it. I had classes with him and often heard him boast about the great scholarship that he had "won." Second, I felt compelled to prove that I could do anything that the male students could do (often collectively). They were not friendly, and I didn't ask them any questions lest they think that I was not capable of doing my own work. In effect, I was isolated. It was me against all of them. In thinking back, I suppose that the worst effect of my undergraduate experiences was the lack of intellectual exchange with my peer group. Finally, when I graduated from Pitt's School of Engineering (the third female and the first black female to do so), I couldn't find out my class rank because girls were not rated in that school.

After working for more than seven years as an aerospace engineer without any particularly illuminating experiences, I applied for graduate school at the University of Cincinnati. One of my last bad experiences came after I had been admitted as a doctoral student in mathematics.

Several of my instructors advised me early in my first term to take the preliminary examinations, which were to be offered soon. I decided to confer first with the professor I anticipated would be my thesis advisor. He warned me that the failure rate on the examinations was about 50% in general and about 98% among "housewives like you." I had never thought of myself as a "housewife," since I was an engineer and did the same things as the men in my office. My would-be advisor suggested that I spend at least a year taking courses from those who would likely prepare the examinations in order to enhance my probability of passing. I followed this advice (much to the consternation of my husband, who said it was the first time he had seen me show a lack of confidence) and wasted a year with the suggested study. This professor soon left the university, and I was fortunate in choosing a thesis advisor who was most supportive.

From my personal experiences, I have concluded that those who are inclined to state that Blacks or women cannot make good mathematicians tend to make it a self-fulfilling prophecy. A woman placed in isolation with little or no intellectual exchange and with her confidence constantly undermined has little chance of succeeding. And these are the conditions that such people tend to create.

I know that with the enlightened attitudes that are beginning to prevail, the set of black women in mathematics—which I am told someone once believed to be the null set—could well become a dense set.

Prejudice and Isolation, or Cooperation and Support?

Malone-Mayes was one of AWM's invited speakers on the occasion of the 100th birthday of the AMS. Some excerpts from that later talk on her experiences in Texas in the '60s follow.[20]

[As] Etta Zuber Falconer [has] pointed out,[21] the black female college students in the early fifties were not encouraged to prepare for academic careers. Since black colleges were so few in number, it was unrealistic to expect a teaching appointment. Consequently, it made little sense to pursue advanced degrees. In those days, we were counselled to prepare for health professions, the ministry, or public school teaching— the few careers which offered an opportunity for livelihood. The available teaching positions at small colleges offered little financial reward. . . .

20. Vivienne Malone-Mayes, "Centennial Reflections on Women in American Mathematics," panel discussion organized by Judy Green and Jeanne LaDuke, AWM *Newsletter* 18(6), 1988, 8–10.

21. As quoted in Vivienne Malone-Mayes, "Lee Lorch at Fisk, a Tribute" *American Mathematical Monthly* 83(1976), 708–711.

[With] a master's degree in mathematics from Fisk University supervised by Lee [Lorch], my salary at the small black college in Waco was $100 per month less than public school teachers with the same credentials . . . .

[My] personal isolation [as I worked on my Ph.D.] at the University of Texas in Austin was absolute and complete, especially during the summer of 1961. At times I felt that I might as well have been taking a correspondence course.

For those who completed degree programs, and for many who quit along the way, the lack of interchange with fellow students was a profound hindrance to academic achievement . . . .

The black, female Ph.D.'s faced much the same isolation even after earning their degrees and finding jobs or university teaching appointments. Though many had 15-hour teaching loads, administrative duties, and isolation on the job, most of them nevertheless succeeded in publishing mathematical research or writing textbooks or otherwise advancing in their academic careers. . . .

This sense of isolation may possibly be one of the reasons why so few black mathematicians today—male and female—participate in the [AMS, the MAA, or the AWM]. At the time of this Centennial Celebration of the AMS in 1988, black female mathematicians have *begun* to be accepted into the mainstream of business and academia. These are remarkable achievements attained only in the last forty years. It is my hope that, when the bicentennial celebration of the [AMS] is held, all problems that accompany racism and sexism will long have been solved.

Not all African-American female graduate students found their experiences as bleak as did Malone-Mayes in the '60s. Janice Brown Walker included these comments in a description of her graduate school experiences at the University of Michigan in the '70s:[22]

[In fall 1971], I was relieved and excited to see more than six other new African-American graduate students there . . . [who] formed a closely knit group that still exists. We were a family. We celebrated successes and shared failures. After the Qualifying Review was passed, you became a member of a team for which quitting was no longer an option. The African-American doctoral students also formed the core of a mathematical society that was organized as a forum for providing support and information to each other, presenting mathematical talks to each other, and interacting socially. This society was named the Ishango

22. Janice B. Walker, "Success at Michigan for African-Americans in Graduate Mathematics Programs," in *You're the Professor, What Next?* ed. Bettye Anne Case, MAA Notes, no. 35, 1994, 103–105.

Mathematics Society. [The Society encouraged cooperation] such as lending books, tutoring, and collaborative studying . . . . A few years later, the new black students could and did come to the more senior black students for assistance.

. . . An equally important reason [for our success] was that we were extremely supportive of each other, both professionally and personally. The sheer number of us attending made it easier to develop a group sense of power, courage, and self-esteem. Also we were warmly accepted and supported by a number of [graduate students and] faculty members. The cultural and ethnic diversity among students and the liberal atmosphere in the general community were important factors for our general comfort. Racial tension was not common. . . .

2. Inside the Academy

Profiles of women academics appear throughout this book—from the early women seeking academic connections profiled in part I, to the first-year assistant professors in part V. In this chapter the emphasis is on women fulfilling the tripartite responsibilities of the professoriate. Mathematical research and effective teaching are central, with service to the department and profession as the third member of the academic triad. On many campuses either research or teaching is weighted most heavily in personnel decisions; the importance of service to the academic endeavor is sometimes undervalued.

The chapter begins with Karen Uhlenbeck's description of the mathematical results presented in her 1988 Noether Lecture. Next in the spotlight, with Uhlenbeck, are three other women with very different mathematical careers as examples of the many who have earned professional honors and awards. Their contributions cover the full spectrum of academic experience, and the organizations that honor them represent multiple sectors. Uhlenbeck received a National Medal of Science for career achievements, including her pioneering efforts in establishing formal mentoring programs for young women mathematicians. Among women mathematicians, only Cathleen Morawetz has also been honored with this medal. Alice Turner Schafer, a teacher, mentor, chair of two mathematics departments, and second president of AWM, has been for over fifty years active in human rights causes. It is not so unusual as it once was for women or for mathematicians to hold academic administrative positions above the departmental level. Even so, the career trajectory of Dolores Richard Spikes is remarkable. President of universities in Louisiana and Maryland, and a strong advocate both for the Historically Black Colleges and Universities (HBCU) and for federal land-grant universities, Spikes has received a multitude of awards. Exposition of mathematical results is another important component of scholarly activity, exemplified in Joan Birman's article that was honored with an MAA prize. Expository writing has applications to all three major areas of academic life: clear explication of prior results advances further research, aids in the education of students, and is a service to the profession. Major lectures at the Joint Mathematics Meetings are often expository talks on important new mathematics. The need to explain mathematical results, not just to other mathematicians but also to the lay public, has become more apparent in recent years.

The chapter continues with historical vignettes. The first two are glimpses into graduate education in an earlier day. Next, two professors, one described as "a role model and mentor to many women in mathematics and science long before such terms were current," are remembered fondly by women they influenced. These relationships are important for the development of mathematicians of either gender, but given the cultural baggage associated with displaying an interest in mathematics, they are especially important for women. The importance of mentoring as a professional responsibility is now recognized by a number of awards, two recently won by women who are authors in this book: Etta Falconer received the 2001 AAAS Lifetime Mentoring Award, and Mary Gray received a 2001 Presidential Award for Excellence in Science, Mathematics, and Engineering Mentoring.

Next a variety of faculty perspectives on concerns within the academy are given. In a 1978 talk, Spikes reflected that she did not expect to progress beyond her master's degree: "That was to be the end of my training because I had never known a Black Ph.D. in mathematics. I did a few years later, but not at that time. Certainly I had known no Black women mathematicians."[23] Today, the need to enhance minority student persistence in graduate education is still vitally important. An article here explores the issue and makes recommendations based on a survey of graduate students.

The balance of efforts and achievements in the areas of research, teaching, and service must be a good fit for the needs and mission of the individual's department and institution. In many academic positions, the credo "publish or perish" is entirely accurate. Women perhaps even more than men may allow self-criticism to interfere with timely submission of results of potential use to other investigators. Marguerite Lehr was a faculty member at Bryn Mawr College at the time of the following anecdote:

> During my first leave I attended Zariski's lectures and I thought I could generalize what he was talking about. I worked very hard, but what I got was a mess. So when the editor of the *American Journal of Mathematics* came to me and said, "Zariski says you have some work I should publish," I responded, "It's . . . a bad lead." Zariski asked to see what I had done, and of course I had to give it to him. Then he said, "You didn't get what you went in after. But you proved some things. Now back off, and state what you proved. You have to publish what you have to keep other people from going down your dead end. But we don't know; maybe someone will see a fork in the road."[24]

Graduate students sometimes question whether research is important for faculty at liberal arts colleges. Lynne Butler answers affirmatively in this

23. Dolores Spikes, "Black Women in Mathematics," AWM *Newsletter* 10(3), 1980, 6–8.

24. Pat Kenschaft, "An Interview with Marguerite Lehr," AWM *Newsletter* 11(4), 1981, 4–7; reprinted as a memorial article in AWM *Newsletter* 18(2), 1988, 9–11, after Lehr's death in December 1987. Lehr was a doctoral student of Charlotte Scott, who was profiled in I.2.

chapter. In the same *Newsletter* forum, Stephanie Singer (faculty member, Haverford College) suggests that "you will be better off at an institution with a strong existing informal network of female faculty." Lisa Traynor (faculty member, Bryn Mawr College) comments, "If research is important to you then it is vital to be in an atmosphere where your colleagues are interested and active in research and where the administration values and expects research" and continues, "Compared to a large school where your service may be satisfied within the department, service at a liberal arts college involves dealing with members across the college community. . . . I am always trying to evaluate if I am correctly balancing the amounts of time I spend on the research, teaching, and service components."[25]

The final issue explored in this chapter is whether or not teaching evaluations are fair to women. Susan Basow, psychologist at Lafayette College, has said "those of us who evaluate female faculty must be alert to the various and subtle ways in which gender bias can affect perceptions and evaluations."[26] Here Neal Koblitz provides his perspective on this topic. The chapter closes with tips from a senior woman mathematician on how to achieve success in the academic sector.

Moment Maps in Stable Bundles

KAREN UHLENBECK

Based on "Moment Maps in Stable Bundles: Where Analysis, Algebra and Topology Meet," AWM *Newsletter* 18(3), 1988, 2. Uhlenbeck is profiled in "Honors and Awards," which follows.

This is a time of finding interrelationships within mathematics. Or, more bluntly, it is fashionable to study mathematical problems encompassing many areas of mathematics. This pleases me. As a student I had many crushes on various mathematics subjects, but sequentially. Now I study them all at once, as the full title of my talk makes clear: "Moment Maps in Stable Bundles: Where Analysis, Algebra and Topology Meet."

The gauge-theoretic study of stable holomorphic bundles on complex Kähler manifolds is a prototypical crossroads of mathematical areas. The subject of stable bundles is treated as an infinite-dimensional example of

---

25. Singer's quotation appears in "Notes on Working at a Liberal Arts College," and Traynor's in "The Compatibility of a Liberal Arts College and Research," both sections of "Research and Teaching at Liberal Arts Colleges," AWM *Newsletter* 26(6), 1996, 6–9.

26. Susan Basow, "Student Ratings of Professors Are Not Gender Blind," AWM *Newsletter* 24(5), 1994, 20–21.

geometric invariant theory (as formulated by Mumford). Symplectic geometry plays an important role. The equations for the moduli space are the Yang–Mills equations (from physics). The group is the same gauge group from Yang–Mills, and the topology on the moduli space is studied via Morse Theory for the Yang–Mills functional. So not only do we find the advertised analysis, algebra, and topology, but geometry and physics to boot! There is probably some number theory hidden somewhere in the second application below.

Moreover, the list of possible applications is formidable. Best known are probably Donaldson's uses of these Yang–Mills moduli spaces to construct invariants of smooth compact four-dimensional manifolds. Nearly as important are the calculations of Atiyah–Bott on the topology of moduli spaces for stable bundles over curves. Corlette has used this machinery to classify flat bundles, and Simpson has employed it to study Hodge Structures. Atiyah and Hitchin have studied the interaction of magnetic monopoles (those things astronomers seem to think may exist). One can always guess this might be useful for string theory, which seems to be able to absorb every kind of mathematics.

This is a lot of mathematics! Our graduate courses are not really designed to teach it, but graduate students seem to manage (better than me in fact) to absorb the necessary ideas. May I encourage more young women to try doing so, and join what is (comparatively) a woman's field. Unfortunately, at this time there is no good introductory survey. The list of references below, none of them particularly expository, includes a number of basic papers by women.

References

Atiyah, Michael, and Raoul Bott. "The Yang–Mills Equations over Riemann Surfaces." *Phil. Trans. Roy. Soc. London A* 308, 1982, 523–615.

Atiyah, Michael, and Nigel Hitchin. *Geometry and Dynamics for Magnetic Monopoles.* Princeton, N.J.: Princeton University Press, 1988.

Corlette, Kevin. "Gauge Theory and Representations of Kähler Groups." *Geometry of Group Representations* (Boulder, Colo., 1987), 107–124, Contemp. Math. 74, Amer. Math. Soc., Providence, R.I., 1988.

Donaldson, Simon K. "Anti-Self-Dual Yang–Mills Connections over Complex Algebraic Surfaces and Stable Vector Bundles." *Proc. London Math. Soc. 50,* 1985, 1–26.

———, "Infinite Determinants, Stable Bundles and Curvatures." *Duke Math. J.* 54, 1987, 231–247.

Hitchin, Nigel. "The Self-Duality Equations on a Riemann Surface." *Proc. London Math. Soc. 55,* 1987, 59–126.

Kempf, George, and Linda Ness. "The Length of Vectors in Representation Spaces." *Algebraic Geometry Proceedings: Copenhagen 1978,* Springer Lecture Notes in Mathematics 732, 1982, 233–243.

Kirwan, Francis. *Cohomology of Quotients in Symplectic and Algebraic Geometry*. Princeton Mathematical Notes 31, 1984.

Kobayashi, Shoshichi. *Differential Geometry of Complex Vector Bundles*. Princeton, N.J.: Princeton University Press, 1987.

Ness, Linda. "A Stratification of the Null Cone via the Moment Map." *Amer. J. Math.* 106, 1984, 1281–1330.

Simpson, Carlos. "Systems of Hodge Bundles and Uniformization." Thesis, Harvard, 1987.

Uhlenbeck, Karen, and Shing-Tung Yau. "On the Existence of Hermitian Yang-Mills Connections." *Comm. in Pure and Appl. Math.* 39, 1986, S257–S293.

Honors and Awards

Uhlenbeck Receives National Medal of Science

Based on "Uhlenbeck Receives National Medal of Science," AWM *Newsletter* 31(1), 2001, 9–10. University of Texas press release.

At the award dinner when Karen Uhlenbeck received her Year 2000 National Medal of Science, President Bill Clinton said of the recipients: "These exceptional scientists and engineers have transformed our world and enhanced our daily lives. Their imagination and ingenuity will continue to inspire future generations of American scientists to remain at the cutting edge of scientific discovery and technological innovation."

Ten of the twelve awardees, including Uhlenbeck, had received NSF support for portions of their academic or research careers. Uhlenbeck has been a MacArthur Fellow, a Woodrow Wilson Fellow, a Sloan Fellow, and a recipient of the Sigma Xi Common Wealth Award for Science and Invention. She is a member of the American Academy of Arts and Sciences and the National Academy of Sciences. Born in Cleveland, she grew up in New Jersey, graduated from the University of Michigan, and earned her Ph.D. from Brandeis University on an NSF graduate fellowship. Since 1987, she has been a professor at the University of Texas at Austin, where she holds a Sid W. Richardson Regents Chair.

Spokesmen for the NSF, which administers the medals for the White House, said Uhlenbeck stands out as one of the founders of geometry based on analytical methods. Cited for her pioneering contributions to global analysis and gauge theory that resulted in advances in mathematical physics and the theory of partial differential equations, she was also recognized for her leadership as a mentor for women and minorities in mathematics education.

"I feel very humble, as many greater scientists have received this award. I hope that my acceptance will serve as encouragement to young women scientists and mathematicians," Uhlenbeck said. "Mathematics is a discipline which takes ideas from all branches of science and extends, constructs and develops further these ideas into a body of results that we usually refer to as theorems. These ideas can be used independently as a language to describe new processes that have nothing to do with the original source," she explained. "For example, I study bubbles; that originates with soap bubbles. But I use them in abstract contexts, where they can be used to investigate the shape of space or to study the structure of martials. A basic idea in high-energy physics is part of the description of color and charm (gauge theory). I study this abstractly, and have found ways to use this in the study of waves and magnetic materials. Other mathematicians have used my work in the study of space-time and in string theory."

Uhlenbeck said her group at UT Austin "specializes in looking for new ideas in mathematics in the work of other sciences. While our primary work is theoretical physics, we have a member who thinks about how DNA coils. I have been fascinated by a number of equations I learned about from physicists who study plasma and fluid flow." She added that "in common with all basic researchers, mathematicians do not expect immediate applications, although we do expect the ideas we develop to be around for centuries. The kinds of mathematics that are used in applications have become diverse, and we don't try hard to second guess what will be useful next year."

Gung and Hu Award Conferred on Schafer

LINDA R. SONS

Based on the citation for the Gung and Hu Award for Distinguished Service presented to Alice Turner Schafer by the MAA at the JMM, Baltimore, January 1998. The citation appeared in the program for the Joint Prize Session and was reprinted in the AWM Newsletter 28(2), 1998, 13–15. Sons is a professor of mathematics, Northern Illinois University.

The curriculum vitae of Alice Turner Schafer lists two specializations: abstract algebra (group theory) and women in mathematics. As early as her high school years, Alice exhibited a love for mathematics and an interest in teaching as a career. As a mathematics educator she championed the full participation of women in mathematics. She has been a strong role model for many women and has worked to establish support groups for women in mathematics, to eliminate barriers women face in their study of mathematics and participation in the mathematics community, and to provide opportunity and encouragement for women in mathematics. She was one of the central figures in the early days of the AWM, through which she has helped to change the place of women in American mathematics. Yet her service goes far beyond her work on behalf of women.

Alice Turner is a native of Virginia, where she spent her school years, earning a B.A. in mathematics from the University of Richmond. Lacking the means to attend graduate school, she taught secondary school mathematics for three years before entering the University of Chicago, where she earned an M.S. and a Ph.D. Her dissertation in projective differential geometry was supervised by E. P. Lane. At the University of Chicago Alice met Richard Schafer, who was seeking a Ph.D. in mathematics. They were married as they completed their degrees. This union has been blessed with two sons and three grandchildren.

The Schafer's marriage was an early example of the "two-body problem" and the "commuter marriage." Alice's first postgraduate position was at Connecticut College, followed by a year at the Johns Hopkins Applied Physics Laboratory. She then held positions at several schools before returning to Connecticut College, where she advanced to full professor. Moving to Wellesley College (by now Richard was at MIT), she soon became department head and the Helen Day Gould Professor of Mathematics, retiring in 1980. Indefatigable, Schafer continued teaching, at Simmons College and in the management program in the Radcliffe College Seminars. Upon Richard's retirement from MIT, they moved to Arlington, Virginia, where Alice became professor of mathematics at Marymount University, retiring once again in 1996.

Throughout her career, Schafer sought to eliminate barriers to women in mathematics and to promote human rights for all mathematicians. She directed the Wellesley Mathematics Project (continued jointly with Wesleyan University) aimed at reducing fear of mathematics for women. She helped to prepare lists of women who were eligible for grants and fellowships, including invited lectureships. Schafer has served on the CBMS Committee on Women in the Mathematical Sciences for six years and has worked for many years for the MAA Women and Mathematics Program. Three times in recent years, through the People-to-People program, she led delegations to China— one concerning women research mathematicians; one, mathematics education; and another, women's issues in mathematics and science.

Schafer is known for her love of people, her boundless energy, and her fierce determination for a just cause. Her lifetime achievements and her pioneering efforts to secure opportunities for all mathematicians make her a most worthy recipient of the MAA Award for Distinguished Service.

Spikes Honored as Administrator and Educator

Dolores Richard Spikes has been honored many times for her achievements as a university administrator. Among these honors are the Thurgood

Marshall Educational Achievement Award (Johnson Publishing Company, Inc., 1989); Outstanding Alumnus of the Century (Southern University, 1990); the Thurgood Marshall Scholarship Fund (TMSF) Education Leadership Award (TMSF, 1996);[27] and an award for excellence in government service (the Maryland Coalition of Women for Responsive Government, 1997).[28] She has been described as a "visionary leader, a dynamic and motivating speaker, and a highly effective, compassionate but firm educational manager."[29]

Spikes has long been passionately concerned with educational equity issues. Her parents, Lawrence Richard and Margaret Patterson Richard, instilled the importance of education in her from a young age; she credits them as the single greatest influence on her life. "My father had a fourth-grade education, but he loved to read. He loved education so much that even after his daughters finished college, he went back to get his GED." Her mother, who had a tenth-grade education, shared these values. "We never talked about whether we were going to college. We always knew we were going, even though my parents didn't know where the money was coming from."[30] Twenty years later, Spikes commented on young African-American women college graduates: "When the lure of a job comes to them upon receipt of their B.S. degree, they are sometimes hesitant even to think about going to graduate school."[31] She herself became the first African-American to receive a mathematics Ph.D. from Louisiana State University.

Spikes taught high school science and then college mathematics. The year 1981 found her as full-time professor of mathematics and part-time assistant to the chancellor at Southern University–Baton Rouge. Moving up the administrative ranks, in 1987 she became president of the Southern University and A&M College System in Louisiana, where she served until she accepted the presidency of the University of Maryland–Eastern Shore in 1997.[32] Having retired from university administration in 2002, she is still in

---

27. TMSF is a national organization that awards merit-based scholarships to students attending public HBCUs. See www.thurgoodmarshallfund.org [accessed 15 November 2003].

28. Jennifer M. Schildroth, "Maryland College Roundup," *The Retriever*, published by The Retriever Weekly, University of Maryland, Baltimore County. UMBC Student Media 1999 [online; accessed 15 November 2003]. Available from the World Wide Web: http://trw.umbc.edu/articles/2307?Newspaper_Session=b353018ff42d5bc6bdde639b092b4598.

29. Louisiana Secretary of State, "Dr. Dolores Margaret Richard Spikes," Distinguished Women of Louisiana Exhibit [online; accessed 9 November 2003]. Available from the World Wide Web: www.sec.state.la.us/archives/women/bio-spikes.htm.

30. From the web page "Dolores Spikes, Mathematician of the African Diaspora," at Scott Williams's website [accessed 15 November 2003]. Available from the World Wide Web: www.math.buffalo.edu/mad/PEEPS/spikes_dolores.html. Quotations cited there were originally from *The Oval Message*, UMES alumni magazine, Fall 1997, 2–5.

31. Dolores Spikes, "Black Women in Mathematics," AWM *Newsletter* 10(3), 1980, 6–8.

32. "Biography—Dolores Margaret Richard Spikes" [accessed 13 November 2003], www.umes.umd.edu/president/presidentbio.html, archived on the World Wide Web: www.adec.edu/clemson/spikes_bio.html.

demand as a speaker on mathematics, education, and African-American issues and continues to serve. For example, as president emeritus of Southern University, she serves on the National Advisory Board of the Kellogg MSI Leadership Fellows Program, designed to train the next generation of senior-level leaders at Minority-Serving Institutions (MSIs) to prepare and retain the talent pool of current and future leaders at these institutions.[33]

Spikes has served on numerous boards and committees focused on the HBCU and federal land-grant universities. "I wanted to work at a historically black college, to give something back and to make a difference in terms of enabling people to get a better education," she has said.[34] She was appointed to the Kellogg Commission on the Future of State and Land-Grant Universities when it was first organized in 1996 and was its vice chair. The reports issued by this project of the National Association of State Universities and Land Grant Colleges (NASULGC) have been highly influential in the academic community, providing detailed and effective suggestions in response to this challenge: "Unless public colleges and universities become the architects of change, they will be its victims. Our key challenge is two-fold. We must maintain our legacy of world-class teaching, research, and public service. At the same time, in a rapidly changing world, we must build on our legacy of responsiveness and relevance."[35]

Chauvenet Prize Awarded to Birman

The MAA has awarded the Chauvenet Prize for excellence in expository writing for many years. Many of AWM's Noether Lectures have featured such exposition on the area of specialization of the speaker, including Joan S. Birman's 1987 lecture on knot theory. (See also plate 36 in the *Complexities* "Photo Album.") She received the 1995 Chauvenet Prize at the 1996 Joint Prize Session, JMM. The citation for her prize read as follows:[36]

> The 1995 Chauvenet Prize is awarded to Joan Birman for her article "New Points of View in Knot Theory," which appeared in the *Bulletin of*

33. The program is a project of the Alliance for Equity in Higher Education that is funded by the W. K. Kellogg Foundation. See www.msi-alliance.org/main.asp?catid=6&subcatid=28 [accessed 15 November 2003].

34. *The Oval Message* (UMES alumni magazine), Fall 1997, 2–5, and Winter 1997, 3, as quoted on the web page "Dolores Spikes," www.africanpubs.com/Apps/bios/0184SpikesDolores.asp?pic=one.

35. National Association of State Universities and Land-Grant Colleges, *Returning to Our Roots: Executive Summaries of the Reports of the Kellogg Commission on the Future of State and Land-Grant Universities* [online]. New York, NASULGC, January 2001 [accessed 15 November 2003]. Available on the World Wide Web: http://www.nasulgc.org/publications/Kellogg/Kellogg2000_Ret Roots_execsum.pdf

36. Based on the "Awards and Honors" section of the AWM *Newsletter* 26(3), 1996, 5–6. The citation originally appeared in the program of the 1996 Joint Prize Sessions, JMM.

the American Mathematical Society 28 (April 1993), pages 253–287. This marvelous article does everything one might want an expository account of a subject to do. It is all of these, at the same time:

- It is an article that one can give to a student who is just about to take a first course in Knot Theory. Birman sets down vividly, precisely, and agreeably the basic definitions, aims, intuitions, and examples in the theory. Her article even provides at one point a sketch, readable by any student, of a proof of an important foundational matter which cannot be found elsewhere.
- Birman's article conveys the marvelously Protean nature of the subject matter, and of its history including a sympathetic recollection of the important ideas occurring in the early papers of Alexander, a discussion of the efforts of classification of knots in the 19th century by the physicist P. G. Tait and others, as well as the more recent startling connection to von Neumann algebras, which are factors of Type II_1 stemming from the work of Vaughan Jones.
- As announced by her title, her article describes the "new points of view." Someone who had no inkling of these new developments, e.g., of the HOMFLY polynomial, the Kauffman polynomial, the quantum group invariants—alias "generalized Jones invariants"—the ideas of Arnold and Vassiliev on the "moduli space" of all knots—including degenerate ones—and the "Vassiliev invariants" that one can deduce from the study of this "moduli space," and the connections between these collections of new invariants, can get a clean explanation of parts of this exciting work, including a sense of the swarm of open problems that remain, from Birman's article.

The instructions for the Selection Committee say that "preference should be given to papers that come within the range of profitable reading for members of the Association." Birman's article is particularly appropriate in this regard. It provides something for everyone, and does so with clarity and spirit.

Birman's response to her citation appears next.

Expository Writing

JOAN S. BIRMAN

Based on the "Awards and Honors" section of the AWM *Newsletter* 26(3), 1996, 6, the text originally appeared in the program of the 1996 Joint Prize Sessions, JMM. Birman is a professor of mathematics, Barnard College, Columbia University.

Much has been written about the pleasure of research in mathematics, but little about the pleasures in expository writing, so that is my theme for this response. We all share a love of teaching, and expository writing is an aspect of teaching, and I had real pleasure writing this article and telling others about a part of mathematics that seems beautiful to me. Even more, teaching is best when there is a response, and I was fortunate to receive a rich mix of responses.

One was from an obviously young student in an unknown college halfway around the world, who introduced himself via e-mail. In my article I had described a homomorphism $\eta: SB_n \to CB_n$ where B_n is a group, SB_n a monoid which contains it, and CB_n its group ring. In the last three lines on page 278 I conjectured that η is injective, i.e. (and here I correct an error in the original text) that $\eta(x) = \eta(y)$ implies that $x = y$. This student understood my question well enough to correct my error without asking and go on to work on the corrected problem, arriving at a reduction that got to the heart of the matter. That was good work, and I told him so. (I wish I could report that he settled it, but mathematics is not that easy—the problem is still open.)

Then there was a junior colleague who, some months earlier, had discussed with me his symptoms of a well-known mathematical illness: he had completed and extended his Ph.D. thesis, and then the ideas stopped coming. But he didn't write to talk about that; instead, he wrote to say that he had read my article and had a question. Eventually that question led to a new project, and the project to a contribution to research, and he was on his way.

In a different direction, I had given (p. 282) the first few terms of an infinite sequence that arose in studying the weight systems that determine Vassiliev invariants. The sequence is impossible to calculate, with existing computers, beyond the ninth term. The first nine terms are 0, 1, 1, 2, 3, 5, 8, 12, 18. The response was from a combinatorist whose hobby is collecting infinite sequences, and after a few exchanges he told me my 9-term sequence matched exactly one sequence in his very large collection, with a reference. That got me very busy trying to make a connection. Alas, I have not (yet) done so, and that problem is still open, but his response opened new doors.

My last example is before all of you, in the citation. I don't know who wrote it, but clearly that person read and understood and learned from and responded to my article, and nothing is more rewarding to a teacher than to have that kind of substantive response. I was very moved by it. Thank you.

Country School to Grad School

BURTON W. JONES AND ROBERT A. ROSENBAUM

Louise Rosenbaum, 1908–1980. Based on "Louise Johnson Rosenbaum," AWM *Newsletter* 12(4), 1982, 16–19. Then, Burton W. Jones (1902–1983) was a professor of mathematics, University of Colorado–Boulder, and Rosenbaum was a professor of mathematics, Wesleyan University. Now Rosenbaum is the chair of the Project to Increase Mastery of Mathematics and Science (PIMMS) housed at Wesleyan.

Laura Louise Johnson was born on January 21, 1908, in Carrollton, Illinois. When she was ten, she and her parents moved to the area of Boulder, Colorado, and lived on a farm northeast of town. Louise had gone to a country school through grade nine, but Latin was not offered there. She took a room in town—at age thirteen—so that she could complete her college-preparatory education.

Expenses were minimized through the vegetables, eggs, and other farm produce that she brought back from weekend visits home. Louise's initial financial resources for her undergraduate career at the University of Colorado consisted solely of a contribution of fifty dollars from her parents. Throughout her four years, she worked as a checker in the university cafeteria; the standard compensation was board for three hours of work per day and room for one hour. As a lover of the outdoors, she often skimped on meals to have time for weekend outings with the Hiking Club. She might have majored in geology if she could have afforded the laboratory and field trip fees, but in those days a female geologist was even rarer than a female mathematician!

Upon graduation from the university in 1928, with election to Phi Beta Kappa, Louise began teaching high school in eastern Colorado, where her teaching duties included physics, civics, and economics as well as the complete curriculum in mathematics. After two years of high school teaching, she returned to Boulder and supported herself with various subfaculty and part-time faculty teaching assignments at the university while doing graduate work in mathematics. She would not know until the last minute what her assignment would be or, indeed, whether she would have a job at all; for administrative reasons, she was always described as a "part-timer" even during those (rare) quarters when she taught seventeen or eighteen hours per week. With scrupulous attention, department head Aubrey J. Kempner saw to it that she received *more* than a full-time salary on such occasions. During one year in the early thirties, Louise commuted a substantial distance to a Civilian Conservation Corps (CCC) installation to teach mathematics.

Louise was very active in the university recreation department. She arranged and guided tours of students into the mountains by bus or back-packing and climbed many of the high peaks of Colorado. She similarly assisted members of the MAA and the AMS and their families at their summer meeting in 1929.

Louise received her master's degree in 1933 under Kempner with a thesis on transfinite numbers. It contains the basic ideas of set theory and the various kinds of simply and well-ordered sets. The fundamental properties of ordinal numbers and Zermelo's axiom are discussed in detail. The list of twelve references ranges from Cantor to Whitehead.

She received her Ph.D. degree in 1939, again working with Kempner.[37] Louise's dissertation was on the Diophantine equation

$$P_n = x(x + 1) \cdots (x + n - 1) = y^k, n \text{ and } k > 1.$$

After a rather extensive review, with some proofs, of solutions of special cases of this equation, she proved several theorems, including two significant results:

1. If $n = 2k$, there are no solutions.
2. The equation has at most one solution if $n = 4, 5, 6,$ or 7 and k is a prime greater than 5.

For the former, a rather involved argument using symmetric functions, based on work of Liouville, Sylvester, and Schur, is given. The latter uses a result of C. L. Siegel[38] and some ideas of S. Narumi.[39] Her thesis, except for the expository material, was published in the *Monthly* 47(1940), 280–289. At the close of her paper, a communication of Paul Erdös, dated February 22, 1940, was mentioned, to the effect that he had proved but not published her first result and that G. Szekeres (no reference given) had proved that the Diophantine equation she considered has no solutions for $n \leq 9$. Erdös visited her and her family on several occasions during and after 1946.

After Louise received her degree, Kempner told her that there was really no future for her in Boulder, since two of the four professorial department members were women. She went to Reed College on a General Education Board Fellowship, where she worked with F. L. Griffin on the preparation of an integrated mathematical curriculum for small liberal arts colleges. The other fellows in the program were Harry Goheen, Robert Rosenbaum,

---

37. Marjorie Louise Heckel Beaty also received her Ph.D. under Kempner's direction that year. Her dissertation was "On the Complex Roots of Algebraic Equations." She went on to the University of South Dakota, where she was a professor emerita at the time this story was written. Beaty died July 2002 in Vermillion, South Dakota, at the age of ninety-six.

38. "Die Gleichung $ax^n - by^n = c^n$," *Math. Ann.* 114(1937), 57–68.

39. "An Extension of a Theorem of Liouville's," *Tôhoku Math. Jour.* 11(1917), 128–142.

and Henry Scheffé. She and Rosenbaum remained at Reed as faculty members and were married in 1942. While her husband served as a naval aviator in the Pacific, Louise continued to teach at Reed, carrying a particularly heavy load in a military premeteorology program. She and Griffin were proud that the Reed unit stood first in the nation on the uniform exams administered to all students in the program.

Despite the duties of a growing family of three sons, there was no semester during which Louise did not teach at least one course, but she often experienced tension between the demands of her career and those at home. Nevertheless, she functioned smoothly in both environments, directing undergraduate theses, serving on major college committees, running a household with minimal help, and joining her husband and sons in skiing and hiking.

In 1953, she gave up her position at Reed to go to Connecticut, where her husband was appointed to the Wesleyan University faculty. Louise held visiting appointments in the mathematics departments of Trinity, Connecticut, and Smith Colleges and a professorship at St. Joseph College in West Hartford, Connecticut. She remained active in mathematics education. She served on a School Mathematics Study Group committee, directed summer institutes for teachers in Connecticut and Oregon, collaborated with her husband on *Bibliography of Mathematics for High School Libraries* (which went through five editions), and wrote the highly successful booklet *Mathematical Induction* for Houghton Mifflin. After a long illness, she died in 1980.

The Real World of the 1930s

DOROTHY L. BERNSTEIN

Based on a talk at the panel "Women Mathematicians before 1950," organized by Patricia Kenschaft, 1978 Summer JMM, as reported in the AWM *Newsletter* 9(4), 1979, 9–11. Bernstein (1914–1988) was then a professor of mathematics, Goucher College. She was the first woman to be president of the MAA, 1979–1980.

I came to Brown in 1935 as one of five or six women among the graduate students in mathematics. Though we felt a bit isolated, our fellow graduate students accepted us readily. But here I became aware for the first time that some people made a distinction between men and women in mathematics. For example, I was assigned to teach a course in remedial algebra at Pembroke College, at that time a separate college for women at Brown, and Hugh Hamilton was assigned to teach the same course to the Brown boys. When we discovered that I had three girls in my class and he had forty-five boys, it seemed natural to both of us to make two classes of twenty-four students each. However, the chairman, C. R. Adams, would not hear of the idea, saying that the Brown boys would not stand being taught by a woman instructor. I pointed out that I had taught boys in Madison the previous year, but nothing was done.

I took my qualifying exams, or prelims, after one year at Brown; they consisted of two full afternoons—about eight hours—of oral examination by the entire Brown mathematics department. It was quite an ordeal, yet I was able to answer everything they asked. I found out later that other graduate students had prelims that lasted only two or two and one-half hours. When I asked Professor Tamarkin about this, he admitted that my exam was extra long for two reasons: one, I was a woman, and two, I had taken most of my courses at a midwestern university. I am not sure which of the two was more prejudicial. (By the way, Brown's attitude toward women was typical of many other universities at that time—my experience is an illustration.)

My first full-time teaching job was as instructor in mathematics at Mount Holyoke College. I cannot resist telling a story connected to this. R. G. D. Richardson was dean of the Brown Graduate School and also secretary of the AMS for many years. In his latter capacity, he was consulted by many people about hiring personnel. I have heard stories, perhaps exaggerated, that he was a one-man employment bureau for mathematicians throughout the country. I do know that when I came to see him about a college teaching job, he took out a map of the United States, covered the region west of the Mississippi and said, "You can't get a job there, because you are a woman." Then he covered the part south of the Ohio River and said, "And you can't get a job there, because you are Jewish." That left the Northeast quadrant. It happened that I heard of a job at Mount Holyoke and, after a visit to South Hadley, I got the position. When I told the dean, he said, "But I had that job all reserved for Hamilton!"

Two Mentors

Violet Bushwick Haas

PAMELA G. COXSON

Based on "In Remembrance of Violet Bushwick Haas (1926–1986)," AWM *Newsletter* 16(4), 1986, 2–3. Then a visiting assistant professor of mathematics, Ohio State University, Coxson is now a mathematics specialist in the Department of Medicine, University of California, San Francisco.

I met Violet Haas in 1983. At age fifty-seven, she approached mathematics with an intensity and enthusiasm that one had to admire. At the time she was a visiting professor at MIT under the NSF Visiting Professorships for Women program. She was extremely pleased to have this opportunity to devote full-time to research. Several research papers were completed in the next couple of months, and she was very active in the control systems community.

In the course of our daily trek across Longfellow Bridge from Boston to the MIT campus, Haas talked about the years when she devoted much of her energy to breaking down or detouring around numerous barriers. Women graduate students were not permitted to teach courses. She pushed very hard to get herself an office, which turned out to be literally a small closet. When she married, rules against nepotism resulted in her taking a position in electrical engineering rather than mathematics. The difficulties of combining work and family responsibilities were compounded by the need to adjust to a different academic environment that was sometimes hostile. She recalled occasions when the entire department put together a grant proposal in which she was the only member excluded. But those days were now past. She had friends and supportive colleagues within her department at Purdue University. In the preceding fifteen years her research output had accelerated. A dozen M.S. and Ph.D. students had studied under her supervision. She was a mathematician/engineer in her prime, not giving any thought to the prospect of retirement.

Over the years, Haas made a substantial commitment to furthering the careers of younger women and to creating a university environment that would be less hostile for them. She was a member of AWM since 1975, actively involved as a coordinator for the Speakers' Bureau. She served for fifteen years as counselor to the Purdue Student Chapter of the Society of Women Engineers and was a member of the IEEE Committee on Professional Opportunities for Women and the ASEE Constituent Committee on Women in Engineering. In 1983 at MIT she confronted the student "tradition" of showing an X-rated film during registration week. There were many others who voiced their objections that year, and in previous years as well, but her letters in the student paper and her appeals to students and administrators at MIT clearly had an impact.

Eight months after leaving MIT in 1984 and one month after I last saw Violet Haas, she suffered a brain tumor that left her unconscious until her death early in 1986. There are many things I wish I could have told her about how much I have appreciated and benefited from her advice and encouragement.

Alice B. Dickinson

JOAN P. HUTCHINSON

Alice B. Dickinson, 1921–1987. Based on "Remembering Alice Dickinson," AWM *Newsletter* 17(6), 1987, 16–17. Then an associate professor of mathematics, Smith College, Hutchinson is now a professor of mathematics and computer science, Macalester College.

Alice B. Dickinson was born in 1921 and earned her Ph.D. from the University of Michigan in 1953. After her death in 1987, I was moved to write some thoughts about her life and the inspiration many of us drew from her. As professor of mathematics and later dean at Smith College, an undergraduate school for women, Dickinson was a role model and mentor to many women in mathematics and science long before such terms were current. She was a warm and unconventional person: she set and followed her own priorities, both professionally and personally. She filled her life with books and music, rather than more usual material possessions. She cared about individuals and affecting their lives positively.

I was an undergraduate at Smith College from 1963 to 1967. As a freshman I was eager to try essentially all the liberal arts before choosing a major. In my second semester I had Dickinson for calculus—literally within days I knew my choice of a major was made. Mathematics with Dickinson became a challenging and irresistible exploration of a new and intricate world. For example, in that class I learned about the Lascaux cave-dweller paintings and the "floating" Babylonian calendar, and how carbon-14 dating and the related mathematics helped to date these. A subsequent differential equations course given by Dickinson was alive with examples of clepsydras (water clocks), tunnels through the earth, vibrating strings and membranes, and bridges in resonance. The notes from this course became her book *Differential Equations: Theory and Use in Time and Motion*. Not only were her class presentations stimulating, but so also were the daily class discussions, based on homework assignments and outside reading. In fact, students were asked to do a lot of independent work on in-depth assignments, on take-home problem sets (rather than tests), and on independent final projects that involved reading, working with newly learned mathematics, and writing up the work in an expository paper. This type of study led me to a senior independent honors project under Dickinson's direction in which I read Bourbaki in French, attended graduate seminars at the University of Massachusetts, and studied the theory of rings of continuous functions, a beautiful meeting ground of algebra, analysis, and topology.

Dickinson's appreciation of mathematics was felt beyond the classroom: she introduced English change ringing to the Smith campus with the installation of a peal of English tower bells and by teaching tower and hand-bell ringing (in English change ringing, permutation groups, subgroups, and cosets are rung in historically prescribed ways). She and other faculty members provided a fascinating learning environment for the women students of Smith College; many have continued in mathematics and related fields. I remember two particular comments of hers that really affected me. First, she suggested that a true test of what one was most interested in was the kind of books one read during summer vacation. Second, she asserted that if I was really serious about mathematics, I should study for a Ph.D. Both comments were surprises to me, but she was right.

Increasing Minority Representation in Mathematics

WILLIAM A. HAWKINS, GLORIA C. HEWITT,
JOHN W. ALEXANDER, AND BETTYE ANNE CASE

Based on *Survey of Minority Graduate Students in U.S. Mathematical Sciences Departments*, February 1997, published by the MAA and NAM with the assistance of the AMS. Hawkins is the director of the Strengthening Underrepresented Minority Mathematicians (SUMMA) Program of the MAA and a professor of mathematics, University of the District of Columbia. Hewitt was then professor and chair, Department of Mathematics, University of Montana, where she is now a professor emeritus. Alexander was then the director of the Board on Mathematical Sciences and is now a professor of mathematics, Miami Dade College.

> Since I came from a small black college, I think the thing that I miss the most is the individual support and oneness with the instructor.
>
> —*African American female, master's level*

> I received a phone call from [a] Professor and Graduate Advisor . . . who stated that if I were to attend the university, I would "flunk out" of the program within the first year. He went on to "warn" me that if I should attend the university, ". . . it would be difficult. . . ." [At] no time was there any mention of mentors, tutors, or program assistance . . . no encouragement! The professor went on to say that he knew I was a Native American and that there was an underrepresentation of mathematicians in my ethnic group, ". . . but let's be realistic." The conversation, in my opinion, was unprofessional and very discouraging.
>
> —*Native American male, master's level*

Anecdotal comments about adjustments and discouragements faced by minority students as they assimilate into the graduate student population abound. The quotations above and those interspersed below are responses to open items on a survey in 1995–1996. Information was collected from which to frame recommendations that could improve the climate for such students and that were later disseminated within the mathematics community. The goal was identification of ways to increase the numbers, and the retention and completion rates, of graduate students from population groups underrepresented in graduate programs. The survey results make clear that efforts of individual faculty members to effect change must have appropriate support from departments, universities, and the professional societies.

At the first stage, departments were surveyed to determine their minority graduate student population and distribute surveys to them.[40] For the second stage, a total of 233 of the identified students responded individually, 43% female and 57% male. The ethnic composition of the sample is 58% African Americans, 35% Hispanics, 4% American Indians or Alaskan Natives, 2% Native Hawaiians or other Pacific Islanders, and 2% of unknown ethnicity. These rounded response percentages by group are similar to those reported by departments for the larger pool of 657 students, and so the survey returns may approximate the actual population of minority graduate students in terms of ethnicity.

The differences in experiences of these students led, as would be expected, to great variability in responses to the survey questions. For example, we hear quite a contrast in these two replies:

> Most professors are so keen to help me in whatever way they can.
>
> —*Pacific Islander male, master's level*

> I don't recall anyone being particularly helpful. I don't believe most people believed in my ability.
>
> —*African-American male, doctoral level*

Some of the major findings of the survey were:

- Almost all respondents were full-time students, with the vast majority receiving teaching assistantships or fellowships.
- 45% entered graduate school because of their interest in mathematics or desire to further their education.

---

40. Responses were received from 267 of 505 surveyed master's-only and doctoral departments. The total number of students identified was 657, with 492 in doctoral departments and 165 in master's-only programs. No gender data were collected from departments.

> [A non-minority professor has been helpful to me] . . . through encouragement to pursue graduate mathematics and through his willingness to mentor me.
>
> —*African American male, doctoral level*

- 36% indicated that family, friends, and professors had helped them most by providing encouragement.

> My undergraduate professors always told me that if I worked hard and was dedicated to my studies that I can and would succeed. I believed them and I learned to believe in myself.
>
> —*Hispanic female, doctoral level*

- 54% said that the person(s) most instrumental in pursuit of their goals had been undergraduate professor(s).
- 46% gave a negative response or none at all when asked to identify organized, planned helpful activities or programs.

> [I miss] . . . other Latinos. No one here has a similar background to mine. I miss speaking Spanish.
>
> —*Hispanic female, master's level*

- 15% reported an adequate support system. 18% responded that one or more of the following were missing from their support system: other minority or female graduate students in their department or specialization, graduate students from their minority group, and interaction with minority faculty or mathematicians outside academia.
- 48% saw themselves eventually as college, university, or research professors.

The report urged departments to use faculty consultants to examine impediments to minority student success and to make specific recommendations on how to remove them. Many of the report's recommendations come directly from student suggestions; individual faculty members and departments should adapt these recommendations to fit their local situations.

1. Grow your own minority graduate students by cultivating the mathematical interest and involvement of minority undergraduates or terminal master's students already in your department.
2. Effectively use minority alumni of the department to familiarize new minority graduate students with the "unwritten rules" of the department. Invite alumni for departmental events, seminars, and conferences, and set aside special times for minority alumni and students to meet and discuss mathematics.

3. Create mechanisms such as bridge programs or summer mathematics institutes for introducing minority students to numerous areas of mathematics that might not have been done as undergraduate work.

> At [a major Southern university] math department each first year grad student has been assigned a faculty mentor to assist with academic concerns and any other concerns that may come up. Note: This person is not intended to be my research advisor. That's another process for later in my academic career.
> —*African-American male, doctoral level*

4. Assign each first-year graduate student a faculty mentor to assist with academic and other concerns and issues that face the new student. This person need not become the student's eventual thesis advisor.
5. Conduct weekly "case studies" seminars for first-year students where faculty present their research. This would be useful for learning what different professors are doing and in finding a thesis advisor.

> I believe I can do it. Not many Mexicans do it. Someone has to open up doors so that more people will be encouraged to do it. . . . I still feel somewhat culturally and socially isolated. I am the only Mexican in . . . the Math Department.
> —*Hispanic male, doctoral level*

6. Use willing minority graduate students to recruit more minority students to mathematics. They can return to their baccalaureate institutions and talk with students about the rewards of choosing mathematics as a life's vocation.
7. Inform minority graduate students of on-campus minority organizations and volunteer opportunities within nearby minority communities. This will assist minority students in adjusting to the culture of the institution and their new surroundings.
8. Sponsor departmental parties, picnics, teas, coffee hours, and get-togethers to help minority and nonminority students and faculty to get acquainted on other levels than just mathematics. This will encourage a welcoming atmosphere within the department.
9. Recruit minority and female faculty for tenure-track positions. Hiring faculty is a powerful indicator of departmental concern with increased minority retention and degree completion.

Implementation of these policies, procedures, and activities will build more minority representation in the professoriate in the future. Additionally, the local climate will improve for all graduate students, not just those in minority classifications.

Research and Teaching at Liberal Arts Colleges

LYNNE M. BUTLER

Based on "Research and Teaching at Liberal Arts Colleges," AWM *Newsletter* 26(6), 1996, 5–6. Butler is a professor of mathematics, Haverford College.

Women graduate students often wonder whether it is possible to do research after joining the faculty at a liberal arts college. It is not only possible, it is essential at the best liberal arts colleges. You will not be hired unless your thesis research is of high quality and your potential for continued growth as a mathematician is apparent. You will not be tenured without steady publication of good papers in refereed journals. You must establish a strong research program and demonstrate that you can maintain it alongside very good teaching at all levels and active service to your department and the college. On the other hand, you do not have to be the best young person in your field or publish many papers every year. But if undergraduates at a research university think you are a great teacher, students at a liberal arts college might find you merely a good one. So before you take a tenure-track job at a liberal arts college, I recommend your research be in full swing, because the demands of becoming a very good teacher at a liberal arts college are considerable.

I am a professor of mathematics at Haverford College, and I cannot imagine a better life. Haverford needs and values what I have worked to become: a strong researcher with broad interests within and beyond mathematics, a devoted teacher and mentor for students at all levels, and an able servant of my department and the college. These three aspects of my life at Haverford enhance each other. Consider these facts: to attract mathematics majors and student research assistants I need to be a stimulating teacher; to teach mathematics courses that range from algebraic topology to linear optimization I need to have mastered and used mathematics far beyond my thesis research; to earn support for mathematics from colleagues in other departments I have to interact productively with them on search committees in related disciplines and interdepartmental curriculum development efforts; to earn respect for my department within the college I need to teach effectively and maintain a strong research program. Haverford does not want me to neglect teaching or service to leave more time for research, nor does Haverford want me to neglect research to devote more time to teaching and service. My life at Haverford is varied and intense.

Students come to Haverford to work closely with faculty. (Every mathematics major writes a senior paper.) Colleagues applaud when your most recent grant is announced at a faculty meeting. (Ten humanities professors came to hear my faculty research talk last spring.) Alumni endow professorships and contribute funds to support faculty research and travel. Haverford appreciates everything I have done and want to continue doing, especially, but not only, my research.

The research universities where I was a student and junior faculty member (Chicago, MIT, and Princeton) provided the ideal environment to develop my mathematical talents, explore my mathematical interests, and establish a strong research program in my field (algebraic and enumerative combinatorics). I was not the very top mathematics student who graduated in 1981 from Chicago, but I pursued both economics and mathematics through first-year graduate courses, led problem sessions for calculus courses, and tutored in a number-theory program for high school students. I was not quite the best of my advisor's students during my five years at MIT, but I studied topology as well as combinatorics, taught calculus several summers in a program for minority students, and learned statistical methods in speech recognition at IBM during a summer internship. I was not the hottest assistant professor at Princeton, but I taught a much wider variety of courses than most (including graduate courses in my field and courses for economics majors, one of which I created), worked as a cryptanalyst several summers, and served productively on the university course of study committee. I took full advantage of a postdoctoral year at the Institute for Mathematics and Its Applications, and I forced myself to apply for regular grant support. I seized every opportunity to publicize my work and participate in activities of the mathematical community.

I always want more time for research, but I do not take time away from teaching or service. So most days I do not have a minute to waste. (I really cannot waste a summer, a fall/winter/spring break, or even a weekend.) I try to be increasingly efficient and energetic. And I find myself able to do more every year. I love working with smart, dedicated, capable people and working on hard, interesting, important problems. Go for it!

Sustaining a Research Program

KAREN BRUCKS, BETTYE ANNE CASE, AND ANNE M. LEGGETT

Based on a talk at the discussion "Finding a Traditional or Nontraditional Job and Growing in It," organized by Krystyna Kuperberg, 1999 Taussky Todd Celebration, as reported in the AWM *Newsletter* 29(6), 1999, 20–21. Brucks is an associate professor of mathematics, University of Wisconsin–Madison.

Spending concentrated time on developing the mature research program essential in most academic career settings is the purpose of postdoctoral appointments whatever their nomenclature (fellowships, named instructorships, visiting assistant professorships, and research scientists are typical titles). A responsibility of faculty in graduate departments is working with postdocs to help them develop independently as researchers, and in some cases also to help them develop teaching programs.

The typical time for this is immediately after award of the doctorate. But there are some postdoctoral opportunities beyond the ones available to new doctorates. The NSF's ADVANCE (Advancement of Women in Academic Science and Engineering Careers) Fellows Award program was designed to increase the representation and advancement of women in academic science and engineering careers. It is intended for those who demonstrate high potential to develop or resume active, full-time, independent academic careers at institutions of higher learning in a science or engineering field (e.g., after a career break due to child care or elder care).

In a panel contribution at the Taussky Todd Celebration, Karen Brucks talked about how to maintain a productive research program using resources in addition to traditional early postdocs and federal grants:

> One issue I want to address is having a productive research program even though there may be few or no colleagues in your department working in your area of research. Travel and bring researchers in to visit you. Plan your summers carefully; do not teach in the summer. Attend workshops and conferences. Give talks! Of course, travel takes money. Your home institution may have funding available for your travel. The AWM administers a travel grant program; the AAUW has many grants and fellowships. Search out funding options. If your institution has a sabbatical program, take one when it becomes available to you.
>
> Currently I am serving on the American Fellowship Panel of the AAUW. This panel makes funding decisions for dissertation fellowships, postdoc grants, and summer grants. Over the past three years, the number of applicants from mathematics has seriously declined. In mathematics we often think of postdoc funding as funding available the first few years past your Ph.D. AAUW thinks differently; their postdoc simply means "past Ph.D.," any time past your Ph.D. They also award summer grants which are targeted toward women at smaller institutions that may have higher teaching loads during the academic year to support summer research. We need to increase the number of applicants for these programs from mathematics!

Are Student Ratings Unfair to Women?

NEAL I. KOBLITZ

Based on "Are Student Ratings Unfair to Women?" AWM *Newsletter* 20(5), 1990, 17–19. Neal Koblitz is a professor of mathematics, University of Washington.

Recently, I asked for information on whether or not student ratings tend to discriminate against women and was pleased to receive a large number of quite varied responses. Some people wrote their general impressions and described their personal experiences. Others generously sent me reprints of papers on the subject, or gave me advice on where to look for more material. To my surprise, it turns out that quite a lot has been written on this question, but not in journals that mathematicians normally read.

I will not attempt a systematic survey of the research and opinions on the subject. For this the reader is referred to the short list of references, which includes the papers which I found to be the most interesting (more extensive lists of papers can be found in their bibliographies). Rather, I will summarize my own conclusions based on the material that was sent to me.

A few of the letters I received and some of the early studies indicate that often women receive equal or higher student rating numbers than men. In many situations students perceive that women instructors tend to be more sensitive to their needs, more concerned and caring, and more dedicated to teaching than male instructors (it also helps if the woman is thought to be lenient)—and as a result reward them with higher ratings. Thus some people conclude that there is little or no discrimination against women in student ratings.

However, a more careful examination of the question shows that the reality is more complex. Note that the traits listed in the last paragraph that may lead to high ratings for women are compatible with sex-stereotyped expectations of women as "mother figures." According to Kierstead et al. [5], "Taken as a whole, [our] results suggest that if female instructors want to obtain high student ratings, they must be not only highly competent with regard to factors directly related to teaching but also careful to act in accordance with traditional sex role expectations. In particular . . . male and female instructors will earn equal student ratings for equal professional work only if the women also display stereotypically feminine behavior."

Thus, the difficulty for women tends to occur in cases where a "get-tough" approach is needed. These situations are much more likely to arise in a math department than, for example, in psychology or sociology, because (1) mathematics departments typically are called upon to perform the role of enforcer of academic standards, with service courses acting as a "weeding out" device for the engineering and science departments, and (2) the discrepancy between students' high school preparation and study habits and the demands of college work is especially glaring in mathematics.

If an instructor feels compelled to put students under pressure (assigning a lot of homework, giving challenging exams), then only the most serious and mature students are at all likely to respond with high ratings at the end of the course. Most students are inclined to "punish" the instructor. There is considerable evidence that the "punishment" is more severe if the instructor is female.

> [According to] Susan Kay's classroom studies . . . male students were far more likely to give lower ratings to those female faculty perceived to be hard graders. . . . This finding is consistent with a series of experiments at the University of Dayton that indicated that college students of both sexes judged female authority figures who engaged in punitive behavior more harshly than they judged punitive males. [7, 484–485]

See also the studies by Kierstead et al. [5] and Bennett [2], which lead to similar conclusions.

Bennett, in particular, found that women will be rated highly only if they are especially accessible to the students and spend a lot of time with them, while men can receive equally high ratings while remaining more aloof. In other words, students tend to allow men but not women to spend most of their time on research and other nonteaching activities without penalizing them in the ratings: "[M]ale instructors are judged independently of students' personal experiences of contact and access, whereas female instructors are judged far more closely in this regard. In this sense women are negatively evaluated when they fail to meet this gender appropriate expectation" [2, 177–178].

An especially interesting study was made in the 1970s by Ellyn Kaschak [4]. Fifty male and fifty female students were given descriptions of the teaching methods and practices of professors in various specialties. In the forms received by half of the students (twenty-five males and twenty-five females), the professors were given names of the opposite gender from the professors in the forms received by the other half of the students. Kaschak found that the male students were biased against women, while the female students were not.

The possibility of sex discrimination is one complex and controversial aspect of the broader question of the validity of student ratings as a measure of teaching effectiveness. It would take us too far afield to discuss some of the other problems identified in the many studies that have been conducted. But it is worth noting that, generally speaking, math departments are usually put at a special disadvantage if administrators and faculty in other departments have excessive confidence in the meaning of student rating numbers and in the value of cross-department comparisons. A larger proportion of our students take courses as requirements rather than electives and view the subject as difficult. This tends to bring down math department ratings across the board and leads to an unjustified belief on campus that the math department has worse teachers than other departments.

People outside of the mathematical sciences often have a naive faith in the value of numbers and are less aware than we are of the pitfalls in taking raw statistics at face value.

> [S]tudent rating scales are a form of measurement and, according to American Psychological Association standards, should be accompanied by information about the meaning, interpretation, and limitations of the scores—yet most student ratings are not accompanied by such information; [in fact,] promotion and tenure decisions are usually made by an array of administrators and faculty committees who are naive about the standard criteria for measurement instruments, and hence do not know how to interpret the results or do not realize their limitations [8, 88].

In practice, the treatment of student ratings by college administrations varies considerably. On the one hand, some institutions have conducted careful studies of the validity of student ratings and have adopted a cautious and sophisticated approach to the subject. At the other extreme, some administrators use student ratings in an unfair and cynical way as a weapon against the faculty, especially the female faculty.

Some Conclusions

1. Student ratings can provide valuable feedback to the instructor her- or himself but cannot be properly understood by someone who is unfamiliar with the nature of the course being rated, the characteristics of the students, and the pedagogical objectives of the instructor.
2. On the student rating forms, questions that are very specific (e.g., "promptness in correcting exams," "availability for office hours") are less likely to invite biased responses than questions of a general nature ("rate the instructor overall").
3. In certain teaching situations that are frequently encountered in math departments (especially in introductory-level and service courses), students tend to discriminate against women instructors on the rating forms.
4. Math departments and administrators have an ethical and legal obligation not to base promotion and salary decisions on data that are biased against women.

References

1. Susan Basow and Nancy Silberg, "Student Evaluations of College Professors: Are Female and Male Professors Rated Differently?" *Journal of Educational Psychology* 79(1987), 308–314.
2. Sheila Kishler Bennett, "Student Perceptions of and Expectations for Male and Female Instructors: Evidence Relating to the Question of Gender Bias in Teaching Evaluation," *Journal of Educational Psychology* 74(1982), 170–179.
3. R. Craig Hogan, "Review of the Literature: The Evaluation of Teaching in Higher Education," Instructional Development Centre, McMaster University, Hamilton, Ontario, 1978.
4. Ellyn Kaschak, "Sex Bias in Student Evaluations of College Professors," *Psychology of Women Quarterly* 2(1978), 235–242.
5. Diane Kierstead, Patti D'Agostino, and Heidi Dill, "Sex Role Stereotyping of College Professors: Bias in Students' Ratings of Instructors," *Journal of Educational Psychology* 80(1988), 342–344.
6. Diane Kierstead et al., "Report of the Course Evaluation Committee," Colby College, Waterville, Maine.
7. Elaine Martin, "Power and Authority in the Classroom: Sexist Stereotypes in Teaching Evaluations," *Journal of Women in Culture and Society* 9(1984), 482–492.
8. Stanley N. Miller, "Student Rating Scales for Tenure and Promotion," *Improving College & University Teaching* 32(1984), 87–90.
9. Rhoda Unger, "Sexism in Teacher Evaluation: The Comparability of Real Life to Laboratory Analogs," *Academic Psychology Bulletin* 1(1979), 163–171.

Rules for Academic Success

AUDREY A. TERRAS

Based on "Five Simple Rules for (Academic) Success (or at Least Survival)," at the panel "How to Be a Successful Woman Mathematician," organized by Chuu-Lian Terng, 1997 JMM, as reported in the AWM *Newsletter* 27(3), 1997, 10–11. Terras is a professor of mathematics, University of California, San Diego.

Here are five simple rules for academic success (or at least survival).

1. Don't give up. This is the most important rule.

Yes, good jobs can be hard to find. The year I got my Ph.D. was 1970 and there were around 2000 math Ph.D.'s and very few jobs. I sent out 100 job applications and got only rejections. Then I sent out another 100. I had a two-body problem. My (now ex-) husband was a mathematician and we looked for jobs in the same department, which made the search much harder. Finally a job came in June, and it was not at one of the top five schools, but at least my spouse and I were in the same department. The next year he had a job on the West Coast while I was on the East Coast. A year later I moved again.

After all that, papers are hard to publish, books even harder. And teaching can be a challenge. Develop a thick skin. Change those things you can but don't apologize for doing your kind of math. If the first referee claims your paper is useless, send it to another journal. If the book reviews aren't glowing, don't brood. There are rival fashions in mathematics, and referees and reviewers may represent one side in a many-sided debate while you are on another side. Proofs are either right or wrong (although standards of proof can change over time), but there are huge differences in taste. The same goes for teaching: you have to try hard to sell math to an increasingly skeptical audience, and evaluations can be surprisingly bad when you think you are giving your all.

Finally, mathematics is hard to do. Progress can be slow, and patience is necessary, but it is important to keep working. When stuck on a problem, start writing up the proofs of the things you have already done. Heisenberg said: "You just have to be able to drill in very hard wood—and keep thinking beyond the point where thinking begins to hurt."

2. Keep learning and teaching.

Keep a notebook (like Ramanujan) with a list of problems you are working on. Keep reading books and papers. Keep writing, including expository papers. Give and go to seminars and colloquia. Give courses on a wide range of subjects. Work hard on making your lectures understandable; give lots of examples, then write up your lecture notes. Become a reviewer for *Math. Reviews* and *Zentralblatt*.

3. Network. Go to meetings, give talks, collaborate.

Take leaves and visit other universities and research institutes. Send out reprints and preprints to lots of people. Seek out supportive and friendly people (not all mathematicians will fall into this category). In particular, keep in touch with your letter writers. Do not get a secretary to mail out old letters. When asking for letters, be sure that you send your letter writers copies of your CV, recent works, and address labels. Not only is this polite, but it helps you to get up-to-date and timely letters. Write up short summaries of your works to help your letter writers cope with a giant stack of papers. Always support other women and all underrepresented groups, making a particular effort to include them in your network. Try to be a good friend, mentor, and advisor to your students and colleagues.

4. Try to do useful mathematics. But do "your kind of mathematics."

Be flexible. Expand your area of expertise. Do beneficial applications if possible. If not, at least connect with other parts of mathematics. As a number theorist, for example, I find it both useful and inspirational to make connections with graph theory, Fourier analysis, and matrix groups. Applications of number theory in computer science and signal and image processing are actually a stimulant to thought.

5. Have a (good) life.

Brains get tired and need rest or change. Often the solution to that problem that has been bugging you for weeks comes during a vacation. Do not let the life take over, however. If you have children, make use of daycare. Many universities will allow you to stop the tenure clock. But don't stop doing math. I find that it is like any kind of sport—you must keep exercising the math muscles. I don't have any children, but this doesn't mean I don't have a life apart from mathematics: I like cats, gardening, music, netsurfing, StarTrek, travel, hiking, reading mysteries and science fiction, photography.

3. Outside the Academy

Many mathematicians, both female and male, spend part or all of their careers working in either government or the private sector. For example, until her recent retirement, Marjorie Stein had been employed for more than twenty-five years by the U.S. Postal Service, working on many interesting problems, calling on both her mathematical knowledge and her analytical skills. She describes some of her early work in this way: "My first project was to analyze the system for allocating transportation costs to classes of mail. Most of the work was 'investigative reporting,' researching what proved to be a very poorly documented data collection system and preparing a report on the findings. Other tasks have included 'in-house' consulting projects such as formulating a model for estimating total piece handlings at post offices, given incoming volumes and percentage flows between handling points."[41]

Although written over a span of twenty-one years, the comments of these women from government and industry retain their original vitality and applicability. Over that time, the work and life patterns they discuss illustrate striking similarities. Moving back and forth between industrial and academic appointments, as shown by the early example of Olga Taussky Todd, is common to many mathematicians. For example, Margaret Wright describes the mathematics related to her work at Bell Labs; she is now a department chair at Courant Institute, although she retains a connection with the scientific computing group at Bell Labs. Sarah Holte, who works in the growing biomedical field, and Barbara Brown Flinn, at the National Security Agency, provide insights into employment options for mathematicians. These three articles are based on talks at the panel "Finding a Traditional or Nontraditional Job and Growing in It," organized by Krystyna Kuperberg at the 1999 Taussky Todd Celebration.

Maria Klawe, Elizabeth Ralston, and Margaret Waid discussed their careers paths at the panel "Non-academic Careers in Mathematics," organized by Patricia Kenschaft at the 1985 Joint Mathematics Meetings. An award citation describing the accomplishments of Anneli Lax, who worked for the publishing arm of a professional society, showcases another employment possibility for the doctoral woman mathematician. Together with stories

---

41. The first woman to complete the requirements for the Ph.D. in mathematics at Princeton University, she received her degree in 1972. Her first job was at the National Bureau of Standards. See Marjorie L. Stein, "The Woman Mathematician in Government: A Personal View," in AWM panel discussion "Women Mathematicians in Business, Industry and Government" (moderated by Lenore Blum), 1976 JMM, AWM *Newsletter* 6(3), 1976, 3–4.

in other chapters, these give an overview of what mathematicians have to offer away from the ivory tower.

The chapter begins with an excerpt from an autobiographical talk by Mina Rees, an accomplished administrator. In contrast to Dolores Spikes (see III.2), Rees began her administrative work not in academia but with the federal government, ending her career as a university administrator. Rees's story describes her government work in applied mathematics during World War II and her subsequent rise to serve as the first head of the postwar Office of Naval Research (ONR). Taussky Todd herself did government work in two countries, for the war effort in the United Kingdom in the 1940s and at the U.S. National Bureau of Standards in the 1950s. Rees and Taussky Todd ended their careers in academe on opposite coasts, Rees distinguishing herself in administration, and Taussky Todd directing numerous theses at CalTech. Their respective career paths have a "back to the future" feeling, given the anecdotal evidence from midcareer women today and recent suggestions in the popular press that everyone should be prepared to change job directions every five to seven years.[42]

Government and Administration

MINA REES

Based on a talk at the panel "Women Mathematicians before 1950," organized by Patricia Kenschaft, 1978 Summer JMM, as reported in the AWM *Newsletter* 9(4), 1979, 15–18. Rees (1902–1997) was then president emeritus of the Graduate School and University Center of the City University of New York. The library there was named the Mina Rees Library in 1985 in her honor. In 1971, she was the first woman president of the AAAS.

My A.B. was from Hunter in 1923, my Ph.D. from Chicago in 1931. After Chicago, in the midst of the depression, I returned to Hunter to resume the usual duties associated with academic life. The decisive event in my life came in 1943, when I accepted a wartime job that introduced a whole new orientation into my career. This was an invitation to join the staff of the Applied Mathematics Panel (AMP) of the Office of Scientific Research and Development (OSRD) when the organization to handle scientific war work was reorganized in 1943. I became technical aide and assistant to Warren Weaver, the chief of the AMP.

---

42. See "Employee Tenure Summary," in "News" [online]. Bureau of Labor Statistics, U.S. Department of Labor, 2002 [accessed 29 June 2003]. Available from World Wide Web: http://www.bls.gov/news.release/tenure.nr0.htm.

Why do I call this change decisive? First, because it greatly broadened my awareness of unfamiliar fields of mathematics and my contacts with mathematicians and, second, because it greatly increased my understanding of the character and activities of many of our major educational institutions and of the structure and operations of the government, including the military establishment. In short, it gave me the kind of experience that made it appropriate for me to be invited to become head of the mathematics research program of the ONR when it was established after the war.

Because of the importance of the AMP in this tale, I think I should say a few words about what it was and what it did. The National Defense Research Committee (NDRC) had been set up to provide scientific assistance to the military even before the United States entered World War II. There was no mathematics division. Weaver was at that time director of the Natural Sciences Division of the Rockefeller Foundation and had been chair of the mathematics department at the University of Wisconsin. He was head of a section of the Fire Control Division, which had been assigned the task of developing an Anti-Aircraft (AA) Director, an essential component in the system that was needed to protect Britain from German bombing. By the end of 1942, that task had been brought to a spectacularly successful conclusion by the production of an AA Director that shot down every single buzz bomb that came over the part of the east coast of Britain to which the director had been assigned for tests. It was at this time that NDRC was reorganized to enable it to perform its task more completely, reflecting what it had learned from its early experience. It was then that Vannevar Bush, who headed the new OSRD of which NDRC became a part, decided to establish an official government body, the AMP, to help with the increasingly complex mathematical problems that were coming to the surface as well as with those other problems that were relatively simple mathematically but needed mathematicians to formulate them adequately.

Weaver set up the AMP, which, like comparable bodies handling other aspects of scientific war work, wrote contracts between the government and various universities to provide, in our case, mathematical services. There were, of course, mathematicians engaged in many parts of the war effort: in the War Policy Committee, in the armed service,[43] on war tasks in industry,[44] in training programs at colleges and universities, and in other divisions of OSRD including the Radiation Lab and the Manhattan Project. AMP was to provide additional mathematical assistance to the military and to other divisions of OSRD when asked, provided that they

43. For example, Herman Goldstine in uniform, E. J. McShane at Aberdeen as a civilian in a government establishment, B. B. Price in an operations research unit attached to the Air Force.
44. For example, at Westinghouse, RCA, and BTL.

considered they had a reasonable chance of doing something useful. The panel set up contracts with numerous academic institutions,[45] as well as with the Mathematical Tables Project (originally established by the WPA). Many of the country's ablest mathematicians were employed under these contracts, and many moved from their home universities in order to participate.[46] The members of the panel itself (all government appointees) included, among others, Thornton Fry, Marston Morse, G. C. Evans, H. P. Robertson, A. H. Taub, Oswald Veblen, and Richard Courant. Courant was responsible for my being invited to join the staff of AMP as a civil servant. It was not because I was an applied mathematician (my degree was in abstract algebra); it was not because I was a woman (there was no equal opportunity then). It was the good old buddy system. Let me explain.

I was at Chicago as a graduate student from 1929 to 1931. In the summer of 1930, I attended my first summer mathematics meeting at Brown University. What I remember most clearly about that meeting was that I had breakfast with Morse and G. D. Birkhoff. I found the experience overwhelming: I was enchanted with mathematicians and, at least partially because of that, with mathematics. After I returned to New York, I continued my attendance at meetings, both winter and summer. And so it came about that I met Courant when he was a visiting AMS lecturer at a meeting, and I continued my acquaintance with him when he came to NYU. Though I was not a research mathematician, and though I soon learned that I couldn't understand much at the meetings, I did find them useful in giving me some idea of the directions mathematical research was taking. (I might add that I understand even less current research now.)

When AMP was being organized, Courant recognized that Weaver would need administrative as well as mathematical help[47] and suggested my name. The job involved contact with the work going on under all contracts and attendance at all AMP meetings. There were trips to military installations (including headquarters in Washington) in the company of appropriate contract personnel, to clarify the problems we had been asked to help solve, to determine whether we could do something useful, and to formulate recommendations to the panel, which met weekly. There were reports on completed projects that were published and circulated to people in this country, England, and Canada with legitimate interests in the

45. For example, Princeton, Columbia, New York University, Berkeley, the Franklin Institute, Brown, the Institute for Advanced Study, Harvard, the Carnegie Institution of Washington, and Northwestern.

46. Among others, there were MacLane, Albert, Courant, Friedrichs, Prager, Leighton, Garrett Birkhoff, Wilks, Mosteller, Abraham Wald, Allen Wallis, and Milton Friedman (the last two operating as statisticians).

47. Mathematical help was being provided by Sam Wilks, Ivan Sokolnikoff, Hal Germond, and, later, by Don Spencer.

problems. Regular reports had to be prepared and distributed, relaying progress, or lack of it, on projects that had been undertaken by the panel but not yet completed. The panel carried on a broadly based and successful effort to bring mathematics to the service of the war effort.

My work with the AMP lasted until 1946, after the war ended. It gave me familiarity with the work of many of America's most able mathematicians, and it gave me considerable understanding of the changes that were occurring in mathematics as a result of experiences in World War II. These included the emergence of mathematical statistics in its great variety, the development of the computer and the need for extensive work in mathematics to insure a sound exploitation of the potential of the computer, the clear opportunity to extend the use of operations research to important new areas, the potential—through the use of computers—for more applicable results in analysis.

When the ONR was established after the war, I was a natural candidate to head the mathematics component, although my being a woman raised some serious doubts. After my confirmation as head, my sex was only one of the problems I had. I very much doubted that the mathematicians would want to receive support for their research from a military organization after the war was over. Initially, this judgment was right, but, as time passed, mathematicians found the program a very desirable one. I found the ONR experience to be exciting. One activity we supported was the newly established National Applied Mathematics Laboratory at the National Bureau of Standards that, as part of its program, brought distinguished foreign mathematicians to this country, sometimes as visitors, sometimes for longer stays. Olga Taussky Todd and Jack Todd were among those who came in those early postwar years. Now, of course, Olga Taussky is one of the American women making important contributions to mathematical research.

I spent seven years in Washington until, in 1953, I left to return to Hunter College as dean. The experience in Washington had given me administrative and academic sophistication that would have been hard to get elsewhere. During my further career, first at Hunter and later as a central administrator of City University of New York (CUNY), the importance of the ONR experience was in the rather intimate knowledge it gave me of the modus operandi and of the ambience of virtually all of the country's leading research universities and many of the liberal arts colleges. It also resulted in the establishment of warm friendships with many mathematicians and university administrators.

My course was set. I was committed to administration, not research, but administration with a heavy orientation toward science. Invitations to serve on national committees continued to come. I was a member (and served as chairman) of the Advisory Committee on Mathematics of the

National Bureau of Standards, of the Advisory Committee for Mathematics of NSF, and of the General Sciences Advisory Board of ODD. I was a trustee of AMS from 1955 to 1959 and a member of the NRC Mathematics Division from 1953 to 1956, where I served on its executive committee for two years. For many years I served on the Board of Directors of AAAS, and was elected to its presidency for 1970, serving as chairman of its board the following year. I was the first woman to hold these posts. In 1958, at the suggestion of NSF, I called the meeting at MIT out of which grew the School Mathematics Study Group. In 1964, by presidential appointment, I became a member of the National Sciences Board.

My stay at Hunter lasted eight years. In 1961 I was invited to develop the Ph.D. programs of the newly established CUNY. The university was based on existing city colleges, including City College, Hunter, Brooklyn, Queens, and Lehman, and resembled a British university, with its many largely autonomous colleges, more than a typical American university with one or two liberal arts colleges. I ended my career in 1972 as president of the Graduate School and University Center of CUNY.

The building of a graduate school that called upon and stimulated the growth of the scholarly and physical resources of so many established liberal arts colleges, and that achieved graduate work acknowledged first-class in a brief period of time, required combining traditional elements of academic structure with often difficult innovations. During this period, when my prime concern was with graduate education, I was active in the work of the Council of Graduate Schools in the United States and was elected chairman of that council in 1970.

I believe it is fair to say that, in the navy and while I was active in graduate education, I had some impact on the growing acceptance of women both in graduate education and in administrative and policy-making posts. In 1972, in a paper presented at a meeting of the American Council on Education, I reported that at CUNY, in contrast to several other universities, there was strong evidence that women's performance as graduate students was about the same as that of men. This was true with respect to all three parameters used to measure this performance in other studies: completion of the first or qualifying examination, completion of all requirements for the degree except the dissertation, and completion of the degree. Moreover, at our graduate school, women's admission to graduate study, access to fellowships, and acceptance on the faculty and in the administration seemed to be substantially without discrimination. At ONR, other women have been appointed to top administrative positions.

I have enjoyed and cherished my associations with mathematicians, but I have made no significant contributions to the corpus of mathematical work. When I was young, my only ambition was to be a research mathematician.

Citations

Rees has stated that among the many honors she has received, she cherishes most the three she received from mathematical groups: resolutions passed by the Council of the AMS and the Institute of Mathematical Statistics on her return to Hunter from ONR, and the first Award for Distinguished Service to Mathematics conferred in January 1962 by the MAA.

From the resolution of the Council of the AMS:

> Needless to say as the purest of all sciences, mathematical research might well have lagged behind [in the large-scale fostering by the U.S. government of fundamental research]. That nothing of the sort happened is beyond any doubt traceable to one person—Mina Rees. Under her guidance, basic research in general, and especially in mathematics, received the most intelligent and whole-hearted support. No greater wisdom and foresight could have been displayed and the whole postwar development of mathematical research in the United States owes an immeasurable debt to the pioneer work of the Office of Naval Research and to the alert, vigorous and farsighted policy conducted by Miss Rees. The influence of these policies has been such that it vitally affected later developments: the activities of Air Force and Ordnance Research, the National Science Foundation itself.

From the resolution of the Institute of Mathematical Statistics:

> Under [Rees's] leadership, the Division of Mathematical Sciences of the Office of Naval Research gave whole-hearted support to basic research, in particular to basic research in mathematical statistics and probability. The whole action was conducted with remarkable foresight and wisdom. . . .

> Mathematical Statistics owes Mina Rees a public "well done."

In retrospect, I don't believe that was ever possible for me, although it is for some women. It is clearly essential that we provide opportunities for women as well as for men to find the satisfactions and the rewards of research careers if their talents and commitments make such careers possible. And other women are now showing the way.

Reference

Judy Green and Jeanne LaDuke, "Mina S. Rees: 1902–1997," AWM *Newsletter* 28(1), 1998, 10–12.

Computer Science

MARIA M. KLAWE

Adapted from the AWM *Newsletter* 15(4), 1985, 7–9. Klawe was then discrete mathematics manager, IBM San José Research Laboratory. After serving as an administrator at the University of British Columbia, she became dean, School of Engineering and Applied Science, and professor of computer science, Princeton University.

I am manager of the discrete mathematics group, which is part of the computer science department in the IBM San José Research Laboratory. Our mission is to do pure research in any one of a number of areas of mathematics, including combinatorics, logic, number theory, operations research, and algebra. The only constraint is that the research should have some connection with computer science. Most often this connection occurs through interaction with the theoretical computer science group in our department. We believe that pure mathematicians have an enormous contribution to make to the field of computer science because so many computer science problems are now quite sophisticated mathematically. Our group of pure mathematicians is particularly fortunate in being exposed to such problems arising in computer science and in having the freedom to explore the mathematics these problems generate.

Our work consists mostly of what one would do in the research environment of a university—choosing one's own problems and, hopefully, solving them, writing them up, and giving talks at conferences. We also, when needed, act as consultants to the rest of IBM in our own particular area of expertise, though that tends to take less than 5% of our time. Another responsibility is to interface with academia, helping with recruiting and stimulating joint research projects. For example, we try to know who the best graduate students are in our areas and to encourage their interest in working for IBM.

How did I end up in this situation? I finished my Ph.D. on amenable semigroups in 1977 and, like many people who received their Ph.D.'s that year, I wrote about eighty letters to universities and received very few offers. Taking the best offer, I arrived at a small university in Michigan. Although there were many good mathematicians in the department, the students were extremely weak. During the eight months that I spent there, I really could not believe that this was why I had gone into mathematics. I was very, very frustrated. (I'm sure that this is a familiar experience to many people.) In eight months I went to eight conferences, just to get away from that environment.

At one of these conferences I met Vasek Chvatal, a combinatorist I'd heard wonderful things about. To my surprise I found that he was teaching in the computer science department at Stanford. While telling him horror stories about my situation, I noticed that he was continually getting phone calls from computer science students finishing their Ph.D.'s, wanting his advice about which job offer to accept—MIT or Bell Labs, and so on. I sat there wondering, "What did I do wrong with my life that has sentenced me to live forever in the middle of nowhere, teaching students who cannot add a half to a third?" I asked Vasek if these students getting such wonderful offers were all so spectacular. He said they were good, but not exceptional. I commented on the unfairness of life, but Vasek replied that if I wanted offers from Bell and MIT, I had to learn some computer science. I was desperate enough to take this seriously and asked which were the best computer science departments. He replied that they were MIT, Stanford, and the University of Toronto. By this time it was March, and Stanford had closed its enrollment. MIT accepted me, but without support. Toronto, perhaps because I'm Canadian, said, "Great, we'd love you to come. We can support you and there's no problem." I went to Toronto.

I spent the next year taking enough graduate courses for a Ph.D. in computer science. When I arrived, I knew no computer science. Before that year I'd never programmed, never read computer science books, and honestly had no idea what computer science involved. On the whole I hated that year, partly because I was taking so many courses, but even more because it was a different culture. I was used to proofs, theorems, and above all, rigor. In computer science there is some of that, but there's a lot of other stuff, too.

By the end of the year I'd learned a great deal and was suddenly employable anywhere. It was a remarkable change. I just could not believe the difference, because in computer science I'd done no research, yet I too could have wonderful offers. I joined the faculty at the University of Toronto and then met someone I wanted to marry, who happened to be working for IBM research. We considered whether he should come to Toronto, or if we should both go to MIT, or if I should go to IBM. In the end I went to IBM, mostly because having made such a change in my area of research, I wanted to have time to concentrate on it and to establish myself in another field.

I found there were many things I loved about being at the IBM San José Research Laboratory. There is an enormous amount of freedom. It is wonderful to be able to concentrate on research. One major objection I had, however, was that I felt the computer science department did not appreciate the potential contributions that mathematicians could make. More than once, when an outstanding mathematician expressed an interest in being hired, we took no action. Finally I decided that instead of fighting to hire individual mathematicians, I would try to convince high-level management to start a mathematics group, thereby establishing the importance of hiring mathematicians into a computer science department once and for all. Last August, that group came into existence. It took a year of my life, but it was worth it.

Let me finish by mentioning the attributes that I think are necessary to work in this kind of group: First of all, you have to be willing to talk with nonmathematicians and be interested in listening to mathematical problems that come from other areas. You have to be very flexible. You have to be willing to learn a lot of things that might not seem interesting on the first round but end up generating good problems. I think you have to be more of a problem solver than a theory builder.

Most of the mathematics I do now is combinatorics. I personally am most satisfied when I solve a problem that I think is genuinely difficult and interesting and whose solution is clever. I'm not saying that I think all mathematicians do or should share my values. Fortunately for the future of mathematics, not everyone is like me—many people have better taste and more interest in deep structures. However, for interfacing between mathematics and an applied area, I think it's useful to have both a love for pure mathematics and a love for solving problems for their own sake.

Aerospace

ELIZABETH RALSTON

Adapted from the AWM *Newsletter* 15(4), 1985, 9–10. Ralston was then a computer scientist with Inference Corporation; she is currently teaching mathematics and physics at Mount Saint Mary's College.

I finished graduate school in 1970, which was just about the time the job market was starting to dry up, although it was probably not as bad then as later. After a one-year job in New York, my husband and I moved to southern California, where I got a job at a state college in the Los Angeles area. Probably because of the job situation at the time, the prevailing attitude in the administration was, "You're lucky to have a job at all, and if you don't like it here you can quit." There were a number of things I didn't like about the job, one of which was the poor students. So after two years, when I got a temporary offer elsewhere, I decided that getting tenure on those terms was probably not worth it, and I did quit. I had a couple more temporary jobs over a period of about three years, by which time the job market in academia had really dried up. At that point I began my search for nonacademic employment.

The first nonacademic job I took was with Aerospace Corporation. This is a not-for-profit corporation that does general systems engineering on Air Force space programs. Basically what this means is that Aerospace does technical studies and provides technical advice for the Air Force. I was referred to Aerospace by a friend of a friend who worked there. In the search for nonacademic employment, it is almost always better to try to circulate your résumé or vita through some personal contact, no matter how tenuous. While I was at Aerospace, probably two-thirds of the people who were hired into the department were hired through some type of personal referral. It's not impossible to get a job another way, but a referral really helps. I think that's particularly true in a field like mathematics, where you have relatively nonspecific job skills and are making a career transition.

My job at Aerospace illustrates another fact of life, which is that in southern California most of the technical jobs are in defense-related industries. Defense contractors or large corporations are the ones that have sufficient funds to pay for someone who (like me at that time) needs training.

You might ask what kind of mathematics I used at Aerospace. I used to tell people being interviewed: "Basically calculus and linear algebra with a smattering of elementary—very elementary—probability and statistics." So I would have to say that, in general, you should probably not go into nonacademic employment if you want to be on the frontiers of mathematical research.

Then why would a Ph.D. in mathematics be hired to do calculus and linear algebra? It may be that most people with undergraduate degrees in mathematics, and certainly most people with undergraduate degrees in engineering, don't really know undergraduate mathematics. So Ph.D. mathematicians have an advantage.

When I first worked at Aerospace, I would be given some specific, semimathematical problems to solve. Since this was basically an engineering company, the problem often had been distilled from some engineering problem and put into mathematical form. What I quickly found out, and what I think most people in this situation find out, was that in most cases it is important to get back to the original problem. In attempting to simplify and extract the mathematical content, people often totally obscure the original problem, and the reformulated problem may have nothing to do with the actual problem.

For example, I was given a problem that basically asked for numbers a and b so that there are no integers n and m such that a times n equals b times m. I said, "This is easy, you just make the ratio of a and b irrational." And then I thought, this can't be the real problem. So I found the person who originated the problem and discovered what it really meant, and we went into an interesting and fruitful collaboration.

Aerospace Corporation was an extremely pleasant place to work. Initially I was in a department that included five or six Ph.D.'s, and this number grew as time went on. In fact, while I was there the number of Ph.D.'s in mathematics who were hired throughout the company seemed to be increasing. I think there may be a couple of reasons for this. One was probably that as Aerospace started to have success with the Ph.D. mathematicians they had hired, they were more willing to take on others. It likely also had something to do with the revival of the aerospace industry and the shortage of trained engineers at a time when mathematicians were available. With Ph.D.'s as colleagues I found the intellectual stimulation on a day-to-day basis a lot greater than I did when I was teaching calculus and precalculus and pre-precalculus to students who had failed in high school.

Gradually the work I did became less specifically technical or mathematical and more in the nature of evaluating outside contractors' proposals of work. Aerospace has a consulting role for the Air Force; it does not actually produce hardware or software. This has both advantages and disadvantages. On the one hand, you get to see large projects from a fairly high level, whereas if you were working for a company building a particular system you would typically be a very small cog in a very large wheel. On the other hand, after working there for several years I began to feel like a kibbitzer on these projects, as I was offering advice and suggestions but never had the responsibility for seeing something through to a conclusion. About two years ago, therefore, I decided I should look for a job change, and I am now working at Inference Corporation.

I must report a rather sad fact from a mathematical point of view. When you start modeling, the success of the model depends far less on the mathematical sophistication of the model than on being able to obtain some reasonable input data for it. I think this is something mathematicians often find when they start working on real problems. You have to make compromises on mathematical purity and concentrate on getting answers to the specific question that has been asked.

Oil Industry

MARGARET WAID

Adapted from the AWM *Newsletter* 15(4), 1985, 11–13. Waid was then manager of production services, NL Sperry-Sun. Now she is retired from her position as head of the Electromechanical Research Group, Halliburton Inc., and is president of a family-owned business, Comet Crawfish Technologies Inc.

I brought my props with me. When I was a university professor, I learned that you have to get the attention of the audience, and you have to get it before you say something important. So today I wore the coveralls and the hard hat I wear to the field when I'm going out on oil rigs, land rigs and offshore rigs, in the Gulf or offshore California or in the North Sea. (In case you don't know, one of the big differences between onshore and offshore rigs is that offshore rigs have bathrooms.)

Now that I have your attention, I want to make my major point: if you decide to work outside academia, you must not regard your company's capital as a substitute for a research grant, and you must not plan on waiting behind your desk for someone else to define the problems and bring them to you so you can apply your mathematics to them. It doesn't work that way. If you decide you want to work in an industry (e.g., the oil services industry, where I work), when you go for an interview, you must make sure the people you are talking with know you are very excited about that industry. You've already done your homework and learned a lot about it, you know what you're getting into, and you're very much interested in getting out into the field and participating in delivering the service, understanding and defining the problems, and then helping solve the problems.

In my current position, I thought, "We really need to get people in here with good mathematical background to do the kind of job that needs to be done." I got approval from upper-level management for the openings and recruited some friends who wanted to leave academia. I knew they could do these jobs, but they absolutely and totally bombed in the interviews. They didn't learn the necessary language and terminology, and they couldn't convince the right people that they would actually go out where the problems are. They couldn't convince them that they had a good feeling for the industry or of what they could do for the bottom line. They didn't do their homework.

I spent about ten years at the University of Delaware working on partial differential equations. Because it followed from my thesis work, which involved fluid flow through porous media, some of my work dealt with problems involving reservoir analysis. I decided to leave a comfortable tenured position and go into the oil field because I was really interested in doing applications, and I didn't have the right people to talk with where I was. I couldn't attack the problems properly because they were not well-defined, and they weren't well-defined because nobody really knew how to define them. I needed to go to the source.

I left academia and spent three years working for Schlumberger as a senior development engineer. One of the things I did there was model tools used for logging wells. Nuclear, electrical, and mechanical tools are used to make measurements concerning the formations, and then the information gathered helps determine where the oil is and how to extract it.

I modeled the response of all tools that involved pressure and flow measurement. I did *not* hide behind my desk, I talked with the engineers who were designing the tools. I wanted to influence their designs because I felt that often a complicated mathematical problem could be eliminated simply by designing the tools properly. The engineers took a lot of educating to understand this. They felt they should design a mechanically sound tool and it was the mathematician's job, or the theoretical modeling engineer's job, to try to figure out how to interpret the response of the tool. I did as much teaching there as I did at the university, but I had a more willing audience in industry because they were ready to listen. They were indeed receptive; they got to the point that whenever they were thinking of a basic design or making a design change, they got in touch with me to find out how this would affect the interpretation. We designed some tools that will now gather the proper information for the mathematical models that have been developed.

After three years at Schlumberger, I became very interested in management. Schlumberger didn't seem too interested in having women as managers. So I joined NL Sperry-Sun with the title of Supervisor of Software and Analysis; I was in charge of all their computer software. This put me in position to get out into the field. (Actually I did that at Schlumberger, too; I designed field tests and went out on lots of rigs.) In the field I discovered that by talking to customers on sales calls and engineers who were actually delivering the services, I could see how to define the problems and what the problems were. I could determine what tools were actually needed and what software was necessary to make the measurements required. I talked quite a lot with the people in research departments at the major oil companies as well as with the engineers designing the tools. You have to go out there and talk with *all* the people involved and look at the entire system. I determine the specifications for the tools so that they actually get the information we need. As a result, good information and proper parameters to fit the fancy reservoir models in use at Exxon, and Shell and Texaco, will be obtained. I was recently promoted to manager of production services, so I manage one of their two main product lines worldwide. I have an R&D department working for me, and an operations department, and people who deliver the services—that is, the salespeople—and I am in a position to get things done.

It is important for women to get into companies where they are not discriminated against and where they can follow career paths that are good for them as individuals. That's what they're entitled to. If I had stayed in academia, I would have stayed at the University of Delaware. It was a nice place to work. But I view academia as being very discriminatory and *not* a good place for women to be. That was one of the reasons

I left. I also left Schlumberger for that reason and went to NL Sperry-Sun. I don't find that situation anywhere here.

If it is not clear that an industry wants or needs you, your expertise and talent, then find one that does! Opportunities for women exist in industries trying to "make money," as opposed to service industries, because decisions are based upon "getting the job done" and the "bottom line." With the economy in the shape it is in, we are talking about *survival* of companies in the marketplace. There is a real shortage of technically qualified middle managers in this country. Upper management wants people who can get the job done. This delivers a great opportunity to technically qualified women with management skills. In this day and age, the losers are people who do not recognize these women, and industries and companies that cannot recognize the talents of the women working for them. So let them go down the tubes!

Publishing

ANNELI LAX, BETTYE ANNE CASE, AND ANNE M. LEGGETT

Based on the "Awards and Honors" section of the AWM *Newsletter* 25(2), 1995, 14–15. The citation and response for the MAA Award for Distinguished Service received by Lax originally appeared in the program of the 1995 Joint Prize Sessions, JMM. Lax (1922–1999) was then a professor emerita in mathematics, NYU.

All Ph.D. mathematicians have written at least a thesis, and many continue to write, producing scholarly research, expository papers and books, articles about mathematics and the profession, and textbooks. A few mathematics doctorates are self-employed as writers or work with publishing firms as writers or with editorial responsibilities, typically working on publications by scientists or about science. One such person was Anneli Lax, who was at the center of the MAA's publication program for thirty-three years, overseeing the New Mathematical Library (NML) series. Lax was honored by the MAA for this work in 1995 at the JMM Joint Prize Session with the Yueh-Gin Gung and Dr. Charles Y. Hu Award for Distinguished Service to Mathematics.

Lax handled every aspect of the NML from acquisition to cover design. The NML series was planned by Lax and the editorial board to "make mathematics accessible to the general reader without sacrificing technical accuracy." The citation for her award says: "No other person in the history of the Association's book publishing effort has played a larger role in developing and nurturing a book series. Some of her admirers have suggested that the NML be retitled as ANML, Anneli's New Mathematical Library."[48]

Lax's interests in language and mathematics led her to develop a combined course of expository writing and mathematical thinking with the writer Erika Duncan. The course, taught at NYU, was so successful that with Ford Foundation support, Lax expanded the curriculum into several junior and senior high schools in New York City.

In response to the award, Professor Lax wrote:

> I am overwhelmed by the unexpected honor of receiving the . . . Award for Distinguished Service to Mathematics. . . . I have the privilege of having known all but four of the thirty-three previous winners of this award and am delighted to be in their company. I had always known that there were various ways of being of service to mathematics, many of which are not officially recognized.
>
> As in the case of the NML, my concern has been access to mathematics, and my efforts have been directed to making sure that our schools do not deprive students of learning how to think for themselves by developing, among other skills, one of their natural talents: looking at the world mathematically. . . . There have been many promising experiments of implementing this agenda: those I have been trying to promote are attention to use of language in all learning, particularly learning mathematics, and developing the art of listening (and reading) so that we can apply this art to looking at our students' emerging ideas as these are voiced in our classes and written in writing assignments we give. . . .
>
> Let us practice what we preach, read and write carefully, avoid trendy slogans, and go beyond mathematical correctness, syntactic correctness, and political correctness in serving our discipline in our individual ways.

National Security Agency

BARBARA BROWN FLINN

Adapted from the AWM *Newsletter* 29(6), 1999, 21–23. Flinn holds the position of mathematician at NSA.

After finishing my Ph.D. at Michigan, I took a postdoc at the University of Texas at Austin (UT). Although there was only one other person in my

---

48. Upon her death in 1999, it was indeed retitled the Anneli Lax New Mathematical Library.

field there, my husband was at Texas A&M, where there was no one in my field, so UT seemed better for my research and a good compromise as a job choice. But UT and A&M are two hours apart, so my husband did a lot of commuting, and we tired of the lifestyle quickly. We both wanted to keep research positions, and a visiting friend suggested NSA as a possible solution. Before we knew it, we had job offers.

We had not researched any other jobs, and we still had academic jobs, but we accepted anyway. I found that time very scary, as I was leaving the only kind of job I had known or prepared for. Fortunately, both my husband and I have been very happy in our careers at NSA. In particular, I have found myself well-suited to applied math and focused problem solving.

Throughout my fourteen years at NSA, I have benefited from a number of development programs. I joined NSA as a member of a three-year intern program, which, through rotational assignments, gave me a broad base of familiarity with the people and problems. I stayed for four years in my first post-intern assignment, where I worked with some fantastic cryptomathematicians on important problems. Eventually I moved on, both to further my own professional development and to let junior colleagues in on the high visibility projects. I left to work on harder, riskier problems and found this to be a tremendous way to learn. Thanks to prodding from my managers, I applied to, and was chosen for, a new in-house development program for senior technical people. This allowed me to study at the Harvard math department for a school year: another great learning experience! At present, I am working in a different area altogether, trying to discover what we mathematicians can do to help solve some of the computer science problems.

My outward-directed activities include teaching in-house, conducting research studies, leading research activities, mentoring—all the things you might expect—and I had two exceptional outreach experiences. First, I spent a summer as a technical director for our Director's Summer Program when it was new. I truly believe this is NSA's best way to connect with the mathematical leaders of the future, and it is amazing to experience the energies of so many exceptionally talented students focused on our problems! The second activity I viewed at first with great skepticism, but now am glad to have participated in: planning the 1993 Women in Mathematics Symposium. Sure, the symposium was a great way for NSA to raise its profile with outside women mathematicians, but I hadn't realized just how wonderful it would be, for both the hosts and guests, to have so many women mathematicians together.

In conclusion, I want to underscore some general points about being a positive force in your workplace. These grow increasingly important as you progress in your career, that is, as you become increasingly influential.

You have a tremendous amount of power, so every now and then, assess how you are using it. I used to pooh-pooh the idea of role models. I reasoned that—since as a child I had never known any grown-up who was remotely like me or like who I wanted to be—role models were not necessary. Now I know better. Accomplished colleagues with whom we can identify inspire us at all stages of our lives. What's more, we are all role models *right now*! No matter how far down you think you are on the food chain, you have accomplished much, and less experienced people are looking to you to see how it's done.

Take opportunities to make an important difference. This could mean saying "yes" when you'd rather say "no" to a task you think is important and needs your leadership. It could also mean saying "no" to something that sounds like fun in order to suggest another candidate who has the skills but needs the chance to shine, stepping aside to give someone junior a chance.

Give extra weight to activities that build community, be it math department, AWM, faculty, etc. This includes organizing or sponsoring conferences. There's something very special, a subliminal confidence boost we all get, in hearing myriad versions of success stories, told by women together, all of whom are a lot like us.

Biomedical Research

SARAH E. HOLTE

Adapted from the AWM *Newsletter* 29(6), 1999, 19–20. Holte is a staff scientist in the Public Health Science Division at the Fred Hutchinson Cancer Research Center, Seattle.

My career as a mathematician began when I wrote a dissertation in point-set topology under the direction of Lew Ward at the University of Oregon. Then I went to the University of Missouri at Rolla, where I had a tenure-track appointment. At this point, I followed pretty much the traditional career path for Ph.D. mathematicians. I taught and did research in topological dynamics.

However, after a few years in this position, I decided that research in abstract mathematics was not for me, and I began to investigate other opportunities for mathematicians. I obtained journals from other professional organizations, many of which have advertisements for positions and postdocs. I was primarily interested in moving into medical research, and after about a year of checking out the options, I found a postdoc in biostatistics at the University of Washington.

The postdoc allowed me to learn statistics and some programming (both areas where I had no experience as a graduate student or assistant professor). It also gave me the opportunity to gain experience and make contacts in the medical research community. After a year and a half on the postdoc, I was offered a permanent position at the Fred Hutchinson Cancer Research Center, where I am currently employed.

My work at Hutchinson involves all sorts of projects where I contribute mathematical and statistical expertise. I've worked on projects on the role of genetics in cancer, the assessment of environmental risk factors for cancer, and mathematical modeling of the carcinogenesis process. Currently I'm working exclusively on HIV research. This involves both mathematical and statistical modeling of HIV infection, as well as offering statistical guidance for large trials of methods to prevent HIV transmission. I also work toward developing improved statistical methods for data analysis.

I've found the biggest difference in working in a nonacademic setting is the pace. Everything in medical science happens extremely fast, at least in HIV research. New results seem to turn the field around on a regular basis, and often the focus of work shifts dramatically as a result. It's both frustrating and exciting. You have to be flexible and work quickly, sometimes letting go of the "but we haven't proved it yet!" ethic we learn in graduate school in mathematics. Another difference is that I work mostly with nonmathematicians, so communication is challenging. It takes a lot of patience to get truly collaborative research results when the collaborators are speaking two different languages, for example, mathematics and biology. But on a regular basis you learn exciting new things in other disciplines.

For people interested in work in the life sciences or medicine, good training in statistics is extremely useful. Statistics is the quantitative language of biology and medicine, and if you have the ability to analyze data, it's an excellent way to get involved with medical research. Mathematics is more outside the mainstream, but I've found that many researchers are interested in exploring mathematical models if you can offer some statistical guidance as well. It's also often the case that "less is more." I rarely need to use extremely sophisticated mathematics. Almost everything I use is part of the undergraduate curriculum in mathematics: differential equations, multivariate calculus, and linear algebra.

I've found my "nontraditional" career to be sometimes frustrating, but overall extremely rewarding. I learn new things every day and feel that I have the opportunity to contribute to advances in HIV science.

Communications Industry

MARGARET H. WRIGHT

Adapted from the AWM *Newsletter* 29(6), 1999, 20–21. Wright was then a Bell Labs Fellow at Bell Laboratories, Lucent Technologies. She is now a professor of computer science, Courant Institute of Mathematical Sciences, New York University.

My career as an applied mathematician has been highly nonlinear. After receiving a B.S. in mathematics and an M.S. in computer science from Stanford, I was not ready to go for a Ph.D.—I had no idea what academic research was, and also I needed to earn some money. So I worked for several years at GTE Sylvania doing numerical analysis and scientific programming. This experience taught me a lot, not necessarily about mathematics, but about some of the ways of the world. It also made me realize that I would not be satisfied in the long term unless I had a job with more individual responsibility. Luckily, by this point I had become very interested in optimization and was eager to begin work on a Ph.D.

Let me first describe where I work. Bell Labs is a large industrial research lab. The "research" part of Bell Labs includes about 1200 Ph.D.'s in different scientific areas; I am in the Computing Sciences Research Center, an organization of about 65 people. It's difficult to give a precise job description of a Bell Labs researcher; one way to think about it is having several careers at the same time.

First, I'm supposed to be a visible scientist in my own field. This means that I do basic research, write papers, and give talks. Bell Labs is a wonderful environment for research because of its openness and collegiality, which have to be experienced to be understood. Second, I was hired because my area of research is seen as important to Lucent Technologies (the parent company, which was part of AT&T until 1996), so I serve as a resource for the company in this area. I'm very happy about this because I like to apply mathematics to real-world problems, and the broader Lucent environment provides plenty of these.

As well as these two core activities, I try to be an active member of the mathematical sciences community. I serve on editorial boards and committees and am active in scientific societies, particularly SIAM (of which I was president in 1995–1996). And although I am still (after all these years!) uncomfortable with the idea of being a "role model," this position is inevitable, whether we like it or not, for all women working in science and engineering. I care a lot about encouraging women (and minorities) to pursue careers in science and engineering, and I give "rah-rah" math and science talks to students of different levels.

Turning now to the theme of growing in a job, I want to mention a few pieces of advice, which I offer with no claims of accuracy. Being a good mathematician is without question hard work for anyone, but there are certain extra, familiar difficulties that women mathematicians tend to encounter. We all have our stories of what I sometimes call the "presumption of incompetence," in which we have to prove that we are good rather than being given the benefit of the doubt. One can argue that this makes us tough, but it is definitely frustrating to have to convince someone that we know what we're talking about when it should be perfectly obvious that we do! However, at least in the foreseeable future, this will continue to be a problem for women. My only advice about this is to be prepared and not to let it get you down except at the level of an occasional irritation. (And remember that it is a good source of anecdotes.)

From the beginning of any job, it's helpful to develop your own individual style. I'm not suggesting striving to be eccentric or peculiar—simply that you should establish a definite identity, so that people will notice and remember you in a good way. My advice on this is, first, to know yourself, and second, to watch others for things you like. Think about what you admire in a speaker, teacher, or colleague; then adapt that person's behavior so that it suits you. Don't be afraid to experiment with something new; this is not a context where you can determine in advance what will work best for you. Style comes with no effort to some people—but for me, it is something I have had to work on (and I'm still working on it).

A related suggestion is that you become familiar with your weaknesses. It is a stereotype, often true, that women as a group lack self-confidence. This is certainly true of me, and of most of my close women colleagues. But lacking confidence (which is usually bad) is not the same thing as knowing your faults (which is good). You need to take a hard, objective look at your research, writing, talks—every aspect of your job—and figure out which things you can do better. This is ideally done with an honest friend who will tell you the truth. We all like it when friends reassure us that everything is fine, but in the long run this is unhelpful if in reality everything is not fine. So, even though it is incredibly difficult to accept criticism, we need to do it. I used to worry because I became upset when people made critical comments to me after I had asked them to be frank, but I now think (rationalize?) that this is natural. What matters is that, once you calm down, you think carefully about what they said. If, being rigorously honest, you find that their criticisms are valid, then you can try to correct the problem—or accept that it is part of you and try to work around it. (And remember to thank the person for being honest.)

Finally, a key part of growing in a job is a highly developed sense of self-awareness. What I mean by this is that you need to keep track of your "professional temperature" and think periodically about whether you are really satisfied with what you are doing. No job is perfect, and all jobs have ups and downs, but you should pay attention to serious, chronic unhappiness. A key part of growing within a job is to understand what things you would like to improve, and then to figure out an action plan to change them. Being in control, in the sense that you know what you want, makes a huge difference to your effectiveness. In the worst case, if you realize after careful thought that, despite your best efforts, your job is simply not right for you—this happened to me at Stanford—then you need to take the stressful and scary step of changing jobs. For this reason you should make sure that at any time your publications and visibility are strong enough that you can change jobs.

4. Having a Life

Graduate students often wonder whether "having a life" is truly compatible with employment as a Ph.D. mathematician. The popular press periodically raises issues that resonate: the study that purported to show that past a certain age a woman is more likely to be killed by a terrorist than to find a husband, reports on the "glass ceiling" and the "mommy track," writings about the "biological clock" and the "opt-out revolution." In "Pathways in Mathematics," which began part II, related social and political realities are considered primarily in terms of their effects on the statistical picture for women in U.S. mathematics. Here, individual women give their perspectives on various aspects of balancing personal life and career, and show that, although perhaps difficult, it is possible to find a satisfying balance.

Louise Hay's autobiographical memoir illustrates the themes of geography, two-career couples, and childbearing in a 1950s setting. These themes continue to be critically important. Susan Landau writes eloquently about ways universities today might make it easier to handle family life. Judy Roitman shows some of the pain involved in changing jobs (whether by choice or by necessity) when the job market is tight. Whether partnered or single, job seekers may worry about geographic isolation, for the sake of both mathematical collaboration and cultural amenities. Although the Internet has somewhat mitigated the geographic difficulties of mathematical joint work, there are still obvious advantages to talking mathematics face-to-face. Margaret Murray, who has received critical acclaim for her writing on women in mathematics, here gives her thoughts on the incidental effects of geography on life and career.

Rebekka Struik, in the early 1970s, collected information from "some couples who . . . decided to accept positions far apart and work out the problems involved."[49] One of her respondents was Bernhard Neumann, who wrote a moving description of his life with Hanna Neumann (a Ph.D. student of Olga Taussky Todd; see I and IV). He explained that for about half of their married life of almost thirty-three years, they were "separated by space."[50] His comment: "Our five children and eleven joint papers testify to the fact that this separation was geographical only; in all aspects that really matter, we were constantly close to each other." Both Judy Green and Marian Pour-El share perspectives on their "commuting marriages"—how to work

---

49. "The Two City Problem," AWM *Newsletter* 4(6), 1974, 8–11.

50. Bernhard H. Neumann (1909–2002) and Hanna Neumann (1914–1971) had long careers together at Australian National University, Canberra, after the years of separation.

out a relationship and how to parent, when the partners are employed in different localities.

One message that comes through clearly is that flexibility and imaginativeness are key. As Hay says in concluding her memoir, "If there is a moral to this tale of how I became a mathematician, it is that sources of inspiration and opportunities to change your life can come unexpectedly and should not be ignored, and that you should not neglect the dictates of your own career, taking some risks if necessary, since you never know what the future will bring."

How I Became a Mathematician

LOUISE HAY

Based on "How I Became a Mathematician (or How It Was in the Bad Old Days)," AWM *Newsletter* 19(5), 1989, 8–10. Hay (1935–1989) was then professor and head of the Department of Mathematics, Statistics, and Computer Science, University of Illinois at Chicago.

During the course of a recent conversation I mentioned a PBS program in which Bill Moyers interviewed a philosopher from UC Berkeley, who put great emphasis on how the Western literary classics (sometimes referred to as those written by "dead white males") should be studied because they affect our lives; I commented that my life had been much more affected by *The Feminine Mystique* by Betty Friedan. My colleague Bhama Srinivasan also reminded me of how influenced I had been by a dinner conversation with the late Hanna Neumann, whom I speak of below. Since the AWM *Newsletter* publishes articles on famous women mathematicians both past and present, perhaps it would be instructive, especially for our younger members, to read what it took for a minor mathematician to escape the spell of the fifties and, through a combination of luck and inspiration, to find a satisfying niche in the mathematical world.

Until high school, I was not particularly mathematically inclined—indeed, I was much better at verbal subjects; combinatorial aspects of numbers and equations have never been my strong point (it is not always recognized that mathematical aptitude comes in many different flavors). I was fortunate in tenth grade, however, to take a geometry class taught by David Rosenbaum. He arranged tutoring opportunities for me, making all financial arrangements with the parents involved, which was extremely helpful since my family's financial position was very precarious. He believed in teaching the logic of the subject rather than just following the theorem-proof format of the text, which of course was straight out of Euclid. He wrote up and distributed notes on logical reasoning, fallacies, etc. and expected the students to understand what they were doing when they wrote up a proof. I found the logical aspects of mathematics much more congenial than the numerical aspects, and when I showed aptitude for this, he suggested I read up on non-Euclidean geometry, to put the subject in a new perspective. He had me get Wolfe's book on non-Euclidean geometry, which I found fascinating and which ultimately was the basis of the project I wrote as a senior for the Westinghouse Science Talent Search, in which I won third prize. This was no doubt instrumental in my getting a large enough scholarship to enable me to attend Swarthmore College, where I majored in mathematics, a decision that I had taken as a result of my geometry class and never reconsidered. The Westinghouse award also led to summer jobs at the National Bureau of Standards, which in turn led to a part-time job at the Moore School of Electrical Engineering, which helped support me through college, and to later jobs in industry.

I heard a few years later that Rosenbaum's reward for his efforts was to be given a solid load of remedial courses. He was a bitter man, presumably based on his experiences in the thirties—when I visited him from graduate school, he told me I would never get anywhere professionally because I was a woman and a Jew; unfortunately, I never got a chance to show him that things had changed.

As was common in those days, I got married at the end of my junior year, though still with the intention of going to graduate school in mathematical logic. It was not easy to find universities with strong programs both in visual perception (for my experimental psychologist husband) and in mathematical logic (those could be counted on the fingers of one hand—logic was not exactly considered respectable in most mathematics departments). We settled on Cornell, where we both received teaching assistantships. My husband went there in the fall of my senior year, while I remained at Swarthmore, an unusual arrangement in those days. I then came up to Ithaca for the spring and worked at General Electric; upon obtaining my degree in June, I went to see the personnel manager about getting my position upgraded. He agreed and told me the salary scale, only to retract it in an embarrassed fashion because it applied only to males! I spent two years at Cornell, at which point my husband, who had entered with his master's degree and now had all the data for his Ph.D. dissertation, decided to take a visiting position at Oberlin (the job market was very tight). Like a good fifties wife, I followed him, first doing a master's thesis to have something to show for my two years. (I must admit I was glad to leave—I lacked confidence that I could finish a Ph.D. at that point.) My thesis advisor, J. Barkley Rosser, was extremely helpful, both in getting a visiting job for me at Oberlin (here the "old-boy network" worked in my favor) and in leaving me with detailed instructions for proving a theorem in infinite-valued logic which would constitute a thesis, while he took off for a trip around the world with his family. As it happened, I found a counterexample to the main lemma, which made the thesis publishable under my own name.

My year at Oberlin was followed by a year working for Cornell Aeronautical Laboratory in Buffalo, next a year teaching part-time at a junior college, and then three years as an instructor at Mount Holyoke College while my husband was on the tenure track at Smith College. By then, *Sputnik* had changed the conditions of academic mathematics in the United States—there was great student demand, a shortage of Ph.D.'s, and therefore jobs for non-Ph.D.'s like myself, though without a real future. By the end of that time, I had built up much more confidence (there is nothing like teaching courses you've never taken to increase your confidence in your ability to do mathematics), but there was no convenient opportunity to return to graduate school.

At this time I became pregnant with my first child, a welcome event after several years of apparent infertility. I had recently taught a course in projective geometry, and I read in Emil Artin's book *Geometric Algebra* the first chapter on coordinatization, which I found incredibly beautiful and which really turned me on to the mathematics again. By good fortune my husband was collaborating on research with a friend, and they decided to spend the summer together at Cornell, which gave me a chance to attend some seminars and get somewhat back into the mathematical swing. (As I recall, logician Graham Higman had just proved the undecidability of the word problem for groups, and we read papers by mathematical couple Hanna and Bernhard Neumann on amalgamation of groups, which were relevant.) I then returned to teaching that fall, having resigned my job effective the second semester due to my expected baby. This is what you did in those days: I never even considered any alternative arrangement.

But several things happened to change my life during that spring of 1963. Betty Friedan's *The Feminine Mystique* was published; when I read it I questioned for the first time the rationale of giving first priority to being a wife and mother and sacrificing a career for myself for the sake of my husband's. The second thing that happened was that Hanna Neumann gave a colloquium at Mount Holyoke to which I was invited, and I happened to sit next to her at the colloquium dinner. I must have mentioned my new baby, and she proceeded to tell me some of her history: how she had interrupted her studies to have two children, but then having been evacuated during the war (as an "enemy alien") from the coast of England to Cambridge while her husband joined the British army, she returned to school, using students as sitters, and finished her Ph.D. By 1963, she had raised five children and become a renowned algebraist. The message that came across to me was that if she could finish a Ph.D. with two children, surely I could do so with one. (My first child was an exceptionally easy and good-natured infant, who did not require constant attention, so that I was somewhat bored staying at home.) Anyway, we went back to Cornell that summer for my husband's research, but halfway through the summer I decided to take the plunge and remain at Cornell for the following academic year to try to finish at least the course and exam requirements for the Ph.D.—it was not clear what I would do regarding a dissertation yet. Those were the halcyon days of post-*Sputnik* funding—lots of grant funding so that I was able to walk in halfway through the summer and be given a research assistantship for the following year. So I found an apartment where my son and I would live while my husband returned to Smith. When I realized what I had committed myself to, I suffered some anxiety and insomnia, which a therapist helped me cope with.

As it turned out, I had a wonderful and productive year. I took three courses, passed my prelims, and under the inspiring tutelage of Anil Nerode, did all the work for a dissertation in recursion theory. (I still remember how I proved my first theorem; it was the weekend Kennedy was assassinated—my husband, who visited me about once a month, canceled his visit that weekend, and I spent the time working instead.) By the end of the year (including both summers) I had all the requirements for the degree except for the final writing up and the thesis defense, and I was pregnant again. The thesis defense was somewhat delayed because it turned out to be twins who were born prematurely, but by the end of the following year I had my degree. I stayed home that year, but with a half-time housekeeper, which enabled me to spend some time each day reading mathematics. Where part of my earlier motivation for returning to work with one child was that I didn't have enough to do, I now had the motivation of having too much to do at home (three children under the age of three!). Finding good and reliable child care was an incredible hassle; we must have gone through at least fifteen sitters, everything from a fourteen-year-old girl to an eighty-five-year-old Italian grandmother and live-in South American and English helpers. One of the hard things to cope with at that time was the climate of social disapproval for working mothers: "You mean you're willing to let someone else raise your children?"

With my Ph.D. as a union card, I was rehired at Mount Holyoke on the tenure track and was an assistant professor for three years, one of which I spent on an NSF Postdoctoral Fellowship at MIT (I would have preferred a job at U Mass., because of the graduate program, but I made the mistake of applying for one, which was the "kiss of death"; these were the days before affirmative action, and most hiring was done through the "old-boy network" via your advisor). As it happened, the following year my marriage broke up, and it *really* made a difference that I had my Ph.D. I could for the first time go on the job market on my own account (and in 1968 the job market was very good); I still shudder to think of what my situation would have been had I not finished my degree. I joined the University of Illinois at Chicago at an excellent salary as an associate professor (thereby avoiding the agonies of waiting for tenure). I remarried, to a very supportive colleague who helped make it possible for me to continue to prove theorems, each of which conveyed to me the peculiar thrill of briefly knowing a sliver of mathematical truth that nobody else knows. And I saw the climate toward women change. Professional women with children no longer have to justify pursuing a career—if anything, it's the other way around. Significantly, when I became department head in 1979, the subject of my sex never became an issue as far as I am aware. (Unfortunately, this may be less true on the national scene—at the annual National Chairmen's Colloquium there are mighty few women to be seen, and tenured women mathematicians seem to be concentrated at relatively few institutions.) But women are no longer automatically expected to follow faithfully in the steps of their spouse, and men even share in child care without calling it "babysitting." It may be instructive to the young women of today to know it wasn't always like that. If there is a moral to this tale of how I became a mathematician, it is that sources of inspiration and opportunities to change your life can come unexpectedly and should not be ignored, and that you should not neglect the dictates of your own career, taking some risks if necessary, since you never know what the future will bring.

As a postscript, I can't resist mentioning my boys, in whom I must admit greater pride than in my theorems. My eldest graduated at the top of his Harvard Law School class and married a Law School classmate. At this writing, the twins are each pursuing doctoral programs. So they seem to have survived the babysitters and other turmoil in my life with reasonable success!

[Ed. note: Hay died not long after writing this article. We are grateful she shared her story so openly with us. I met all the boys and their families at her memorial service in Oak Park, Illinois. They were indeed fine young men with whom I shared loving memories of their mother. *AML*]

Is Geography Destiny?

MARGARET A. M. MURRAY

Based on "Is Geography Destiny? One Perspective," AWM *Newsletter* 23(3), 1993, 10–11. Murray was then an associate professor of mathematics, Virginia Polytechnic University, where she was promoted to professor. She is an M.F.A. candidate in nonfiction writing and a graduate instructor in English at the University of Iowa.

In the fall of 1982, I was a graduate student at Yale University, completing a dissertation in harmonic analysis and preparing to go on the job market. During the late 1970s and early 1980s, the academic job market—indeed, the whole U.S. economy—had been unsteady and unpredictable. In part to pay my way through school, and in part to gain a variety of work experience, I had held a succession of summer jobs in business and industry during my undergraduate and graduate days, preparing for the possibility that my career would *not*, in fact, be an academic one. As late as summer of 1982, I was not at all certain that I would even apply for academic jobs. But come the fall, emboldened by some successes I had recently had in research and energized by a semester in the classroom, I decided to apply for tenure-track jobs in mathematics.

The job market turned out to be much better than I had feared it would be; I received a handful of offers. In the end, it came down to a choice between two universities: one public, the other private; one rural, the other urban; one a technically oriented land-grant institution, the other a comprehensive university with a more traditional orientation toward the liberal arts and sciences.

I chose to come to Virginia Tech, Virginia's land-grant university, located in Blacksburg, in the southwestern part of the state—within a five-hour drive of Washington, North Carolina universities and research institutes, and Knoxville, but somewhat off the beaten track. It was not, in the opinion of most people who knew me (my adviser, for example!), the obvious choice. But I'd made the choice with my eyes open. Although I was a city kid, I had tired of living in dangerous urban areas and welcomed the slower pace and relatively safer conditions of small-town life. Although my own formal education was undertaken in private schools, my populist and egalitarian instincts drew me toward the idea of teaching at a public institution. And I was eager to join what appeared to be a friendly, congenial department where one could find a good many analysts, though none working in harmonic analysis per se; I felt that I could easily find my niche there.

Ten years later, I am still at Virginia Tech, though the path that took me from that day to this was not at all what I expected. I have been pleased with the relaxed pace, and lower prices, of small-town life, but I have missed the stimulation of the big city much more than I had imagined I would. I have been pleased with my decision to work for a public institution, but have discovered some of the limitations of state universities, particularly in difficult budgetary times. And it has not always been easy to combat the sense of professional isolation I've felt as "the only" harmonic analyst in my department.

Back in 1983, my rationale for seeking a tenure-track job was that I wanted to be "settled," but in the end, research trips to the University of Texas at Austin and a visiting position in Minnesota constituted the "moral equivalent" of a postdoc. (Of course, I have had the security of knowing that I have a home—and a job—to return to, here in the mountains of southwestern Virginia. This is not the situation of most academic postdocs, who trade security for the opportunity to concentrate on research.) Tenure (which I received in 1990) afforded me the opportunity to stop and catch my breath, and to take a look around me at my life and career, at my place in the university and the academic community more generally.

In today's straitened job market, it is rare that new Ph.D.'s have the luxury of choosing among a variety of job offers when they emerge into their professional careers. On the other hand, many of the geographic challenges that I have met (by choice) during my career are in fact the same geographic challenges that new Ph.D.'s who are lucky enough to land stable academic employment face when they finally settle into the job they so desperately sought.

Unlike many other women mathematicians, I have been able to pursue my career up to this point without worrying about the "two-body problem." This has given me a certain freedom to make my career up as I go along. This is not to say, however, that personal considerations have played no part in my decision-making process. It is very important to me personally to feel a part of the community in which I live; having a "home base" is and continues to be a high priority. Perhaps because I had prior experience in nonacademic employment, and because I have had teaching experience in a fairly wide variety of institutions (including a graduate school summer spent teaching in a community college), I have generally taken a broad view of what my "career" could be. When I was at Yale, there was a tendency among graduate students to visualize their future careers as being, in essence, carbon copies of their advisers': lives spent devoted to research, first and foremost, and secondarily to classroom teaching. But there was ample evidence, even as early as the late 1970s and early 1980s, that higher education would undergo a kind of "restructuring" just as American industries had been doing for some time. I have never had any reason to expect that my life in academics would be even remotely like the life my professors have had. I have considered it my good fortune that I have been able to pursue my "first choice" of career, in teaching and research, but I think I would have found a modus vivendi if such a career had proved impossible. Given the rapid pace of change in the American economy, it's difficult to predict just how much longer the academic enterprise will continue in the form in which we've come to know it. I give thanks for the chance I have to contribute by teaching and learning, but I know better than to be carried away by the thought that this opportunity will continue indefinitely—tenure or not.

Which brings me, at last, to the question of my title: is geography destiny? Are mathematicians, women in particular, constrained by their geographic circumstances? To some extent, yes—all the choices that we make in this life constrain us to one degree or another. But the degree of the constraint is as much determined by the limits of our imagination as by the reality of our circumstances. In my own career, it would have been easy to develop a sense of geographic isolation. The best way to avoid feeling constrained by geography is, first and foremost, to adopt and maintain a flexible attitude toward the meaning of the word "career." I have never taken a narrow view of the word, but rather have tried to think of my own as "vocation," so that *how* one lives is as important as the particular job one pursues or the place in which one pursues it.

Is geography destiny? The answer for me is no. In your case, the answer is up to you.

Making a Choice

JUDITH ROITMAN

The job market for Ph.D. mathematicians was bad when Roitman sought her first and, as she describes here, her second positions. Based on a talk at the panel "Choosing Our Lives" that she moderated, 1977 JMM, as reported in the AWM *Newsletter* 7(4), 1977, 7–8. The next academic year, she was a member of the School of Mathematics at the Institute for Advanced Study in the fall, and joined the University of Kansas (where she is now a professor of mathematics) on the tenure track in the spring. Her first job was as assistant professor of mathematics.

Recently I decided to leave my first job and look for another one. I want to compare this experience with what I went through when I found that first job. The difference between the two times and the two states of mind is great, and I've learned a few things in the process.

The main thing I've gained is a sense of my own worth and integrity. When I first got my degree, I was filled with ideas of my own incompetence. After I passed my qualifying exams, I didn't really believe I had passed them. I felt my degree was somehow just sleight of hand. In some ways I was totally passive, but I sent out 300 applications and drove the secretaries crazy.

I took the first job I was offered because it was a job and the possibilities of tenure looked good on paper; everybody told me to take it. When I went for my interview, I knew the place was wrong for me. There was nothing I could put my finger on, but it didn't fit me and I didn't fit it. However, I didn't have the courage and the sense of myself to say, "This isn't what I want."

It's important to have a sense of yourself as a mathematician, to look honestly at who you are, whatever your primary interests may be, because there may be strong counterpressures. There's the status/prestige thing—being a researcher at MIT is supposed to be better than teaching disadvantaged kids in elementary school—so people who really want to teach feel pressured into taking research-oriented jobs. Then there's the reverse pressure, as in my first job, which pushed me away from research and toward teaching. On paper, it looked like there was time for research, for hanging out at MIT and Harvard, but the unspoken understanding was that you shouldn't put that stuff first, not even a close second.

So almost the minute I got there, I realized it wasn't what I wanted. When the time came for my contract to be renewed, I asked that it not be. I decided to leave academics if I couldn't have what I wanted and went on the job market. I sent out my applications with a sinking stomach and a sad heart. My future was a tremendous black wall called June, and I didn't know what was on the other side. My friends who were hunting for jobs didn't appear as upset as I felt. I thought something must be wrong with me to feel as bad as I did, and I was also feeling bad for feeling bad.

I was greatly cheered by a letter from a friend of mine. I had written him to check on my job prospects at his university, which I didn't think were very good. He wrote back confirming my suspicions, but added that he was sorry to hear I was going through the demoralizing and dehumanizing situation of looking for a job in the current market. This was wonderful to hear. It reminded me that of course I was feeling bad, there would be something wrong with me if I didn't feel bad. I immediately felt much better.

I've talked to many of my friends who are looking for jobs, and we have great solidarity. We tell each other about openings, and we're honest with each other about our interviews. We're not cutting each other's throats, we're all in it together. One of my friends told me how angry he was. He said, "I wrote a really good thesis. I do pretty good research, I teach well, I work hard. I do everything I'm supposed to do, and there is no guarantee of making a living, no guarantee I can keep my family together."

I decided to think seriously about what to do if I didn't get a job. Untrained for a nonacademic career, I felt that my other options were less of career than of place. My friendships and relationships are extremely important to me, so I decided that if I don't get a job I like, I'll go back to Berkeley where many of my friends are, where I still feel part of the community, and then see what happens.

This decision was important. My world is not going to fall in, and I am independent of the job market. I don't have to take any job offered to me simply because it involves mathematics. I have a viable option.

So here I am, possibly on the brink of leaving the mathematical community. Some people in this position work like mad, but research takes a lot of energy, and I can't throw myself into something that may fizzle out soon. If I get the kind of job I want, then I'll go back to mathematics with the same care I had before. But I don't worry about it now.

I know what I want and won't take anything less. Now when I go for interviews, I'm interviewing them much more than they're interviewing me. They have my vita, and they have my letters. They know who I am. What am I going to tell them that they don't already know? But I have a right to ask them all sorts of questions, and I don't have to pretend I like everything they say. I have the freedom to say, "I'm not interested in your job, thank you." Saying this doesn't hurt me, because there's no point in taking a job in which I'm unhappy. I didn't know that three years ago.

Universities and the Two-Body Problem

SUSAN LANDAU

Based on "A Study in Complexity: Universities and the Two-Body Problem," Computing Research Association *Newsletter*, March 1994, 4, reprinted in the AWM *Newsletter* 24(2), 1994, 12–14. Then a research associate professor of computer science, University of Massachusetts, Amherst, Landau is now a senior staff engineer, Sun Microsystems, Inc.

Around the time I was finishing my Ph.D., I received a warmly encouraging letter from a senior faculty member at a major research university, urging me to apply. I did. Nothing happened. What had made my application look so bad? I had my answer a few months later. We met at a conference, and he said, "Sorry, we didn't interview you. We knew you were part of a two-body problem, and we only had one job."[51]

All my job hunting since my Ph.D. has been complicated by the fact that I am married to a fellow computer scientist. A job for one is complicated; for two, the predicament exponentiates. Yet the reality is I am far from a singular point in being married to a scientist. A recent article in *Science*[52] states that 69% of married women physicists are married to scientists, as are 80% of married women mathematicians and one-third of married women chemists. Chairs and deans are not discussing an isolated phenomenon when they say, "We wanted to hire [a female scientist], but she was married to [a male scientist], and there wasn't a position for him."

I do not think universities are using this situation in bad faith as a way to avoid hiring women. I think departments, chairs, and deans view each

---

51. In fact, discrimination on the basis of marital status is illegal, but that is *not* the point of this article.
52. Ann Gibbons, "Key Issue: Two-Career Science Marriage," *Science* 255(1992), 1380–1381.

occurrence individually. But a recent report from the University of Michigan points out that "female faculty seem to benefit from career services even more than men, because women, based on our experience and interviews, often have a spouse or partner in a position equal to or higher than their own. Almost all female faculty recruited by Engineering have a partner with a Ph.D."[53]

A number of universities and colleges across the country have begun to find imaginative institutional responses to the two-body problem. These responses are instructive to describe, for they show different ways to handle what seems to be viewed by many as an insoluble problem. In limiting my discussion mainly to university responses to dual-career married academic couples, I do not mean to diminish the difficulties faced by other types of dual-career couples. But solving this one aspect of the problem for married female scientists will help with an important part of a complicated picture.

At the University of Wisconsin, Madison, the Spousal Hire Program is run by the provost's office, which provides funding for one-third of a position for up to three years. The spousal hire must meet one or more of the following criteria: (1) the department, as judged by the dean, is in need of expansion, (2) a strong case can be made for continued employment after the initial three years, (3) the hire will enhance faculty diversity, or (4) the spouse has a record of receiving research grants providing some salary support.

Both Oregon State University and the University of Nebraska at Lincoln have spousal fellowship programs. The dollar amount is low—$12,000 at Oregon State and $15,000 at Nebraska—but gives the spouse time to find employment in the area. The fellow is employed for one year in an appropriate department. At Oregon one-third of the funding comes from the department and two-thirds from the provost's office, while at Nebraska all the funding is from the provost's office. This investment should be contrasted with the University of Michigan's estimate from its College of Engineering that "the cost per faculty position of . . . recruiting efforts is about $20,000."[54] Of course that number does not take into account the intangible costs when a top recruit leaves.

The University of Michigan has been concerned for quite some time with the issue of dual-career couples, for the unhappiness of an underemployed spouse has a strong effect on faculty retention. Here the focus is on dual-career couples where one member is nonacademic. The Office of the Assistant Vice President for Academic Affairs–Personnel has organized job search workshops for spouses and partners and aided these spouses/partners by arranging interviews and contacts, often *before* the recruited faculty

53. The University of Michigan, the Dual Career Project.
54. Michigan, Dual Career Project.

member has decided whether to accept an offer. There is no formal program in place for dual-career academic couples, although in practice the university has aided several such couples through tenure-track or tenured appointments. The existence of a high-level administrator with official concerns for this problem makes it somewhat easier to effect solutions.

For several years now, Kenyon College, the College of Wooster, and Dennison College in Ohio have run joint job advertisements, listing all positions available at the three institutions. Whatever complications exist when, for example, the art department wants someone, but astronomy has the other member of the couple as second- or third-choice candidate, are exacerbated when the spouses/partners are applying to different institutions. But the provost at Kenyon said that the colleges are deeply committed to doing this and will work it out. This year nearby Otterbein College has joined the advertising. Recently Bates College, Colby College, and Bowdoin College, all within an hour of each other in Maine, have begun similar advertising.

Other arrangements have been developed by an energetic chair, an imaginative dean, or a thoughtful candidate. The two examples that follow involve couples where both are computer scientists.

Several years ago the Iowa State computer science department had a first-choice candidate, but she was married, and the department had authorization for a single slot. The chair went to the dean and the provost for an additional position, which they agreed to because of the institution's commitment to affirmative action. But funding was tight, and it wasn't clear when the position would become available. So the department made an offer, *in writing*, to hire the spouse when the *next* position became available to the department and committed to hiring the spouse within three academic years. The chair urged the husband to accept the offer, saying he would do all he could to expedite matters. The husband signed the contract and joined the faculty two years later.

Another pair approached the University of Waterloo. One of them was already there; the couple asked to be appointed to one and a half positions, three-quarters each. Because fractional load appointments were allowed by Waterloo's faculty handbook, that part of the arrangement was easy to implement. The additional half position was much simpler to negotiate than a full position would have been.

These approaches require changes in hiring criteria. The biology department's hires are affected by computer science, the physics department's hires by history. Already it is not strictly true that departments hire "the best candidate." Departments look for the best candidate in a certain area, or the best candidate that several areas can agree upon, or even the best candidate in an area not planned to be filled that year, if the candidate is sufficiently outstanding that a case can be made to the dean. Candidates are judged by multidimensional criteria, and rarely is one candidate best by all measures.

The programs above add another dimension to the picture. They don't answer all questions, and they certainly don't solve the problem for small, isolated colleges. But these suggestions can stimulate deans and chairs into searching for solutions. Viewing the problem of the two-career academic couple as the problem of the individuals involved is too narrow, as this complication affects so many women scientists.

We should get across to our women students—and their husbands—that being a scientist does not mean forswearing other parts of life. Being a scientist may be complicated and require compromise, but it is also rich and rewarding. You do not have to choose between a career as a serious scientist and marriage. It is possible to have both.

Lifestyle Discussions

After the article above appeared in the AWM *Newsletter*, an article by Mary Beth Ruskai pointed out that single people also have problems finding appropriate employment. Ruskai writes,

> The limited statistics available actually suggest that single women have even more difficulty[55] with some aspects of career advancement! For example, one study[56] looked at the percentage of 1970–74 Ph.D.'s who were tenured by 1979. The results were 66% for married men with children, 53% for single men, 51% for married men without children, 51% for married women with children, 41% for married women without children, and only 37% for single women without children.

Ruskai concludes by suggesting that real progress would require career opportunities to be available "irrespective of sex, marital status, family obligations, race, ethnicity, etc."[57] A letter to the editor, also written in response to Landau's article, pointed out that "the group most heavily impacted by the problem [is] lesbians or gay men in committed relationships with other academics."[58] In response to these comments and others, Landau explained the parameters under which her article was written and continued: "But having said that . . . it is sometimes very easy to have blinders on, and I apologize for my narrowness of vision."[59]

55. Lynne Billard, "The Past, Present and Future of Academic Women in the Mathematical Sciences," AMS *Notices* 38(1991), 707–714; footnote in original.

56. Nancy F. Ahern and Elizabeth L. Scott, *Career Outcomes in a Matched Sample of Men and Women Ph.D.'s: An Analytical Report* (Washington, D.C.: National Academy Press, 1981), as cited in Billard; footnote in original.

57. Mary Beth Ruskai, "Myths about the Role of Marital Status in Career Advancement," AWM *Newsletter* 24(3), 1994, 9–11.

58. James E. Humphreys, "Letter to the Editor," AWM *Newsletter* 24(3), 1994, 7–8.

59. Susan Landau, "Letter to the Editor," AWM *Newsletter* 24(4), 1994, 6.

The Two-City Existence

JUDY GREEN

Based on a talk at the panel "Choosing Our Lives," moderated by Judith Roitman, 1977 JMM, as reported in the AWM *Newsletter* 7(4), 1977, 2–3. Green was then an assistant professor of mathematics, Rutgers University, Camden. For seventeen years, she commuted on a weekly basis from her home in the suburbs of Washington, D.C., to New Jersey. Now she is a professor of mathematics at Marymount University.

I chose a two-city existence when I couldn't get a job in the city where my husband had one. He is employed in a large metropolitan area, but there wasn't an academic job there for me when I got my degree. So I applied all over the country, mainly trying to get a job within commuting distance, and I am now commuting 150 miles. I leave Monday mornings and come back Friday afternoons. The department has been nice to me by arranging my schedule so my first class on Monday is not early and my last class on Friday is not late.

I made the choice to commute because I knew I would be unhappy unemployed. I knew that about myself—if I was not going to school, I was teaching. I was very happy teaching, and I wanted to continue in the academic profession.

The actual mechanics of who takes care of the kids and similar things are somewhat unusual, since my husband is the one who takes care of them. He has an academic job with a low teaching load, so we're lucky. The first year, when the kids were young, we had babysitters come in after school and on evenings when my husband wanted to go out. The second year their standard babysitters happened to be going off to college, and they didn't want to break in new babysitters, so they said they would have babysitters only in the evenings. By the third year, when the youngest one was seven and the older one was nine and a half, they said: no more babysitters. So they don't have babysitters. They're very independent children because of this, and I think it's good for them.

However, I think the mechanics aren't really relevant. I also don't think that the fact that my husband is in academic life is what makes it work, although it does make it easier. What I think was more important was how I reacted to other people. A lot of people thought what I was doing was just awful. I had young children when I started, and there was much how-can-you-do-that-to-the-children type of thing. I insisted that relatives not be openly critical in front of the children—that they should be supportive because it would not work unless they were supportive, and I wanted it to work.

I'm very happy with the arrangement. I like my job very much, and I wouldn't leave it to take a job near my husband's employer unless I liked it as much. I would do this again; I would probably even do it with a longer commute.

Spatial Separation in Family Life

MARIAN BOYKAN POUR-EL

Based on "Spatial Separation in Family Life: A Mathematician's Choice," in *Mathematics Tomorrow*, Lynn Arthur Steen, ed., © 1981 Springer-Verlag New York Inc., reprinted by permission in the AWM *Newsletter* 12(2), 1982, 4–9. Pour-El was then a professor of mathematics, University of Minnesota, where she is now a professor emerita.

For approximately twenty-three years, my husband and I have alternated periods of "living apart" with "living together" in order to pursue our careers. During that time we raised a daughter from a two-year-old toddler to a twenty-five-year-old college graduate. Contrary to the prevailing attitude, our family believes that our life was richer for having lived apart than it would have been had we lived together continuously. Furthermore our experiences indicate that continuous physical nearness need not always be a primary consideration in family life. If a mother's love for her grown children can extend undiminished over the miles, why cannot other aspects of family life survive and flourish under similar conditions?

Our lifestyle evolved slowly, beginning as a temporary response to our professional needs. Although we had already committed ourselves to a two-career family before marriage, we never considered the possibility that we would not live together continuously. Our lifestyle, once begun, progressed on its own as we gained experience and understanding. Eventually it included four types of arrangements: long-term separation, short-term separation, commuting, and living together continuously. It all began almost by chance in 1958, shortly after I obtained my Ph.D. from Harvard University. I received an offer of a tenure-track assistant professorship at Pennsylvania State University. As Penn State had some very fine mathematical logicians and no nepotism rule, I could not refuse. My husband, who had not quite completed his Ph.D., would remain in Berkeley. My daughter and I would move to Pennsylvania. Except for one brief visit of three weeks, we would live apart for an academic year, a long-term separation.

We planned our move and time apart very carefully, and with particular concern for our two-and-one-half-year-old daughter. It all seemed safe and assured, if Spartan. Unfortunately, Murphy's Law took over; my daughter contracted bronchitis. Although she managed to catnap between bouts of coughing day and night, I could not. I was all alone, teaching, doing research, homemaking, and taking care of a sick child, with little sleep. The first attempt at our lifestyle seemed doomed to failure. However we had journeyed too far to turn back. After I found an adequate babysitter, no easy task in central Pennsylvania, life could once again proceed on a more even keel.

My husband's single visit during that academic year was awaited with great anticipation. When he arrived, the interlude of togetherness was more than we had hoped for. The bond between us had strengthened, our discussions were deeper and more meaningful. We parted with sorrow, but also with joy at the discovery of our strengthened ties. This was a glimpse of a new lifestyle: the yearning when apart, the delight when together.

My husband returned as planned in the summer with his Ph.D. and a position at Penn State. We remained together in Pennsylvania for three years.

The next separation was short-term, begun when I received a fellowship to the Institute for Advanced Study in Princeton. Since the Institute was associated in my mind with Einstein, Gödel, and other mathematical luminaries, I could not refuse the offer. My husband, Akiva, would remain in Pennsylvania, and my daughter Ina and I would live in New Jersey. My husband would visit us on weekends.

Again we planned carefully. Finding babysitters for Ina for the after-school hours, now that she was six, would be easy in Princeton. The weekends during which Akiva visited were very satisfying. All of our activities were family affairs, none of the "you do this while I do that" arrangement. The intervals of separation provided enough stimulus so that strong bonds were forged during the periods of togetherness.

Our commuting experience was initiated by Akiva a year later. During my second year at the Institute, he developed a severe back ailment. After being treated to no avail by several doctors in central Pennsylvania, he came east to join me and to secure adequate medical service. When he accepted a position at the Veteran's Hospital in Philadelphia, he began commuting. Although we found this to be a pleasant experience, it did not have the positive effect of even our earlier short-term separation.

While in Princeton, we began to search for a place where we could live together continuously. We still had not realized that the patterns of alternate separation and togetherness were to continue throughout our lives. But we had already become quite experienced, both mentally and physically, in undertaking such an arrangement.

More than twenty years ago we began this lifestyle. Today I am again amused when told that what we did is incompatible with successful family life. Ours is richer for having lived apart. It is my hope that those who are beginning to practice this lifestyle, perhaps with considerable trepidation, will be somewhat reassured by our experiences.

Tenure Track, Mommy Track

SUSAN LANDAU

Based on "Tenure Track, Mommy Track," AWM *Newsletter* 21(3), 1991, 12–14. At the time she wrote the article in 1988, Landau was an assistant professor of computer science, Wesleyan College; when it appeared, she was a research associate professor of computer science, University of Massachusetts, Amherst. Now she is a senior staff engineer for Sun Microsystems, Inc.

My husband and I married while I was a graduate student in computer science at MIT. "Don't have children until you finish," cautioned a friend, the wife of a history professor. I nodded easily. I was then twenty-five. At twenty-eight I completed my doctoral thesis. "Don't have children until you get tenure," warned a member of the faculty. I was leaving to become an assistant professor at Wesleyan University. This time the nod didn't come so easily. Tenure is typically a seven-year process, and my husband and I wanted a family. I didn't want to wait until I was thirty-five to begin one.

Choosing which came first was not hard for me. The security of tenure was important, but children were more so. If I had tenure at thirty-five, but was then unable to have children, the pain would have been unbearable. I knew I could handle the opposite situation. I had my first child at thirty-one, my second at thirty-three. At thirty-four I have my family even if I don't have academic permanence.

All along I felt that the choices were more mine than my husband's. We both raise the children, but I'm the one who's pregnant. I have the fuzzy brain for nine months; I'm the one who can't go off to conferences during the late months of pregnancy and the early months of nursing. My work suffers, my energy flags, my batteries fade. I lost about two years of research during the first five years after my Ph.D. (What I've gained is immeasurable—but not the subject of this essay.) So I get 51% of the vote. As it turns out, we both voted for children first, tenure second, so it was no contest. But there's a price I may yet pay in my career.

In the new professional world of recent years, many women face the hard choice between career and family. That decision is particularly sharply etched in academia: the typical Ph.D. recipient requires five to seven years of graduate study, and a tenure decision generally comes seven years after that. The years between the degree and tenure are the years one proves oneself: as a scholar, a professional colleague, a teacher. They are not the years for distractions, the languor of pregnancy, the time-consuming demands of infants and young children.

"Are you a serious scholar?" says academe. "Publish (or perish). Lecture. Go to conferences. Are you a concerned professor? Advise students. Serve on university committees. Establish yourself as a teacher and a researcher."

Tenure is a seal of approval, the university's vote of confidence in a professor's abilities and direction. Having tenure, a scholar can take the long view and tackle problems that may take years to come to fruition. Those first years after the Ph.D. are crucial for developing momentum and establishing one's professional reputation. It's also the time many of us want to have children.

I chose to—and was lucky. I didn't know I'd be in a state of torpor for nine months of pregnancy, but I also didn't expect the burst of creative energy that followed the birth of each child. That energy more than made up for those lost nine months. Each academic mother has a different experience, but all of us face the ticking of those simultaneous clocks of tenure and the childbearing years.

Academia doesn't help us. Few universities have maternity leave, and those that do ignore what happens next. For example, my university has an excellent maternity policy (one semester's leave at two-thirds salary), but no day care facilities, despite over a decade of lobbying by male and female faculty.[60] Thus my kids' center is forty-five minutes away. I can't attend late afternoon colloquia or faculty meetings. Last year my husband and I were both invited to spend our sabbaticals at a university where we would have great research opportunities. Lack of day care there meant we couldn't go.

American business and industry aren't much different. In general, maternity leaves are inadequate and on-site day care is rare.

Fifteen years of affirmative action haven't substantially improved things. To talk about women in academia is to talk about tenths: nationally one-tenth of all full professors are women, less than one-tenth of the tenured faculty at the most prestigious institutions (the Ivies, MIT, Stanford) are women, only one-tenth of the current Ph.D. recipients in science are women. It is hard to hire women—there are so few qualified—but then the universities do little to keep us.

---

60. On-site day care has been available at Wesleyan since 1989, when the university entered into an ongoing relationship with the Neighborhood Preschool to run the program in facilities provided by Wesleyan.

The lack of women sends a discouraging message to our brightest students, male and female. Few women go on to pursue graduate degrees, and fewer still to teach. This percolation effect extends down the line, and at less prestigious institutions, there is a similar lack of women. By example, or lack thereof, universities and colleges are telling their students that women do not succeed as scholars. We are effectively eliminating half the research talent this nation has to offer.

I didn't meet a female mathematician until I was twenty-five. Despite the smoke signals, I was convinced women could be mathematicians. (I can thank a sixth-grade math teacher, male, for that.) When I decided to become a professor, it was because I loved mathematics. I wasn't married, wasn't thinking of children or timing or any of the issues that are now so crucial. Had I been, my decision might have been different.

There's a touch of the priesthood in the academic world, a sense that a scholar should not be distracted by the mundane tasks of day-to-day living. I used to have great stretches of time to work. Now I have research thoughts while making peanut butter and jelly sandwiches. For sure it's impossible to write down ideas while reading *Curious George* to a two-year-old. On the other hand, as my husband was leaving graduate school for his first job, his thesis advisor told him, "You may wonder how a professor gets any research done when one has to teach, advise students, serve on committees, referee papers, write letters of recommendation, and interview prospective faculty. Well, I take long showers."

The tenure process was established in an era when men had professions and women had babies. Women now have professions as well as babies, but the academic world hasn't changed. My two maternity leaves in two years seemed like a lot to several of my colleagues. ("You shouldn't vote on this," complained one, "you're never here.") I see it as two maternity leaves over a lifetime. Even if a faculty member chooses to work half-time for ten years, that still leaves thirty years for full-time scholarship and teaching.

There are any number of complex reasons why women have not reached the top echelons in a variety of sectors. This is a simple, avoidable one. Fellowships, maternity leaves, on-site day care can make a huge difference. Universities can afford to be farsighted; they should be leading society on this one. As long as they make it difficult for us to be professors and mothers, they are engaging in a policy that effectively keeps a significant segment of women off the faculty.

Small changes can make a great deal of difference. Universities have flexibility, and they can use it without sacrificing standards. A few have adopted a "stop-the-clock" policy: if a woman takes time out—a semester, a year—for maternity leave, the tenure clock is set back that semester or year. Others allow a temporarily reduced teaching load, but at a reduced salary. This allows faculty members to concentrate on research and babies at a crucial time. Some fellowships exist that free women from teaching duties.

These solutions are not without problems. A delay on tenure creates pressure because it extends the probationary period. Colleagues who are sympathetic to lowered teaching loads because of professional commitments often look askance at those who request it for personal reasons. Many untenured women cannot afford to risk the option. Fellowships are few and far between. But these changes are a start.

They helped me. My university's generous maternity policy gave me time after childbirth to catch up on the research that I had been unable to do while pregnant, and a government fellowship has given me more time during the years when my children are young. I am one of the lucky ones. Many women are not, and they leave—or don't enter—academe. Solutions cost money. So does the lack of solutions, but this doesn't show up on the universities' balance sheets. Instead, we, as a nation, are paying with a growing shortage of scholars and researchers.

IV. CELEBRATION

The well-established mathematicians in part IV portray doing mathematics—the excitement of creating new mathematics—in the context of living a life and having a career. The research process is illuminated by their vibrant discussions of theorems or technical results. Many aspects of women's careers in mathematics are examined. These talks were plenary addresses delivered to a general conference audience of women and men mathematicians at all career stages, with an intense focus on the doctoral students and recent Ph.D.'s (together, the Workshoppers) attending.

These papers are part of the archival record of the Olga Taussky Todd Celebration of Careers for Women in Mathematics; see also III.2 and III.3. The Celebration follows the precedent of earlier conferences loosely organized around the theme of a woman mathematician's work; previous conferences honored Emmy Noether, Sonia Kovalevsky, and Julia Robinson. Historical information to put the life and work of the namesake in perspective and mathematical talks to showcase the work of active researchers (not necessarily in the same field as the woman honored) are important components of these meetings.

In this part, other conference activities are also described, along with career suggestions gleaned by Workshoppers. The spirit of the event is conveyed in pictures included in the "*Complexities* Photo Album." Expository papers by Christa Binder and Helene Shapiro on the work of Taussky Todd bracketed the plenary talks at the conference. Binder's account concentrates on early work before Taussky Todd's forced emigration from Vienna, while Shapiro focuses on the mathematics of her later years in California as a much-loved advisor of doctoral students. Chandler Davis's memoir in I.1 proceeds from her early days and flight from Austria in 1934, on to war work in England, then to employment at the U.S. National Bureau of Standards and finally her lengthy return to academia. Taussky Todd's story illustrates in sharp relief how the political and social environment in which a woman lives may affect her career as a mathematician. Her ability to turn a situation to her advantage, to move in mathematical directions based on circumstance, may presage the directions of future careers in mathematics for both women and men.

Cathleen Morawetz tells of her early acquaintanceship with Taussky Todd as she recounts the beginnings of her own mathematical investigations that led to an illustrious career. Earlier, Morawetz's work and that of Taussky Todd had been tied to the World War II effort. Later, in 1968,

Morawetz interviewed Taussky Todd and numerous other women for an AMS committee survey. The results showed that few of the women interviewed had ladder professorial academic positions; Morawetz did, but Taussky Todd did not until age sixty-five.

During the summers of her undergraduate years, plenary speaker Evelyn Boyd Granville also did war work. In 1949, she became the second African-American woman known to have been awarded a Ph.D. in mathematics. She went to industry a few years after receiving her Yale degree; she had found her academic employment opportunities limited, but prospects for minority mathematicians in government and private industry had, in her words, "evolved as a result of the personnel needs dictated by the war." In midcareer at a time of greater opportunity, Granville came back to academe and turned her energy to mathematics education, continuing this involvement during her retirement. She gives her perspective on the current educational crisis in the United States and urges her fellow mathematicians to raise their voices and be heard in the ongoing debate on this issue.

Part IV concludes with three midcareer investigators describing applications problems that arose in different employment sectors. Fern Hunt has spent most of her career in government labs; Diane Lambert works in the communications industry; Linda Petzold is a professor. Hunt, who has an impressive body of work in mathematical biology, writes about her recent work in materials science at the National Institute of Standards and Technology, a shift in research focus to meet the needs of her employer. Lambert, at Lucent Technologies, explains some of the statistics used in her research. Petzold blends problems from her previous experience in industrial settings into her research in academe, where she works in a college of engineering. Each woman discusses her career pathway in the context of her research. Petzold's exhortation to the beginning mathematicians at the conference is a fitting chapter's end for all who find inspiration in these pages: "Go for it."

Problems, Including Mathematical Problems, from My Early Years

CATHLEEN SYNGE MORAWETZ

Morawetz is a professor emerita of mathematics, Courant Institute of Mathematical Sciences, New York University.

It seems impossible to recall when I first met Olga Taussky-Todd. She was an old Göttingen friend of both Courant and Friedrichs, and she spent a year with John at the fledgling Courant Institute in the early 1950s, but I feel sure I met her before then. I also heard about her from my parents, who had known Olga's Irishman, John Todd, for a long time.

I recall two conversations with Olga. In the first one, very long ago, I asked Olga about her war work on flutter problems (flutter is caused by the resonance interaction between an airfoil and the gas flowing past it). But she was no longer interested in the subject, and my interest was also mostly gone. By then, algebra was her main concern, and her great arena.

Our later talk took place around 1968. I had just been appointed to the AMS Committee on Women. I used a trip west to ask all the women mathematicians I knew who were roughly my age about their careers and positions. It was a sobering mission as none of the women, except the statistician Betty Scott, had the standard academic job I had.

Olga insisted our talk was off the record, and so it shall essentially remain. However, it was an opportunity for her to put away her wonderful smile and air her complaints. Her greatest difficulties had come from being both Jewish and a woman. Her early year in Bryn Mawr had been difficult, and not having a regular position at Caltech rankled within her. But her beloved work in mathematics saved her.

I shall talk now about my own early years in mathematics. I worked on my first scientific problem during 1943–1944. I was measuring the muzzle velocity of shells for the Inspection Board of the United Kingdom and Canada. I was there because I had wanted to be a radar officer in the Navy, as my male fellow students became, but I was told I would have to start with boot-training, scrubbing floors, etc. (That made me angry!)

I found an error in a graph that was used whenever the weather was bad. Fixing it up was great fun. I will not bore you with the details, but it showed me that a little bit of undergraduate mechanics could get you something useful. Well, useful, although maybe the old graph had pragmatic information that I had not used. I owe it to my boss, Malcolm Macphail, a very able physicist, for making me write it up.

At the Inspection Board, I had my first exposure to visible prejudice—against my French-Canadian coworkers. So I retained a lifelong sympathy for their nationalist reaction as I did for my parents' Irish nationalism, that is, nationalism tempered by understanding and goodwill.

My second problem in mathematics was a master's thesis at MIT in elasticity. It was in fact a rather unmanageable problem in flutter theory. I did not really understand it or find it interesting. I am glad to see that master's theses in math are mostly replaced today by learning more mathematics.

The atmosphere at MIT, though essentially male, was not at all hostile to women. I was there because Caltech took no women and Harvard only put them through via Radcliffe, which I found offensive. However, I did find at MIT, as I have found almost everywhere, men whose attitude toward professional women is best told through a story of today:

A few weeks ago I was traveling in a remote part in the west of Ireland, where I met a country man. He asked me if it was true that I "was in the higher education." I replied that I was a professor of mathematics, to which he responded, "It would not do a man to make some kind of failure and you around."

On to my third mathematical problem. I married and moved to New York in 1946 to the group working with Courant, Friedrichs, and Stoker. They had worked together on military problems during the war. For better or worse, since the time of Archimedes's catapult, the military has been the source of very interesting problems in mechanics, and I was engaged to work for and be supported by an ONR contract. The job, shades of Olga's past at Göttingen, was to fix the English and not tamper with the mathematics of Courant and Friedrichs's great book, *Supersonic Flow and Shock Waves*,[1] which was being made out of a previously classified set of notes, written during the war. This had been, I now think, the greatest wartime contribution of that group. Except for another year at MIT, I spent the rest of my career at NYU as Courant's group evolved.

Friedrichs gave me, from time to time, spin-off problems, but mostly I edited. I learned fluid dynamics, at the same time complaining about the obscurity of Friedrichs's writing and the shortcuts Courant introduced to make the book more readable.

My first "published" contribution was an appendix to a paper using stationary phase. I learned stationary phase that way. For me, doing, not reading, has been the best way to learn. The paper was by Friedrichs and Hans Lewy on flow near a dock [1]. My job was to find the *second* term in an asymptotic expansion. The first was of course found by Friedrichs and Lewy, and, I might add, no estimate of the error was made by me or anyone else.

---

1. Published first by Interscience in 1948 and reissued by Springer-Verlag 1977.

I found that I had some skill and got great pleasure out of the manipulation of the necessary formulas. Friedrichs urged me to be a coauthor but out of some perverse vanity I declined. This was 1948, and the paper was published in one of the first issues of *Communications*.[2] I was duly credited in the text.

By then I had one child. Courant had a big ONR grant. He generously allowed me to work part-time. No timetable was mentioned except that I should take my orals as soon as possible, which I did. I started to work on a thesis in quantum mechanics, Friedrichs's lifelong love, but by that time I had one-and-a-half children, and Friedrichs and Courant persuaded me to convert one of my Navy problems into a thesis.

The thesis was a study of the stability of a spherical implosion under a perturbation of the initial conditions, while preserving spherical symmetry. To me, the application was to collapsing stars, and in fact the interest today would again be in the stars. The military application to igniting the first nuclear bombs was unknown to Friedrichs and to me. It was not made public until the spy trial of the Rosenbergs in 1952.

I never published my thesis because it was disappointingly incomplete. I know now that almost every thesis has unfinished business, and one should think twice before *not* publishing one's thesis. The underlying problem is a perturbation of a particular solution of Euler's equations that is spherically symmetric. Thus the equations are for u radial velocity, ρ density, and p pressure, respectively. The radial distance is r. The equations are

$$\rho_t + u\rho_r + \rho\left(u_r + \frac{2u}{r}\right) = 0 \qquad \text{Mass}$$

$$u_t + uu_r + \frac{1}{\rho}p_r = 0 \qquad \text{Momentum}$$

$$(p\rho^{-\gamma})_t + u(p\rho^{-\gamma})_r = 0 \qquad \text{Entropy}$$

The exponent γ is the ratio of the specific heats. The special solutions are of the form $u = r/tU(\eta)$ etc. with $\eta = r^{-\lambda}t$. The pictures are in figures 1a and 1b.

The problem reduces to solving a single autonomous ordinary differential equation:

$$\frac{dC^2}{dU} = \frac{B(U,C)}{A(U,C)}$$

2. *Comm. in Appl. Math.*, which became *Comm. in Pure and Appl. Math.*

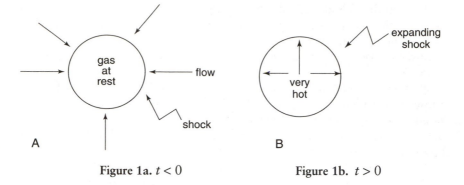

Figure 1a. $t < 0$ Figure 1b. $t > 0$

C is essentially the sound speed. There are lots of lovely singular points where $A = B = 0$. The parameter λ has to be chosen so that a solution exists which represents an implosion.

This solution was found during World War II by v. Neumann and others at Los Alamos, G. I. Taylor in Britain, and, of all things, by G. Guderley in Germany, where it was published during the war in the open literature [2]. This literature was combed by scientists after the war, and the result was described in Courant and Friedrichs's book.

To look at stability we perturb around this solution and look for complex solutions of the resulting linear equation that grow exponentially in time, like exp (k, t) with $Re\ k > 0$. I succeeded in reducing the problem to studying a second-order ordinary differential equation (o.d.e.) with real coefficients depending on η; i.e., $r^{-\lambda}t$ and the complex parameter k. It was singular where $\eta = 0$; i.e., $t = 0$, the moment of reflection. I had to prove there were no eigenvalues with $Re\ k > 0$. But k appeared in an odd non-standard way *and*, even worse, the equation was singular at $\eta = 0$ in a manner not considered in the then standard works of Langer on o.d.e. Langer's results worked for "something" less than zero. I needed equals zero. Torture! I tried hand computation on a Marchand calculator—too time-consuming and inaccurate. Later asymptotic theory for such equations was worked out, but too late for me. I never looked back. What was really wrong? (1) The problem had too many cliffhangers for a thesis. (2) I had too many duties as a mother of two. But I got my degree or, as Courant liked to say, my union card.

The next problems were similar: singular o.d.e. with a parameter. But this time they were solvable. I was at MIT working for C. C. Lin and paid for part-time from a NASA grant. I wrote my first two papers, and they were published [3; 4].

Even before this, Courant invited me to come back to his institute, again to be supported by ONR. For the next six years, I worked on an

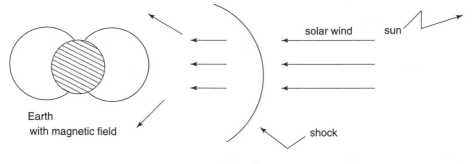

Figure 2.

assortment of problems, mostly on transonic flow. Everything worked well, more or less. And I had two more children.

Courant, around 1956 I think, proposed that I join Harold Grad's group in magneto hydrodynamics (MHD) and plasma physics. He enticed me with the prospect of helping to solve the world's energy problem through thermonuclear fusion. Still an open question to say the least! But it was a golden lode for problems, from nonlinear hyperbolic partial differential equations in MHD to statistical mechanics without collisions. I nibbled away for twelve years, but the rewards were all modest. As an example, here in figure 2 is a plasma problem, connected actually to the solar wind as well as thermonuclear fusion. An ionized gas or plasma, the solar wind flows toward earth. The magnetic field of the earth repels the wind. Because of the high speed of the wind, a shock, like the bow shock in front of a supersonic airplane, is formed between the earth and the sun. *Question*: What is the mechanism for the shock?

In gas dynamics you argue that the shock (jump in flow quantities) is smoothed out by adding viscous terms to the questions. In figure 3, a one-dimensional model, the width of the transition goes to infinity with the viscosity or mean free path. I do not know what the mechanism is for the solar wind, but for the controlled thermonuclear fusion setup for making energy it is enormous. If there is going to be a realistic shock limit, there has to be a better mechanism. I did create a crude model, but this is still today an interesting unsolved problem involving asymptotics and statistical mechanics and with lots of room for new ideas.

Aside from untreatable mathematics problems, and that's the way most problems from physics are, the problem of those years was that I was moving forward professionally only very slowly. I had successfully avoided teaching. I considered that I could not raise four children, do research, and teach, so I settled for the first two. I hoped to get a regular position when my youngest child went to school. In 1957, six years after

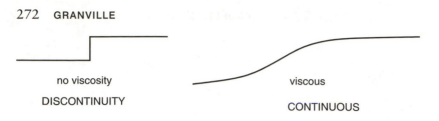

no viscosity

DISCONTINUITY

viscous

CONTINUOUS

Figure 3.

my Ph.D., Courant offered me an assistant professorship in the Institute. I accepted knowing that I might not have a second chance. So in 1957 I stopped being a hybrid, part-time research worker cum postdoc and became a regular faculty member.

Those six early years gave me a very special start. I wish something like those opportunities were more available to women—as I like to call them, those valiant women—who have to choose between having children and risking tenure, or getting tenure and risking not having children.

References

1. Kurt O. Friedrichs and Hans Lewy, "The Dock Problem," *Comm. on Applied Mathematics* 1(2), 1948, 135–148.
2. G. Guderley, "Starke kugelige und zylindrische Verdichtungstösse in der Nähe des Kugelmittelpunktes bzw. der Zylinderachse," *Luftfahrtforschung* 19(9), 1942, 302–312.
3. Cathleen S. Morawetz, "Asymptotic Solutions of the Stability Equations of a Compressible Fluid," *Journal of Math. Physics* 33(1954), 1–26.
4. ———, "The Eigenvalues of Some Stability Problems Involving Viscosity," *Journal of Rational Mechanics and Analysis* 1(1952), 579–603.

Looking Back . . . Looking Ahead

EVELYN BOYD GRANVILLE

Granville is a professor emerita of mathematics, California State University, Los Angeles. She is an educational consultant and serves on corporate boards.

For the past two years I have been traveling in the eastern part of the state of Texas visiting middle schools as part of a campaign sponsored by Dow Chemical Company to encourage more students to consider and prepare for careers in science, engineering, technology, and business. I will have more to say later about my experiences during my travels to schools in the Houston, Dallas, Beaumont, East Texas, and Shreveport, Louisiana, areas.

Ordinarily when I am introduced as a speaker, my date of birth is given. I have no objection to this because it isn't hard to figure out that I have been around for a while if you know when I finished high school or college. At one school—I believe we were in Houston—a question was posed to me in the question and answer period that let me know that anyone born in the 1920s is very, very old in the eyes of a youngster. The question was—"Did you have to study by candlelight?" I suspect in the eyes of that young boy I had to be in the category of a great-grandmother, born long before any modern conveniences were available. I assured the youngster that my generation enjoyed not only electricity, but also indoor plumbing—at least in big cities.

So I will tell you that I was born on May 1, 1924, in the nation's capital, Washington, D.C., and I received my early education in the public schools of that city. The current educational system of the capital of this nation has earned the reputation of being one of the worst school systems in the country. This is a real tragedy, because the rest of the world looks to this country as the leader of the free world in so many respects.

Early this year I was interviewed by a columnist for the *Washington Post*; his main question for me was, "What was significant about the school system as it existed in your day that could produce the graduates who brought national recognition for excellence to the colored school system of the District of Columbia?" When I was growing up in the nation's capital there were two school systems—one for white students and one for colored students. Although the systems were separate, the colored system was in no way inferior to its counterpart. The system achieved a national reputation for excellence because teachers and administrators were well trained in their subject areas and were dedicated to providing the kind of education that students needed to be able to compete in a larger community. Job opportunities for persons of color were limited, and teaching was one of the few well-paying professions open to Negroes. As a result, our schools attracted the very best—the cream of the crop. Our teachers received their training in some of the finest institutions of higher learning in this country. Among the many outstanding teachers I had were two great math teachers, one a graduate of the University of Pennsylvania, the other a graduate of Yale University. Although schools were segregated, many public institutions were open to us. The city public libraries, the U.S. Library of Congress, federal museums and art galleries, and other national institutions provided us with the means to expand our knowledge beyond the classroom.

My generation benefited by being brought up in a culture that prized education and stressed education as a means of improving the quality of one's life. We studied our past history; we learned about the difficulties encountered by others who preceded us when they sought to become educated. We were made to understand why slaves were forbidden to learn to read and write—it was impressed upon us that knowledge is power, and to be truly free you must gain knowledge. We listened and we learned. And we succeeded. I described to the newspaper columnist the educational climate in the nation's capital as it existed when I was getting my early education, and ended by expressing my dismay that too many students today either are not told the message or choose to tune out when the message is presented to them. How can we turn this around? More about this later when we look ahead. Right now, I am still looking back.

Washington, D.C., had three high schools for colored students, Cardoza High for business students, Armstrong High for those who wanted technical training, and Dunbar High for college-bound students. Our teachers knew that we would have to compete with students trained in the finest public and private high schools in the country, so our curriculum was designed to provide us with the academic preparation required by the best institutions of higher learning. It included three years of mathematics, history, English literature, composition, Latin, a modern foreign language, biology, chemistry, physics, and another course that I treasure to this day. We were required to take a course in formal grammar; this course has benefited me all my life. It was a tradition to encourage Dunbar students to attend Ivy League colleges in the Northeast. Boys went to Yale, Harvard, Dartmouth, and Amherst; girls went to Vassar, Mount Holyoke, Wellesley, Radcliffe, and Smith. I was encouraged to consider attending one of these schools, even though by today's standards my family would have been considered low-income, far from middle-class status. My mother worked in a nonprofessional job as a stamp examiner at the U.S. Bureau of Engraving and Printing; my father held a variety of jobs in his lifetime, all nonprofessional. However, my mother supported my applications to Smith and Mount Holyoke, and, although I did not receive scholarship aid for my freshman year, my mother with financial help from her sister sent me off to Smith College. I chose Smith over Mount Holyoke due to the encouragement of two alumnae, Otelia Cromwell, a colored graduate of Smith in 1900, and her niece, Adelaide Cromwell, Smith 1940.

I entered Smith in the fall of 1941. Often I am asked how I felt to be one of fewer than five or six colored students on a campus of 2000. I never encountered racism from either students or faculty. I never felt different because of my race, my public school background, or the fact that in comparison to most of the students I was definitely financially disadvantaged. Academically I was on a par with other students, and I was able to compete. Dunbar High School had prepared me well, and that was all that mattered.

The United States entered World War II on December 7, 1941, a few months after I started college. This war brought about changes that affected me personally, and changes that brought new freedoms and new opportunities to people of color. At the start of the war the armed services were segregated; by the end of the war all services were integrated. At the start of the war people of color had only limited career choices; by the end of the war there were expanded opportunities for employment in many fields, especially for persons trained in engineering, technology, and the sciences. During the summers of 1942, 1943, and 1944, I was employed at the National Bureau of Standards, where I helped engineers with computations related to defense projects. We did not have modern-day electronic computers; we had a few electric calculators, Marchands and Fridens. Often I had to use a hand-cranked mechanical calculator when others beat me to the electric ones. But I was happy to have this work because it meant funds toward expenses for the next school year. I lived in a co-op house after my first year at Smith. Students who lived in co-op houses received a deduction on room and board because they did work in the dorm in place of hired help. When the war started most of the hired help left for jobs in defense industries and, as a result, hired help was in short supply, and all students had chores to perform in the dorms.

World War II created new opportunities for women and minorities. We hate to talk about benefits that accrued to women and minorities as a result of such a horrible, devastating war, but the aftermath cannot be ignored, because it is a fact of life. At the beginning of the war very few government agencies in this country hired minorities in significant numbers, and certainly not in professional positions. By the end of the war minorities could be found in professional positions with important decision-making responsibilities. It is a sad fact that it took a devastating war to provide people of color with equal opportunities.

When I entered Smith I had planned to major in French and become a French teacher. However, I soon realized my real talents were in mathematics and science. I concentrated my studies in math, physics, and astronomy. I was fascinated by the study of astronomy. To think that from the surface of the earth we could discern what was happening and what had happened millions of miles away in the universe was mind-boggling. I toyed with the idea of picking astronomy as my major but abandoned the idea when I contemplated the few jobs available in the field and the isolated places where astronomers worked. Had I known then that astronomers would be in great demand for space exploration in the future, my decision might have been different. But this was in the early '40s, when the only space travel we read about was in the exploits of Flash Gordon, a comic strip character.

I graduated from Smith in 1945 and received scholarships from both Smith and Yale to begin graduate study. I had applied to both Yale and Michigan but chose Yale because of the financial help I received. I did not know that another African-American woman was attending the University of Michigan about the same time and that we would become the second and third African-American women to receive doctorates in mathematics in 1949. Unfortunately, I never had the pleasure of meeting Marjorie Lee Browne.

My Yale experience was a fulfilling one. Again, I experienced no problems because of my color. I was treated by students and faculty as just another graduate student. I was fortunate to receive scholarship and fellowship aid throughout my four years at Yale. In 1949 I was awarded the doctorate in mathematics with Einar Hille, a former president of the AMS, as my advisor.

The prospect of living in New York City and having the theater, art galleries and a NYC lifestyle was alluring, and so I accepted a position as research assistant at the NYU Institute for the Mathematical Sciences, now called the Courant Institute of Mathematical Sciences. During the year I spent at the Institute with scholars such as Richard Courant, Fritz John, Peter Lax, and Cathleen Morawetz, I gained much more academically than I could possibly contribute to this outstanding mathematical community. On the personal side I found the city to be much less appealing than I had imagined. Good housing was scarce and expensive; New Yorkers were abrupt and unfriendly; the pace was too fast for me. I quickly lost my enthusiasm for living in this city.

So when I received an offer to teach at Fisk University in Nashville, Tennessee, I readily seized the opportunity to get out of New York. I had never traveled south of Virginia, and I welcomed the opportunity to see what life was like on a campus in the South, even though segregation was still the way of life in the southern states. Fisk was an oasis in the midst of a very unfriendly environment. Our lives were centered on the campus, and as long as we stayed on the campus, we were shielded from southern attitudes and mores. Nevertheless, I enjoyed my two years at Fisk and made some lifelong friends there. Lee Lorch was one of them. Two young ladies who took classes from me went on to receive doctorates in mathematics—Etta Zuber Falconer and Vivienne Malone Mayes. I credit Lorch with inspiring them to pursue advanced study. Yet both have said that my having the advanced degree showed them that it could be done.

Since World War II opened up jobs in government and private industry for women and minorities, in 1952 I accepted a position at the National Bureau of Standards in Washington, D.C., in a division that later split off to become the Diamond Ordnance Fuze Laboratories of the Department of the Army. I left academia and started on a career in government and private industry that I could not have envisioned when I left Dunbar High School as a shy seventeen-year-old. I had seen professional women solely in the position of teachers; these were my role models. Now, I began to see women as mathematicians, engineers, chemists, biologists, computer programmers. A different world had evolved as a result of the personnel needs dictated by the war and, subsequently, by the cold war and the growth of the electronic computer industry.

After four years in government, I joined IBM in 1956 and was introduced to the world of computer technology. When IBM contracted with NASA to produce computer programs for the U.S. space program, I was assigned to the IBM Space Computing Center to join a group of IBM mathematicians and scientists working with NASA scientists to formulate calculations for computer programs to track vehicles in space. When I left IBM in 1960 to move to the West Coast, I continued to work in the field of trajectory analysis, first at Space Technology Laboratories and later at North American Aviation. I returned to IBM in 1964 to join the Federal Systems Division to work on a variety of projects for defense contractors. If you think that I made a lot of moves in a short span of time, you are absolutely correct. The demand for mathematicians and scientists was high, and it was not unusual for a former colleague or a friend to lure you to his or her company by dangling an offer of more interesting projects, greater benefits, and/or higher pay.

In 1967 a request from IBM to transfer me back to the East Coast led me to look for a more permanent position. Therefore, I sought a teaching position and was hired at California State University, Los Angeles (CSULA), where I taught for seventeen years before retiring in 1984. When I started teaching there, the so-called new math curriculum had already appeared in many schools across the country, and the debate about the value of this new approach had already begun. After teaching a class for prospective elementary teachers and using the new ideas in the class, I was won over to this new approach. I developed a great interest in standards for the mathematics curriculum for the elementary grades and standards for the preparation of mathematics teachers, an interest that continues to this day.

I moved to Texas when I retired from CSULA. My husband is a native of Texas, and his desire was to return to his home state for retirement. It wasn't hard to get me to agree to the move. We were both tired of the Los Angeles smog, traffic, high cost of living, and the burgeoning population. A home on sixteen acres in East Texas sounded like "heaven on earth." I enjoyed retirement for a few months and then started looking for something to do.

My first postretirement job was a disaster! I must tell about this experience. After learning of my background in mathematics and computer programming, a trustee of the school district in which we resided invited me to join the faculty at the junior high school to teach a newly mandated computer literacy class for eighth graders. It sounded like this would be a lot of fun when I accepted the offer. I soon learned a lot about eighth graders: they are not disciplined college students, and it takes a different type of skill to keep them in line. I quickly learned that I was out of my league. Before the fall semester ended, I parted company with the school district with the blessings of the district superintendent and school administrators. I vowed never to try public school teaching again.

Following this debacle I taught math and computer science at Texas College in nearby Tyler, Texas, for three years before deciding in 1988 to really give retirement a try. I managed to stay retired for two whole years before accepting a position as visiting professor at the University of Texas at Tyler in 1990. I remained there for seven years, before retiring for the third time in 1997.

When I retired again, everyone wanted to know what I would do next. I had no definite idea, though I knew I would certainly have to find something stimulating to do. The problem was solved when I received an invitation to participate in the Dow Chemical campaign. This company, like other private industries and government agencies, is concerned that our schools are not producing enough competent, well-trained people to fill the jobs that exist today and jobs that will be created in the future.

I was recruited to bring the message that the study of mathematics provides students with the skills needed to pursue these careers. I tell students what employers look for when they are hiring. They want workers with skills in oral and written communication, facility with mathematics, and knowledge of a variety of problem-solving strategies. My message to the students is—"You are our future"; "Private industries and government agencies need you"; "Math opens doors." I urge the students first to master algebra and geometry and then to include advanced math topics as they continue their studies. I cite a few horror stories of what happens to students who avoid the study of algebra beyond the rudimentary first course.

As I talk with the students, there is a concern in the back of my mind that I do not express to my audiences. Do we have in the elementary, intermediate, middle, and senior high schools of this country enough teachers adequately trained in mathematics to guarantee that every child has access to a quality education in mathematics? The answer is *no*. Over the past seven years I have participated in several Eisenhower workshops offered to improve the teaching of mathematics at the elementary and middle school levels. In these workshops math concepts for the school curriculum are reviewed, and the participants (all classroom teachers, some with considerable years of service) are introduced to manipulatives and computer technology that will aid the teacher in presenting the concepts in the classroom. It is quite evident that adequate training in mathematics was not provided when these teachers were being trained for the profession. The participants acknowledge that their preparation has been poor, and they are eager to learn and to take back to other teachers what they have gained from this experience. But only a small fraction of the teachers who need this professional development can be reached by programs like the Eisenhower program.

Every week you can find an article in a newspaper, a weekly magazine, or a professional journal about the state of public education in this country. Syndicated columnists, news reporters, politicians, professional educators, and laymen report on flaws in the system and give their advice on how to correct the situation. You have heard the suggestions. The answer lies in home schooling, or vouchers, or charter schools, or smaller classes; some even propose that we junk the whole public school system and privatize the system. I look back to my own education in the segregated, colored public school system of the District of Columbia, and I know why I was able to compete and succeed in an environment that did not want to see persons of my skin tone succeed. I was privileged to have competent, well-trained teachers who wanted to teach, who appreciated the value of a good education, and who cared enough to see that their students got the education they needed to go as far as their talents could take them.

Can we produce teachers of this caliber today? I think we can. Dow Chemical Company wants students to get the training in mathematics that will prepare them to qualify for demanding jobs in private industry and government. Can we produce competent teachers of mathematics for our public schools? I think we can.

We need to establish national standards for the math curriculum for grades K–12 and national standards for teacher preparation in mathematics. The Council of Chief State School Officers reported in 1997 that twenty-four states had no math requirement at all for prospective elementary school teachers. In many states the requirement is minimal—one course in mathematics offered in the mathematics department and one course in methods offered in the school of education. As I have visited schools over the past year and a half, I have heard algebra teachers complain that they cannot begin to teach algebra because students are weak in the fundamentals of math. Can you teach factorization of polynomials if students do not understand factorization of whole numbers? Can you teach algebraic manipulation of rational functions if students do not understand operations with fractions? We need national standards so that every child will reach the next level of education fully prepared for this level. To the list of individual rights—life, liberty, and the pursuit of happiness—we must add the right to receive a first-rate education in a first-rate public school delivered by competent, well-trained teachers. It is in our national interest that every child has the tools to survive and thrive in the technological world in which we live. With the adoption of national standards a child in the fourth grade in Watts in Los Angeles would be at the same level as a child in the fourth grade in Beverly Hills. There is a practical need for national standards. We have a very mobile population; families move often, and children transfer from one school system to another. Children should not be disadvantaged when families relocate by finding, at the same grade level, differing expectations and subject matter coverage.

So, what will happen if we upgrade preservice training and tighten standards for teacher preparation? Short term, fewer students may want to enter the teaching profession, but if teachers get higher status, the long-term effect may be to attract more well-qualified people. We cannot put inadequately trained personnel in the classroom and expect the children to get the skills needed in this high-tech world. What do we do to attract prospective teachers at all levels of public education? What can be done to prevent teachers from leaving the teaching field after serving a few years? One report states that 50% of new teachers will have left the profession at the end of five years.

We are indeed faced with a crisis. I believe the problem can be solved if we adopt reasonable steps.

- Reduce the number of classes taught each day to allow time for evaluation and lesson preparation. Limit class size.
- Provide certified teacher aides to assist with record keeping, drill, study and lunch periods, and help for individual students both in the classroom and after school.
- Set competitive starting salaries and provide generous step increases based on years of service in order to compete with job opportunities in government and private industry.
- Provide district-paid professional development classes. These may or may not be in the district, but teachers should *not* have to pay for their own continuing education out of their paychecks.
- Provide a salary plan to keep good teachers in the classroom. Veteran teachers should not have to become administrators to reach top pay.
- Extend the school year. A three-month vacation is no longer needed for students and teachers to help on the family farm. Students forget a lot over the summer, and teachers spend valuable time at the beginning of the school year reviewing the previous year's work.
- Promote partnerships between schools and private industry to provide after-school practical experience for students, mentors, and summer intern programs. The more incentives a student has to learn, the easier the job becomes for the classroom teacher.

- Provide alternative schools for disruptive students. Set zero tolerance for bad behavior.
- Provide early childhood education so that children enter first grade ready for first-grade work.
- Educate the public to the advantages of spending tax dollars to develop first-rate educational systems rather than spending tax dollars on detention homes and prisons.
- Keep politics out of the debate on how to improve the public school systems of this country.
- Tailor high schools to the needs of the students served. Offer one course of study for those who plan to attend college, another for those who prefer technical training, and a third course of study for those who want to develop business skills.

As mathematicians we must raise our voices and be heard in this debate over how best to salvage a good educational system from what we have today. We are what we are today because we were fortunate to receive a first-rate education that enabled us to earn advanced degrees. We have a duty to provide a climate in education in which every child has the opportunity to learn. We have a duty as responsible citizens to champion quality education not only in mathematics, but also in all fields. Thereby, we will be protecting ourselves and the well-being of all the citizens of this country.

Looking back, I can see a lot of changes for the good that have occurred since I first entered school at the tender age of four years. I am grateful to those whose energies and sacrifices made me what I am today. Looking ahead, I see a brighter future possible if each one of us assumes the responsibility to make this a better world where all have the chance to learn and achieve.

Olga Taussky and Class Field Theory

CHRISTA BINDER

Binder heads the research unit for history of mathematics at the Institut für Analysis und Technische Mathematik, TU, Vienna.

1. Fräulein Doktor Taußky in Vienna and Göttingen

Born August 30, 1906, in Olmütz (Moravia), Olga Taussky attended school in Vienna and Linz, where she received her *Matura* in 1925. From 1925 to 1930 she studied mathematics at the University of Vienna; her principal teachers there were Philipp Furtwängler and Wilhelm Wirtinger.

Her other teachers included Hans Hahn, Walther Mayer, Edmund Helly, and Karl Menger; she also studied chemistry and logic, the latter mainly with Moritz Schlick (*Wiener Kreis*). In the autumn of 1927 she began work on her thesis there, although she spent one semester of 1929 in Zürich, where she took courses from Speiser, Pólya, and Fueter. Taussky received her promotion to doctor of philosophy on March 7, 1930.

During the next few years, Taussky gave a number of talks at professional meetings and met many other mathematicians interested in the same subjects. On May 28, 1930, Taussky spoke on "Axiomatik des metrischen Zwischenbegriffs" (Axiomatics of the Metric Concept of Betweenness) in a seminar called the Mathematische Kolloquium run by Menger. On May 30, 1930, at a meeting of the Vienna Mathematical Society, she discussed some of her thesis work, "Über eine Verschärfung des Hauptidealsatzes" (On a Sharpening of the Principal Ideal Theorem). At a meeting of the German Mathematics Society in Königsberg, Taussky gave two talks. The first one, on September 4, 1930, was again about her stronger version of the Principal Ideal Theorem; the second talk on September 6, 1930, concerned "Eine metrische Geometrie in Gruppen" (A Metric Geometry in Groups). Helmut Hasse, David Hilbert, Emmy Noether, Arnold Scholz (from Freiburg), and other well-known mathematicians of that time also participated in this meeting.

Taussky remained in Vienna as an unpaid assistant for the academic year 1930–31. During this time (January 22, 1931), Kurt Gödel presented his famous paper "Über Vollständigkeit und Widerspruchsfreiheit" (On Completeness and Consistency), which appeared in the journal *Ergebnisse eines Math. Koll.* (There is a discussion of this paper in Menger [10].) On March 4, 1931, Taussky spoke on "Über ähnliche Abbildungen von Gruppen" (On Similar Transformations of Groups) in the Kolloquium.

At a meeting of the German Mathematics Society in Bad Elster, held September 14, 1931, Taussky gave another talk, "Zur Theorie des Klassenkörpers" (On the Theory of Class Fields). Other participants at this meeting included Ludwig Holzer, Nikolaus Hofreiter, Kurt Mahler, and Richard Courant. Olga's next talk, "Zur Axiomatik von Gruppen" (On the Axiomatics of Groups) (in the Math. Koll.) was on October 27, 1931.

Taussky spent the academic year 1931–1932 in Göttingen as coeditor, with Helmut Ulm and Wilhelm Magnus, of *Hilbert's Collected Works*, volume 1. There she met Emmy Noether, Emil Artin, Richard Courant, Hermann Weyl, Edmund Landau, and O. Veblen, among others.

Taussky, who attended the 1932 International Congress of Mathematics in Zürich, continued to have an unofficial position at the University of Vienna as assistant to Hahn, Menger, and Furtwängler. On April 27, 1934, she gave a talk, "Abstrakte Körper und Metrik. 1. Mitteilung: Endliche Mengen und Körperpotenzen" (Abstract Fields and Metric, First Part: Finite

Sets and Powers of Fields). In autumn 1934, Taussky left Vienna and Europe to take a position at Bryn Mawr College in Pennsylvania. The remainder of her academic career was spent in Great Britain and the United States.

2. Class Field Theory

We turn now to the development of class field theory.

Gauss's Results. Two hundred years ago, Carl Friedrich Gauss wrote his *Disquisitiones arithmeticae*. Although Diophantus, Fermat, Euler, and Lagrange had worked on the same topics earlier, it was this work that brought number theory to the center of mathematics. In particular we are interested in Gauss's results concerning quadratic forms and the class number.

First recall that it is exactly those odd prime numbers p which are congruent to 1 mod 4 that can be represented as $x^2 + y^2$, for some integers x and y. In order to generalize this fact, Gauss studied general quadratic forms $ax^2 + bxy + cy^2$. He defined two quadratic forms to be *equivalent* provided the coefficients of the first are the result of a transformation of the coefficients of the other by a matrix of determinant one. (The representation behavior is invariant under such transformations.) For each possible *discriminant* D (here $D := 4ac - b^2$), he tried to determine the *class number* $h(D)$, that is, the number of different equivalence classes with discriminant D. Gauss calculated all discriminants with class number 1 and found that the largest is 163. Further developments in this theory are summarized by Cohen ([1], p. 227):

> A deep theorem, due to Baker and Stark (1967), proves that there exist only 13 values of D for which $h(D) = 1$. The first 9 correspond to field discriminants, and are 3, 4, 7, 8, 11, 19, 43, 67, and 163 (or the negatives, depending on the definition of D). There are 4 more values of D corresponding to non-maximal orders: 12 and 27 (in the field $\mathbb{Q}(\sqrt{-3})$), 16 (in the field $\mathbb{Q}(\sqrt{-4})$), and 28 (in the field $\mathbb{Q}(\sqrt{-7})$).

> Although it has been known since Siegel that $h(D)$ tends to infinity with D, and even as fast as $|D|^{(1/2)-\varepsilon}$ for any $\varepsilon > 0$, this result is ineffective, and the explicit determination of all D with a given class number is difficult. I have just stated that the class number 1 problem was solved only in 1967. The class number 2 problem was solved jointly by Baker and Stark in 1969: $D = 427$ is the largest discriminant (in absolute value) with class number 2. The general class number problem was solved in principle by Goldfeld, Gross, and Zagier in 1983, who obtained an effective lower bound on $h(D)$. However, the problem still needs some cleaning up, and, to my knowledge, only class numbers 3 and 4 have been explicitly finished.

Hilbert's Contributions. One hundred years after Gauss's *Disquisitiones*—many mathematicians, definitions, theorems, conjectures, and generalizations later—Hilbert was to write his famous *Zahlbericht*, which summarized the development of number theory and presented it in a unified form. This monumental work would play a role similar to that of *Disquisitiones*, that is, it would be the basis for future students of algebraic number theory, including Taussky. Her work editing Hilbert's collected works will be described later, so a few words on the content of the *Zahlbericht* are well-deserved here. Hilbert put together everything known at that time on number fields, Galois number fields, quadratic number fields, cyclic fields, and Kummer fields. His main goal in the book was to study the reciprocity law and Dedekind Zeta functions for these fields and to give a unified exposition of the theory, including his own unpublished investigations. Hilbert's own main results appear in his later papers, primarily in "Über die Theorie der relativ-Abelschen Körper" (On the Theory of Relatively Abelian Fields), *Acta Math.* 26 (1902), 99–132. In his *Zahlbericht*, Hilbert shows how to eliminate the restriction that the field has an odd class number; he considers relatively quadratic extensions of a number field. The main point is the theory of class fields, called unramified relatively abelian fields in Hilbert's terminology. He could treat them only for field extensions of degree two, but he conjectured, with admirable foresight, the whole complex of theorems now known as the principal theorems of class field theory. He also conjectured that the results might still be valid if "relatively abelian field" were replaced by "general algebraic number field."

After Hilbert. In the intervening years, nearly all of Hilbert's conjectures have been shown to be true. In particular, Furtwängler in 1906 showed that the results hold for unramified relatively abelian fields, and then Takagi showed them for fields with arbitrary discriminant.

The main points of the theory are the definition of the class field, its connection to the ideal group (every relatively-abelian field K over k is a class field for an ideal group H of k), the existence of infinitely many prime ideals, and the reciprocity law.

The next step was that Furtwängler removed, as Hilbert suggested, the restriction to relatively abelian for prime exponents. At the same time, Artin proved the same theorems using other more elegant methods. By 1927, just one theorem remained to be proven; that was the Principal Ideal Theorem, which states that in the maximal abelian unramified class field K of an algebraic field k every ideal of k becomes a principal ideal. Finally, in 1928, Furtwängler succeeded in proving this theorem—a major triumph! But his proof was clumsy and gave few clues for real understanding.

That was the situation when Taussky started to work in this field. Her work on this topic is mentioned in Cohn's descriptions of later developments ([2], p. 178):

If $K^{(0)}$ is a field and $K^{(1)}$ its Hilbert class field, then all ideals of $K^{(0)}$ become principal in $K^{(1)}$. Yet $K^{(1)}$ need not have class number 1. The best we can do generally is to say, for example, that if the class group of $K^{(0)}$ is cyclic, then $K^{(1)}$ has either class number equal to 1 or relatively prime to $K^{(0)}$ (see Furtwängler 1916, O. Taussky 1937). Thus other nonprincipal classes may arise in the class field $K^{(1)}$. We then construct $K^{(2)}$, the Hilbert class field of $K^{(1)}$, but again $K^{(2)}$ need not be a principal ideal domain. It can be shown that $K^{(2)}$ is unramified over $K^{(0)}$ but not abelian. We might hope that on continuing the succession of a so-called "class field tower"

$$K^{(0)} \subseteq K^{(1)} \subseteq K^{(2)} \subseteq K^{(3)} \subseteq \ldots$$

we should finally achieve a $K^{(n)}$ of class number 1 (so that $K^{(n+1)} = K^{(n)}$). Unfortunately, not every such tower terminates. This was shown in 1964 by Golod and Shafarevich. The trick in obtaining an infinite tower is to have a sufficiently large number of ramified primes in $K^{(0)}$, for instance $K^{(0)}$ might be $Q(\sqrt{-2 \cdot 3 \cdot 5 \cdot 7 \cdot 11 \cdot 13})$.

More recent developments are discussed in [5], p. 30:

A central problem of field theory consists in obtaining a survey of all Galois extensions L of a given field K. Since the law by which these extensions are built up over K must be hidden solely in the inner structure of the ground field K, one wants to classify the extensions L/K by objects which are directly associated to the ground field K. Now class field theory solves this problem for the abelian extensions of K, in that it establishes a one-to-one correspondence between these extensions and certain subgroups of the group A_K.

Neukirch comments on the earlier history too, and he gives the final version of the Principal Ideal Theorem ([5], pp. 107–110):

Theorem (Principal ideal theorem): Every ideal K becomes a principal ideal in the Hilbert class field (the ray class field with modulus 1). . . .

A problem closely related to the Principal Ideal Theorem, and first raised by Furtwängler, is whether the class field tower

$$K = K_0 \subseteq K_1 \subseteq K_2 \subseteq K_3 \subseteq \ldots$$

(where each K_{i+1} is the Hilbert class field of K_i) stops after finitely many steps. A positive answer to this question would have the following interesting consequence: If $K_i = K_{i+1}$ for sufficiently large i, then K_i has class number 1, i.e. every number field K would be contained in a canonical solvable finite extension field in which all ideals are principal. This problem resisted solution for a long time.

3. Taussky: 1925 to 1934

We return to Olga Taussky and her study in Vienna. Taussky was fortunate that upon entering the university in 1925 she knew immediately who her teacher would be. She knew her main subject, and for this there was a unique choice: Philipp Furtwängler, the famous number theorist from Germany. He had started with geodesy during World War I and had studied Hilbert's work in class field theory on his own. He had proved—and in a few cases disproved—Hilbert's conjectures in this field. (He never met Hilbert personally.) Furtwängler was paralyzed and could not write on the blackboard. Nevertheless, during Taussky's first year he gave a course on elementary number theory, and, in her second year, he gave a two-hour-a-week seminar on algebraic number theory which included some of his work in class field theory. After that, Taussky asked Furtwängler for a theme for her thesis. He decided that it should be in class field theory. To be given such a difficult and important topic was a great honor for a young woman in her third year of study; this choice of topic had a major influence on her future career. She had a wide variety of literature to study, but it was difficult for Taussky to work in this field with little direct encouragement from her teacher. It turned out that Emil Artin had developed an ingenious method for translating the still unsolved Principal Ideal Theorem into a statement on finite non-abelian groups. Furtwängler told her about that, but the knowledge did not help her frustration. At about that time, Furtwängler proved that Artin's group theoretic statement was true, thus proving the principal ideal theorem—a tremendous achievement! When later this proof was labeled "ugly" by the mathematical world, Taussky rushed to defend her teacher—a first proof of a major theorem is seldom the most elegant one.

After this success, Furtwängler was ready to assign a more specific thesis problem for Taussky: apply Artin's method to a problem concerning odd prime numbers. Furtwängler had solved it for the prime number 2, but did not show his solution to Taussky, knowing that his method did not work for larger primes. After some struggle she solved it for 3 and discovered that every prime number behaves differently, depending on $p - 2$. The situation is very complicated; the non-abelian cases are, even now, not fully understood. Nowadays the number-theoretic side of the theory has less appeal than in the thirties. The problems and ideas are now part of algebraic geometry, homology, and the geometry of numbers.

Furtwängler himself changed his interest to the geometry of numbers. Since binary quadratic forms can be represented by two-dimensional lattices, with generalizations to higher dimensions, the connection to the original class number problem is obvious.

Taussky's third year was spent working hard on her thesis, taking courses and seminars, and earning some money through private tutoring. By her fourth year, 1928–1929, her thesis was nearly finished, and for the first time in many years she had time to relax. She could not take her final oral examinations yet—the earliest possibility was in the tenth semester. At the invitation of her uncle, she took advantage of this situation to spend one semester in Zürich, where she attended courses by Speiser, Fueter, M. Gut, Plancherel, and Pólya. Also, Pólya gave her good advice on how to lecture.

Back in Vienna she received her doctorate on March 7, 1930.

Shortly after getting her degree she presented her results to the Vienna Mathematics Society on May 30, 1930, "Über eine Verschärfung des Hauptidealsatzes" (On a Sharpening of the Principal Ideal Theorem). She was a regular participant in Menger's Mathematische Kolloquium, a small circle of gifted young mathematicians, including Gödel and Nöbeling.

In September 1930 the whole group around Menger traveled to Königsberg to attend the meeting of the German Mathematics Society, traditionally a "market" for young mathematicians, and an opportunity to present themselves and their results to the leading German mathematicians. Taussky gave two talks. Although she was nervous, she seems to have been a big success. Emmy Noether and Helmut Hasse started a lively discussion after her talk on her thesis, both having worked in the same field, but their discussion was more abstract and algebraic. Taussky did not understand this aspect of the problem at that time. She also met Arnold Scholz, who had similar interests, and they started a collaboration, culminating in 1934 in a major article with many interesting ideas.

It was difficult to find a position in Vienna. A large part of the population was jobless—and this general situation was mirrored at the Institute of Mathematics. There were no available paid positions as assistants; neither Helly, nor Mayer, nor Gödel ever had a paid position at the university. Taussky decided to stay at the Institute anyway. While there, she assisted her teacher Furtwängler, took part in Hahn's seminar and in Menger's colloquium, and earned some money doing private tutoring. Together with Hahn, she wrote a review of van der Waerden's *Modern Algebra*, which was soon to become a classic. She was one of the first to hear Gödel's theorem on completeness and consistency (January 22, 1931), later to be called his greatest achievement. She, herself, discovered other areas of mathematics as she solved some problems in group theory.

The next meeting of the German Mathematics Society, in September 1931 in Bad Elster, brought the success Taussky desired. Again she spoke on class field theory, thus presenting herself as a specialist in this field. It happened that Richard Courant was looking for such a person to edit Hilbert's number theoretic papers (Vol. 1 of his *Collected Works*). This permitted Taussky to spend the academic year 1931–1932 in the mecca of the mathematical world—Göttingen. Together with Wilhelm Magnus and Helmut Ulm, she worked through Hilbert's papers. It turned out to be a bigger task than expected, since they found many small errors and some omissions and false conjectures. The main problem was that the *Zahlbericht* was considered by some to be outdated and in need of rewriting using modern methods. But it was decided to simply make some small changes, sometimes with remarks from the editors. Taussky kept in touch with the leading specialists, asking for their comments, and worked carefully. For Hilbert, she was even too careful, since her meticulous work delayed the edition, and it turned out that the book was not finished in time for its presentation on Hilbert's seventieth birthday. (By the way, there is a new edition of Hilbert's *Zahlbericht*, published by Springer, in the series *Classics of Mathematics*.)

In spite of the time-consuming editorial work, Taussky tried to learn as much as possible from the inspiring Göttingen atmosphere. She heard lectures by Artin on class field theory—her mimeographed version of his lectures has been much in demand (they are translated and published in [2]). She also attended a seminar by Emmy Noether on the same theme—arranged especially for her! And she met many famous mathematicians such as Hermann Weyl, the "Noether boys" (the group of young algebraists around Emmy Noether, including van der Waerden, Deuring, Artin, . . .), and guests from the United States, such as Veblen.

Although Taussky doesn't mention it in her autobiography, we can assume that the change in the political situation must already have been felt in Germany. Only one year later, Emmy Noether (and many others) had to leave Germany.

In the summer of 1932, Taussky and Noether traveled together to the International Congress of Mathematicians in Zürich. Noether even accepted some advice from Taussky on her lecture—to include some examples to illustrate the abstract theory. In Zürich, Taussky met the whole group from Vienna, including Menger, Franz Alt, Hahn, and Wirtinger.

In autumn 1932, Taussky was back in Vienna, still without a paid position. Hahn and Menger were helpful; they managed to provide a small assistantship for her in the year 1933–1934, paid out of the so-called public lectures fund.

Taussky reported in her autobiography that after the excitement of Göttingen, she felt lonely and bored in Vienna. But considering the interesting developments that happened there, most mathematicians would not sympathize with her feelings. Topology, dimension theory, logic, and economics were developing rapidly there, and so was number theory. Furtwängler was still working and had some pupils such as Nikolaus Hofreiter (who filled the only available precious position as assistant). But Furtwängler could not spend as much time with Taussky as she would have liked and—from the point of view of Göttingen—he was considered old-fashioned, and his interest changed more and more to the geometry of numbers. Taussky worked a lot with Hans Hahn in his seminar, where she was responsible for a group of students (including Auguste Kraus, later called Auguste Dick, who even later became a good friend and biographer of Emmy Noether), and she guided two of the thesis students of Hahn while Hahn was in the hospital (he died on July 24, 1934).

Taussky's mother and younger sister (a pharmacist) also lived in Vienna at that time, and so Taussky was able to invite colleagues and guests to her home for tea. Gödel was a regular guest, as were mathematicians from Japan such as Takagi (who had worked with Furtwängler) and his pupils; the lecturers at Menger's Kolloquium also attended.

Taussky continued to work on her paper with Scholz, communicating by letter and working together when attending some conferences.

The general situation got worse. The library of the Mathematical Institute at the corner Strudlhofgasse-Boltzmanngasse became the place to discuss not only mathematics, but also the job market. A major part of the population had no jobs, and there was no possibility of a real job in mathematics. Gödel was never paid in any way in Vienna; Helly was working for an insurance company; Mayer was the heir to a coffee house (and thus considered to be rich)—he later became an assistant to Einstein. Franz Alt, a pupil of Menger, looked for years before he finally found a place in economics at the Institute for Higher Studies—a foundation of Oskar Morgenstern. Alt then left Vienna around 1939; he later became a distinguished computer scientist, and much later Taussky and Alt worked together at the U.S. Bureau of Standards.

Taussky applied to Girton College in Cambridge for a stipend. While Girton was considering her application, she was offered (through her Göttingen acquaintance Veblen) a one-year position at Bryn Mawr (where Emmy Noether had found refuge from Nazi Germany). The Girton stipend also materialized but could be postponed. So Taussky took some quick lessons in English and left for Bryn Mawr in autumn 1934.

That turned out to be her departure from the German-speaking world for a long time.

Much later, she often returned to Europe to attend meetings, to be a guest professor in Vienna, and to receive honors. She was elected corresponding member of the Austrian Academy of Sciences, she received the honorary Cross of Austria, and, in 1980, her doctorate degree was renewed after fifty years.

4. Prominent Mathematical Acquaintances of Taussky

Philipp Furtwängler (April 21, 1869, Elze–May 19, 1940, Vienna). Furtwängler studied mathematics in Göttingen from 1889 until 1894, mainly with Felix Klein with whom he wrote a thesis on ternary cubic forms. In 1897, he became assistant at the Institute for Geodesy in Darmstadt; then he had a succession of positions at Potsdam, Bonn, and Aachen until 1912, when he accepted the chair of mathematics at the University of Vienna as successor to Franz Mertens. During his first years in academia, he worked on the problem of measuring the gravity and the size of the earth—with remarkable success (he even wrote the article on cartography for the encyclopedia), but his great love, number theory, always had a prominent place in his research. Upon his arrival in Vienna, he again made number theory and, more specifically, class field theory, his main goal.

Kurt Gödel (April 28, 1906–January 14, 1978). Taussky and Gödel had many things in common; first, they both came from Mähren (Moravia), Taussky from Olmütz, Gödel from Brno. He was just one year older, both went to school in Austria (he in Vienna, she in Linz), and both studied mathematics at the University of Vienna (he from 1923 on, Taussky from 1925). Taussky concentrated her studies from the start on number theory and the lectures of Furtwängler. Gödel, though also fascinated by number theory, especially elementary number theory, was almost lured into class field theory, but he got caught by logic. His genius was soon recognized by Hans Hahn. From the beginning, Gödel was one of the most prominent members of Karl Menger's Mathematische Kolloquium. In his thesis of 1929, "Die Vollständigkeit des Logikkalküls—On the completeness of logic calculus," he solved Hilbert's second problem. On January 22, 1931, Gödel spoke on incompleteness and consistency; this was the first presentation of his famous work known as the theorem of incompleteness. Later (in 1933) it became the topic for his *Habilitation*. It is recorded that Taussky took part in the discussion after this talk. Despite never having a paid position at Vienna, Gödel became *Privatdozent* in 1933. In 1939 he finally left Austria for Princeton.

Karl Menger (January 13, 1902–1985). Only a few years older than Taussky, Karl Menger, the son of a famous economist, was considered to be a prodigy because he solved a major problem in dimension theory in his first year of studying mathematics. He continued his career in the Netherlands with Brouwer and came back to Vienna as *Extraordinarius* in 1927. Menger was the typical modern professor: open, with many friends in the whole world, energetic, full of problems, and ready to discuss any part of mathematics. He was soon surrounded by a group of young people—many of them later famous (besides Menger himself, they included his mentor Hahn, Gödel, Nöbeling, Wald, Alt, and various guests from Poland). They met regularly in the Mathematischen Kolloquium—on invitation only. That group had separated from the famous Vienna Circle, the group around Schlick, who studied Wittgenstein and mathematical language, along with philosophy and logic—a bit too far away from mathematics (Taussky also attended some of the meetings of the Vienna Circle). The meetings of the Kolloquium were a good opportunity for Taussky to learn about other areas of mathematics, and she soon contributed some results in group theory. Menger left Vienna in 1936, and the Kolloquium died soon afterward.

References

Class Field Theory

1. H. Cohen, "Elliptic Curves," in *From Number Theory to Physics*, ed. M. Waldschmidt, P. Moussa, J.-M. Luck, C. Itzkyson (New York: Springer-Verlag, 1992), 212–237.
2. H. Cohn, *A Classical Invitation to Algebraic Numbers and Class Fields* (with two appendices by Olga Taussky), University, (New York: Springer-Verlag, 1978).
3. H. Hasse, "Bericht über neuere Untersuchungen und Probleme aus der Theorie der algebraischen Zahlkörper I," *Jber. DMV* 35 (1925), 1–55.
4. H. Kisilevsky, "Olga Taussky-Todd's work in class field theory," *Pacific J. Math.* 181 (1997), 219–224.
5. Jürgen Neukirch, *Class Field Theory*, Grundlehren der mathematischen Wissenschaften 280 (New York: Springer-Verlag, 1986).
6. M. Waldschmidt, P. Moussa, J.-M. Luck, and C. Itzkyson, eds., *From Number Theory to Physics* (New York: Springer-Verlag, 1992).

Other References

7. C. Binder, "Olga Taussky-Todd—eine Mathematikerin aus Österreich wird Frau des Jahres," in *Jenseits von Kunst*, ed. Peter Weibel (Vienna: Passagen-Verlag, 1997), 311.

8. E. Hlawka, "Olga Taussky-Todd, 1906–1995," *Monatshefte für Math.* 123 (1997), 189–201.

9. E. Hlawka, "Renewal of the Doctorate of Olga Taussky-Todd," *Mathematical Intelligencer* 19(1) (1997), 18–20.

10. K. Menger, *Ergebnisse eines Mathematischen Kolloquiums*, ed. E. Dierker and K. Sigmund (Vienna: Springer-Verlag, 1998).

11. O. Taussky-Todd, "An Autobiographical Essay," in *Mathematical People*, ed. D. J. Albers and G. L. Alexanderson (Boston: Birkhäuser, 1985), 309–336.

12. O. Taussky-Todd, "Remembrances of Kurt Gödel," in *Gödel Remembered, Salzburg 10–12 July 1983*, ed. P. Weingartner and L. Schmetterer (Naples: Bibliopolis, 1987), 29–41.

13. O. Taussky-Todd, "Zeitzeugin," *Vertriebene Vernunft II: Emigration und Exil Österreichischer Wissenschaft*, ed. Friedrich Stadler (Vienna: Jugend und Volk, 1988), 132–134.

14. O. Taussky, "Recollections of Hans Hahn," in *Gesammelte Abhandlungen Hans Hahn III*, ed. L. Schmetterer and K. Sigmund (Vienna: Springer-Verlag, 1997), 570–572.

Numbers, Matrices, and Commutativity: A Review of Some of Olga Taussky Todd's Work in Matrix Theory

HELENE SHAPIRO

Shapiro is a professor in the Department of Mathematics and Statistics at Swarthmore College.

1. Introduction

Olga Taussky Todd received her Ph.D. in 1930 from the University of Vienna. Her advisor was Philip Furtwängler, and she wrote a thesis in class field theory: "Über eine Verschärfung des Hauptidealsatzes" (On a Sharpening of the Principal Ideal Theorem). This work was published in 1932 [OTT 9].[3]

She spent a year in Göttingen, where she was one of the editors of Volume 1 (number theory) of Hilbert's collected works and was also an assistant to Courant. She then returned to Vienna for a few years, after which she received the Alfred Yarrow Science Fellowship at Girton College, Cambridge. She was also offered a scholarship to Bryn Mawr college and spent a year there when Emmy Noether, whom she had also known in Göttingen, was there. After a year in Bryn Mawr, she returned to England to finish her

---

3. Citations in this form refer to the list of Olga Taussky Todd's publications in the *Linear Algebra and Its Applications* Special Issue (see [7, 67–83]).

fellowship at Girton College, and also had a lecturer position at the University of London. During this time her mathematical interests were number theory, class field theory and related problems in group theory, and topological algebra, a new field then. Some of her work in class field theory and group theory is discussed in articles by Hershey Kisilevsky [4] and Tom Laffey [6]. This article will focus on some of her work in matrix theory.

During her time in England, Olga met and married John Todd, whose specialty was numerical analysis. From 1943 to 1946 she worked in London at the Ministry of Aircraft Production as a member of the "flutter group"; this was her war job. The Todds then came to the United States in the late forties and spent the next decade at the National Bureau of Standards. Olga's work during the war years and at the NBS drew her attention to problems in matrix theory, and over the next fifty years she became one of the most prominent workers in this area. In 1957, the Todds took positions in the mathematics department at the California Institute of Technology.

Olga Taussky Todd had a long and productive mathematical career. *Mathematical Reviews* (MR) lists over 170 publications under her name, and there are about 20 articles which predate MR, as well as about twenty research problems and solutions. MR also lists her as the reviewer for 177 articles. Philip Hanlon (Olga's last Ph.D. student) has collected some insightful statistics about Olga Taussky Todd's work in [5]. Using the MR classification scheme, he has compiled the following breakdown of her publications (see table 7). Many of her papers were short; almost half of her publications were one to five pages in length, with another quarter in the six- to ten-page length. The new discoveries presented in these short, concise papers were frequently further developed and generalized by others. She was generous in sharing her ideas and enthusiasm, and collaborated with many colleagues. Her coauthors included Arnold Scholz, John Todd, Theodore S. Motzkin, T. Kato, Helmut Wielandt, Hans Zassenhaus, E. C. Dade, Morris Newman, Dennis Estes, and Robert Guralnick.

Beyond the direct effect of her own mathematical work, she had a great influence by encouraging and inspiring others. Richard Varga wrote the following in his preface to the 1976 special issue of *Linear Algebra and Its Applications* dedicated to Olga Taussky Todd [1].

As anyone who has had the pleasure of working with her knows, Olga's enthusiasm for mathematics is unbounded and infectious. She has collaborated with some of the most eminent mathematicians of the century, such as Professors Courant, Hilbert, Motzkin, and Ostrowski, and she has had in addition ten Ph.D. students. Because of the breadth of her contributions to mathematics, it would be difficult to single out one contribution for which she is most famous. No one in our field has given more encouragement to young mathematicians beginning their research

TABLE 7. Subjects of Olga Taussky Todd's Publications

Number theory	21%
Linear algebra	37%
Associative algebras	1%
History	10%
Field theory	6%
Group theory	4%
Numerical analysis	3%
Other	9%

careers than she has. I owe much to her for personally introducing me to the beautifully elegant Perron-Frobenius theory of non-negative matrices, as well as to eigenvalue estimation via Gersgorin circles.

In a *SIAM News* obituary [3], Varga said:

Olga was always interested in students, and she brought out the very best in them and in her post-doctoral fellows, a group that included Morris Newman and Alan Hoffman.

In his piece in the *LAA* special issue [7, 13], Alan Hoffman writes about Olga at the National Bureau of Standards:

She took upon herself the role of guide, mentor, and cheerleader for the young Ph.D.'s hired into the Division. She suggested problems, encouraged our careers, sympathized with our failures and beamed at our successes. Everything I know about the care and feeding of junior mathematicians I learned from Olga.

Number theory was Olga Taussky Todd's first mathematical love; here is a quote from her autobiography [OTT 211, 317]:

I recall that at the height of my desperation over my thesis problem, one of the *Privatdozents*, Walter Mayer, asked me how I was getting on. I mentioned that I saw no progress. To this he replied, "Remember that you are not married to Furtwängler." I understood this. He was in search of thesis students and would gladly have given me a problem and helped me with it. But his subject was n dimensional differential geometry—he became an assistant of Einstein later—and I was married to number theory.

However, her work during the war years and then at the National Bureau of Standards drew her in other directions. Taussky writes in her

"Torchbearer" article [OTT 225], "Still, matrix theory reached me only slowly. Since my main subject was number theory, I did not look for matrix theory. It somehow looked for me." This was a period when matrix theory was enjoying a renaissance, with greatly increased interest in both pure and applied problems which involved matrices and linear algebra. The development of computers sparked new interest in numerical methods, and analysis of matrix computations. Classical results were rediscovered, examined in greater detail, generalized, and eventually grew into whole areas of research. Olga Taussky Todd played an important role in these developments, through both her own work and the stimulus she had on many others. Hans Schneider gives an insightful analysis of Olga Taussky Todd's influence on the development of matrix theory in [2].

Olga Taussky Todd was interested in many problems in matrix theory; her great insight, gift for observing beautiful and important connections, and her enthusiasm sparked further work in many areas. Her own "Torchbearer" article and her autobiography give firsthand accounts of her life and work, and the LAA Special Issue [7] includes a list of publications about her on pages 59–61. The remaining sections of this article discuss some of the areas in which she worked.

2. Integral Matrices and the Latimer-MacDuffee Correspondence

The study of matrices of rational integers is a vast and rich field, blending together number theory and matrix theory. Quoting from Taussky's article [OTT 95], based on an address delivered at a 1959 AMS meeting, "This subject is very vast and very old." It involves topics such as the arithmetic theory of quadratic forms, the unimodular group, ideals in algebraic number fields, norms, and discriminants.

Olga Taussky Todd wrote a number of papers dealing with subjects such as ideal matrices, connections between integer matrices and norms and discriminants in algebraic number fields, composition of quadratic forms, and sums of squares. Of particular fascination to her was a result of Latimer and MacDuffee, showing a correspondence between the ideal classes in an algebraic number field and similarity classes of integer matrices, under the action of the unimodular group. She seemed particularly pleased with her paper "On a Theorem of Latimer and MacDuffee" [OTT 45], in which she gave a simple proof of the Latimer-MacDuffee correspondence for the case when the polynomial is irreducible. Thirty-five years later she was still inspiring interest in this subject; the Latimer-MacDuffee correspondence was extended to more general rings by Dennis Estes and Robert Guralnick [8], two colleagues who were once postdocs at Caltech.

For an introduction to the use of matrices in algebraic number theory, and a reference list including much of Taussky's work in this field, see [OTT 177].

3. Gershgorin Circles

In almost any problem involving a matrix, be it theoretical or applied, the eigenvalues of the matrix are of great importance. It is not so easy to find the eigenvalues of a general matrix, and much of numerical linear algebra is devoted to devising, analyzing, and comparing algorithms for computing eigenvalues. Theorems which give information about the location of the eigenvalues of a matrix are thus of great interest. The Gershgorin circles theorem is perhaps the best known of these theorems, and justifiably so: it is a beautiful result which is easy to state, easy to apply, and can be proved by completely elementary means.

Theorem (Gershgorin): Let $A = (a_{ij})$ be an $n \times n$ matrix of complex numbers. For each row i of A, let R_i denote the sum of the absolute values of the off-diagonalizable entries of row i, thus, $R_i = \sum_{j \neq i} |a_{ij}|$. Let C_i denote the closed disk $|z - a_{ii}| \leq R_i$. Then the union of the n disks C_i contains the eigenvalues of A.

Here is an example. Let

$$A = \begin{pmatrix} 3 & .2 & -2 & .3 \\ 0 & i & 1 & 1 \\ -3 & 2 & -11 & 1 \\ .1 & .2 & .3 & 7 \end{pmatrix}.$$

The centers of the four circles are the diagonal entries 3, i, -11 and 7, with radii $R_1 = 2.5$, $R_2 = 2$, $R_3 = 6$, and $R_4 = .6$, respectively.

Proof. Let λ be an eigenvalue of A with associated eigenvector $x = (x_1, x_2, \cdots, x_n)$. Choose p such that $|x_j| \leq |x_p|$ for all j. From the p'th coordinate of the equation $Ax = \lambda x$ we get $\sum_{j=1}^{n} a_{pj} x_j = \lambda x_p$. Rearranging this equation gives $(\lambda - a_{pp}) x_p = \sum_{j \neq p} a_{pj} x_j$. Now use the triangle inequality, and the fact that $|x_j| \leq |x_p|$ for all j, to get $|\lambda - a_{pp}| |x_p| \leq \sum_{j \neq p} |a_{pj}| |x_j| \leq \sum_{j \neq p} |a_{pj}| |x_p| = R_p |x_p|$ and thus $|\lambda - a_{pp}| \leq R_p$. So λ lies in the disk C_p.

Taussky's 1947 report [OTT 38], written for the Aeronautical Research Council, describes the Gershgorin method and applies it to several examples which occurred in work concerning flutter of airplane wings.

By using a continuity argument, one can show that if k of the circles form a region which is disconnected from the remaining circles, then that region must contain k of the eigenvalues (where we must count eigenvalues according to multiplicities). In particular, any isolated disk is guaranteed to contain a single eigenvalue.

Now, by applying the Gershgorin method to $S^{-1}AS$, for various choices of S, one might hope to get better estimates of the eigenvalues. In particular, it is interesting to investigate what happens when S is diagonal, for then it is easy to compute $S^{-1}AS$. If $S = diag(d_1, \cdots, d_n)$, the ij entry of $S^{-1}AS$ is $(d_j/d_i)a_{ij}$; note the diagonal entries are unchanged. So the circles for $S^{-1}AS$ have the same centers as the circles for A, but the radii change. If $d_p > d_s > 0$, for all S, then the radius of the p'th circle will shrink—of course the other circles may get larger. However, if one has an isolated disk, one might hope that a good choice of diagonal similarity will shrink that disk down, thus yielding better information about the location of that eigenvalue.

Schneider's paper [2, 204] quotes a letter from Olga describing how she first learned of the Gershgorin theorem. She says there:

> I immediately tinkered with the theorem, applying it to a very nasty looking 6×6 matrix of complex elements given with many decimals, revealing off hand nothing about its stability . . . I am enclosing xerox copies of the circles. Number I gives the 6 circles, one containing points with large negative real part; Number II comes after a diagonal similarity which already excludes a large part of the negative real axis; Number III then shows what happens if we expand the shrunk circle again and shrink the other 5. This seemed great fun. I then realized an optimum for this process must be possible; this was carried out much later by Henrici, Jack (1965), Varga (1965), etc. At the time I wrote a report for the Aeronautics Research Council (1947) which contains most of what is in the monthly article, apart from the equality case, which I carried out under prodding from G. B. Price. It was he, who in 1947 pushed me into writing the article.

The "monthly article" mentioned is the now well-known "A Recurring Theorem on Determinants," which appeared in the *American Mathematical Monthly* in 1949 [OTT 47]. This article begins:

> This note concerns a theorem on determinants of which proofs are being published again and again; on the other hand, the theorem is not as well known as it deserves to be. The theorem has arisen in many varied connections as is indicated by the titles of the papers quoted. Although it can be proved in a very simple manner, some of the proofs

that have been given are very complicated. The theorem deals with determinants of matrices with a "dominant" main diagonal. Such matrices are particularly useful.

The term "recurring theorem" actually refers not to the Gershgorin theorem, but to the following equivalent result about matrices with a dominant main diagonal.

Theorem: Let A be an n × n matrix and suppose $|a_{ii}| > R_i$ for all i. Then $\det(A) \neq 0$.

Taussky had first encountered this theorem as a student [2, 204]:

> First of all, I seemed unusually interested in Vienna in the "recurring theorem" as soon as Furtwängler got to it in his course on algebraic number theory, when proving Dirichlet's unit theorem. He proved it by induction and it is really quite an interesting proof, but I did not like it then.

The two theorems are equivalent—one can easily prove one from the other and vice versa. Taussky's "recurring theorem" article also clarified what happens when the strict inequalities are replaced by "greater than or equal to," pointing out the role played by the irreducibility of the matrix in this case.

Theorem: Let A be an n × n matrix and suppose $|a_{ii}| \geq R_i$ for all i with equality in at most n − 1 cases. Assume further that the matrix cannot be transformed to a matrix of the form $\begin{pmatrix} P & U \\ 0 & Q \end{pmatrix}$ by a permutation similarity. (P and Q are square and 0 is a block of zeroes.) Then $\det(A) \neq 0$.

As a consequence of this, it follows that a boundary point of one of the Gershgorin circles can only be an eigenvalue if it is also on the boundary of the other n − 1 circles.

Much work has been done on the Gershgorin circles and diagonal dominance. These ideas have been generalized in many ways, and thirty years later Richard Varga found it was time to write another paper about recurring theorems on diagonal dominance [14]. A few of the papers on these subjects are listed in the references [9–16].

Now it is not clear to me exactly what new result Olga contributed to this subject—I once asked her about the Gershgorin theorem and she said that was her husband's work and told me to ask him. Schneider [2, 202–203] credits her with clarifying the role of irreducibility in analyzing the case where $|a_{ii}| \geq R_i$. She also wrote a paper establishing some new results about the Gershgorin circles for 2 × 2 matrices [OTT 43]. Schneider [2, 207] says this about the role of her paper in the *Monthly*:

A renaissance must start with a renewed and novel appreciation of the classics. This was provided by Olga Taussky's (1949) note and that is the significance in the development of matrix theory of that short contribution in a journal largely devoted to exposition at the college level.

4. The Lyapunov Theorem

We turn now to another theorem about the location of the eigenvalues of a matrix, the famous Lyapunov theorem. If we have a linear system of differential equations, represented in matrix form as $dx/dt = Ax$, then we want to know how solutions behave as $t \to \infty$. Do they approach the fixed point, or do they "blow up"? The eigenvalues of A determine this. The system will be stable if and only if all of the eigenvalues of A have negative real parts. It is a bit more convenient to deal with matrices which are positive stable.

Definition. An $n \times n$ complex matrix A is said to be *positive stable* if all of its eigenvalues have positive real part.

In geometric language, A is positive stable if the eigenvalues all lie in the right half plane. Here now is the matrix version of the Lyapunov theorem. We state two equivalent versions.

Lyapunov Theorem I. The matrix A is positive stable if and only if there is a positive definite Hermitian matrix H such that $HA + A^*H = I$.

Lyapunov Theorem II. The matrix A is positive stable if and only if for every positive definite Hermitian matrix K, there is a positive definite Hermitian matrix H such that $HA + A^*H = K$.

Version II is obtained from I by writing $K = P^*P$; then we have:

$$HA + A^*H = P^*P$$
$$P^{-*}HAP^{-1} + P^{-*}A^*HP^{-1} = I$$
$$(P^{-*}HP^{-1})(PAP^{-1}) + (P^{-*}A^*P^*)(P^{-*}HP^{-1}) = I$$

Note that A is positive stable if and only if PAP^{-1} is positive stable, and H is positive definite if and only if $P^{-*}HP^{-1}$ is positive definite.

To help understand this theorem, first consider the special case where A is diagonal:

$$A = diag(\lambda_1, \lambda_2, \cdots, \lambda_n). \text{ Let } H = I \text{ and note that}$$
$$A + A^* = 2\, diag(Re(\lambda_1), Re(\lambda_2), \cdots, Re(\lambda_n)).$$

Thus, for a diagonal matrix A, we see $A + A^*$ is positive definite if and only if A is positive stable.

For a general matrix A, set $R = (A + A^*)/2$ and $S = (A - A^*)/2i$. Then R and S are Hermitian and we have the decomposition $A = R + iS$. We think of $R = Re(A)$ as being the "real part" of A and iS as the "imaginary part." Observe that if $Ax = \lambda x$, and $|x| = 1$, then $x^*Ax = x^*\lambda x = \lambda = x^*Rx + ix^*Sx$. So, $Re(\lambda) = x^*Rx$. Hence, if R, the real part of A, is positive definite, then A is positive stable. However, A can be positive stable without R being positive definite; the situation is a bit more subtle. Note that $HA + A^*H = 2\,Re(HA)$. We can now rephrase the Lyapunov theorem as follows.

Theorem: The matrix A is positive stable if and only if there exists a positive definite Hermitian matrix H such that $Re(HA)$ is positive definite.

Taussky's work [OTT numbers: 98, 101, 106, 110, 120] on the Lyapunov theorem generalized it in several ways and developed connections between the Lyapunov theorem and work of Stein. Ostrowski and Schneider further developed the subject in [17], which contains the result known as the "inertia theorem." For a square complex matrix A, let $\pi(A)$ denote the number of eigenvalues which have positive real part, $\nu(A)$ the number which have negative real part, and $\delta(A)$ the number which have zero real part. The triple of numbers $In(A) = (\pi(A), \nu(A), \delta(A))$ is called the *inertia of A*. When A is Hermitian, this reduces to the more familiar idea of inertia (number of positive, negative, and zero eigenvalues). The Lewis and Taussky paper [OTT 98] contains the following result.

Theorem: Assume the real parts of the characteristic roots of A are all nonzero. Then there exists a positive definite Hermitian matrix H such that $In(HA + A^*H) = In(A)$.

The key idea in the proof is to put A into a Jordan canonical form with the usual "1's" on the superdiagonal replaced by epsilons. In her 1961 paper, "A Generalization of a Theorem of Lyapunov" [OTT 106], Taussky proves the following:

Theorem: Let A be an $n \times n$ complex matrix. Assume that $\alpha_i + \overline{\alpha_k} \neq 0$ for all eigenvalues α_i, α_k. Then the Hermitian matrix H which is a solution to $HA + A^*H = I$ is nonsingular, and the number of positive eigenvalues of H equals the number of eigenvalues of A which have positive real parts.

Finally, here is the inertia theorem of Ostrowski and Schneider [17].

Inertia Theorem: There exists a Hermitian matrix H such that $Re(HA)$ is positive definite if and only if $\delta(A) = 0$. In this case, $In(A) = In(H)$.

Ostrowski and Schneider also study the case where $\delta(A) \neq 0$. On pp. 72–73, Ostrowski and Schneider say:

Recently, investigations into the behavior of economic systems depending on a finite number of parameters have led Arrow and McManus to consider the problem from a different point of view, introducing the concept of the S-stability. The equivalence of Lyapunov's theorem with some of the results of Arrow and McManus was noticed by Olga Taussky, who let us see her then unpublished manuscript, and thereby sparked this investigation.

"Inertia theory" has since developed into a whole subfield of linear algebra. Much work on ideas related to Lyapunov's theorem has been done by many workers in the field, including Alexander Ostrowski, Hans Schneider, David Carlson, Richard Hill, Bryan Cain, and two of Olga's students, Raphael Loewy and Charles Johnson [17–31]. See [29] and [31] for more references on this topic.

5. Generalized Commutativity, Commutators, and Cramped Matrices

Matrix multiplication is not commutative. When two matrices A and B do commute, this is a very strong statement about the pair. The term "generalized commutativity" refers to problems in which we slightly weaken the assumption $AB = BA$ and investigate the consequences.

We may rewrite the equation $AB = BA$ as $AB - BA = 0$, or $[A, B] = 0$, where $[A, B] = AB - BA$ is the additive, or Lie, commutator of A and B. We might now try to find generalizations of the equation $[A, B] = 0$, or describe ways in which $[A, B]$ is "close to" zero. For example, we might think of a matrix as being close to zero if it is nilpotent (all eigenvalues are zero), or if many of its eigenvalues are zero. It is also fruitful to look at "higher order" commutators, such as $[A, [A, B]]$, and to study the map $T_A(X) = AX - XA$. This is a linear transformation on the space of $n \times n$ matrices X. If the eigenvalues of A are $\alpha_1, \alpha_2, \cdots, \alpha_n$, it turns out that the eigenvalues of T_A are the n^2 numbers $\alpha_i - \alpha_j$.

Let A and B denote $n \times n$ matrices over a field F. We look at the iterated commutators:

$$B_1 = [A, B] = T_A(B)$$

$$B_2 = [A, B_1] = T_A(B_1) = T_A^2(B)$$

$$\vdots$$

$$B_i = [A, B_{i-1}] = T_A^i(B)$$

In [OTT 82], Kato and Taussky observed that for 2×2 matrices, $B_3 = (\alpha_1 - \alpha_2)^2 B_1$, where α_1 and α_2 are the eigenvalues of A. Taussky and Wielandt proved a more general result covering the $n \times n$ case in [OTT 109]. Taussky's student Fergus Gaines did further work on such commutator relations in his 1966 Ph.D. thesis [38; 39; 48].

Now again consider the equation $AB = BA$, and suppose A and B are nonsingular. Then we can rewrite the equation as $ABA^{-1}B^{-1} = I$ and look for ways to generalize this. For example, Frobenius [32; 33] proved the following result for unitary matrices A and B where B is "cramped": The eigenvalues of a unitary matrix are complex numbers of modulus 1, hence they correspond to points on the unit circle. We say a unitary matrix is *cramped* if its eigenvalues lie on an arc which is less than π; that is, less than a semicircle.

Theorem (Frobenius [33]): Let A and B be $n \times n$, unitary matrices. Let $C = ABA^{-1} B^{-1}$. Assume that B is cramped and that $AC = CA$. Then $AB = BA$.

Taussky [OTT 102] studied the structure of unitary matrices A and B under the assumption that $AC = CA$, but without additional assumptions on B; see also the paper of Zassenhaus [35]. Marcus and Thompson [37] later generalized these results to the case of normal matrices; for generalizations to nonnormal matrices and more references see [40]. While on the subject of commutators, we should mention the theorem of Shoda, which was a special favorite of Olga's [OTT 225]. Shoda's theorem says that over an algebraically closed field, any matrix of determinant 1 is the multiplicative commutator of two matrices [34]. Robert Thompson, one of Taussky's students, studied this result for general fields [36].

6. Properties P and L

This section also deals with topics belonging to the area of generalized commutativity. For a survey of results in this area, see Taussky's article "Commutativity in Finite Matrices" [OTT 85].

We begin with a basic result of Frobenius.

Theorem (Frobenius [41]): Suppose $A_1, \cdots, A_t$ are $n \times n$ matrices over the field F which commute; i.e. $A_i A_j = A_j A_i$ for all i and j. Assume the field F contains the eigenvalues of these matrices. Then, there is a nonsingular matrix S such that all of the matrices $S^{-1} A_i S$ are upper triangular.

This is proved by induction on n; the key step is to show that the commutativity of the matrices implies that they have a common eigenvector. If the field F is the complex numbers, then one can use a unitary similarity for the S.

Definition. We say $A_1, \cdots, A_t$ are *simultaneously triangularizable* if there is a nonsingular matrix S such that all of the matrices $S^{-1}A_i S$ are upper triangular.

Thus, commuting matrices are simultaneously triangularizable. However, triangular matrices need not commute, so the converse is not true.

The challenge then is to examine this property of "simultaneously trian-gularizable" and see exactly what it does tell us about the set of matrices. Now let's think about the eigenvalues. If $A_1, \cdots, A_t$ are all in triangular form, then the diagonal entries of each matrix are the eigenvalues. Let α_{ij} denote the i'th eigenvalue of the matrix A_j. When two triangular matrices are added, we add the diagonal entries; when they are multiplied, corre-sponding diagonal entries are multiplied. Let $p(X_1, X_2, \cdots, X_t)$ denote a polynomial in the noncommuting variables $X_1, X_2, \cdots, X_t$. Then the i'th diagonal entry of $p(A_1, A_2, \cdots, A_t)$ is $p(\alpha_{i1}, \alpha_{i2}, \cdots, \alpha_{it})$. Thus, if A_1, $A_2, \cdots, A_t$ can be simultaneously triangularized, then there is an ordering of the eigenvalues of the A_j's such that for any polynomial p, the eigenval-ues of $p(A_1, A_2, \cdots, A_t)$ will be $p(\alpha_{i1}, \alpha_{i2}, \cdots, \alpha_{it})$. This is what we call property P. It turns out that a set of matrices have property P if and only if they are simultaneously triangularizable [42; 43; 46].

Theorem (McCoy [42]): Let $A_1, A_2, \cdots, A_t$ be $n \times n$ matrices over a field F which contains the eigenvalues of all of the matrices. Then the fol-lowing are equivalent.

1. $A_1, A_2, \cdots, A_t$ are simultaneously triangularizable.
2. $A_1, A_2, \cdots, A_t$ have property P.
3. For each pair i and j, and each polynomial p, the matrix $p(A_1, A_2, \cdots, A_t)(A_iA_j - A_jA_i)$ is nilpotent.

(*So $A_iA_j - A_jA_i$ is in the radical of the algebra generated by $A_1, A_2, \cdots, A_t$.*)

We have seen that (1) implies (2). It is easy to show (2) implies (3); the real heart of the result is that (3) implies (1). In this theorem, the field F needs to contain the eigenvalues of the matrices. If this is not true, then a "block triangular" version of the theorem holds, as shown by Fergus Gaines [48, 49]. See also [51–53].

The McCoy theorem tells us that property P is a rather strong condi-tion: sets of matrices with property P can be simultaneously triangular-ized. M. Kac suggested investigating property L: we say A and B have property L if there is an ordering, $\alpha_1, \cdots, \alpha_n$, of the eigenvalues of A and $\beta_1, \cdots, \beta_n$, of the eigenvalues of B such that for any x and y, the matrix $xA + yB$ has eigenvalues $x\alpha_i + y\beta_i$. Motzkin and Taussky studied the L-property in [OTT 56; OTT 69]. The first of these papers [OTT 56] con-tains a proof of a conjecture of Kac.

Theorem (Motzkin, Taussky [OTT 56]): If A and B are Hermitian matrices with property L, then $AB = BA$.

Wiegmann [44] and Wielandt [45] later generalized this result to a pair of normal matrices. In their second paper [OTT 69], Motzkin and Taussky studied the characteristic curve associated with the polynomial $\det(zI - xA + yB)$, proving several significant results about the pair A, B, the pencil $xA + yB$ and the L-property. This work involves methods of algebraic geometry applied to the polynomial $\det(zI - xA + yB)$. Here are some of the main results from [OTT 69].

Theorem (Motzkin, Taussky [OTT 69]): Let A and B be $n \times n$ matrices over an algebrically closed field F of characteristic p. Assume all the matrices in the pencil $xA + yB$ are diagonalizable, and, for $p \neq 0$, assume $n \leq p$, or that A and B have property L. Then A and B commute.

Theorem (Motzkin, Taussky [OTT 69]): The variety C of commuting pairs of $n \times n$ matrices over an algebraically closed field F is irreducible.

Generalizations and other work related to these results have been done by Wales and Zassenhaus [50], Gerstenhaber [47], Laffey, Friedland [54], Guralnick [55], and others. Laffey discusses this in his article [6] in the LAA memorial issue.

We conclude with a look at the proofs of the first two theorems from the first Motzkin, Taussky paper [OTT 56] on the L property. These arguments illustrate the sort of techniques that are often useful in generalized commutativity problems.

Theorem (Motzkin, Taussky [OTT 56]): Let A and B be $n \times n$ matrices with property L and assume A is diagonalizable. Let $\alpha_1, \alpha_2, \cdots \alpha_t$, be the distinct eigenvalues of A and let m_i be the multiplicity of α_i. Let β_1, $\beta_2, \cdots \beta_n$ be the corresponding eigenvalues of B. Let $A' = P^{-1} AP$ be in Jordan form and let $B' = P^{-1} BP$. Write

$$B' = \begin{pmatrix} B_{11} & B_{12} & \cdots & B_{1t} \\ B_{21} & B_{22} & \cdots & B_{2t} \\ \vdots & \vdots & \ddots & \vdots \\ B_{t1} & B_{t2} & \cdots & B_{tt} \end{pmatrix}$$

where B_{ij} is size $m_i \times m_j$. Then $\det (zI - B') = \prod_{i=1}^{t} \det(zI - B_{ii})$ and $\sum b'_{ik} b'_{ki} = 0$, where the sum is over all $i < k$ with (i, k) outside of every B_{jj}.

Outline of proof.

Property L is not affected by a translation so we may assume $\alpha_1 = 0$. The matrix A' is diagonal and we write $A' = \begin{pmatrix} 0 & 0 \\ 0 & A_{22} \end{pmatrix}$ and $B' = \begin{pmatrix} B_{11} & C_{12} \\ C_{21} & C_{22} \end{pmatrix}$ where A_{22} has size $(n - m_1) \times (n - m_1)$, the zeros indicate blocks of zeros and B' is partitioned conformally with A. Note that A_{22} will be nonsingular. Now consider the polynomial $\det (zI - xA' - B') = \det \begin{pmatrix} zI - B_{11} & -C_{12} \\ -C_{21} & zI - xA_{22} - C_{22} \end{pmatrix}$

and note that the coefficient of x^{n-m_1} is $\det(zI - B_{11}) \det(-A_{22})$. However, since A and B have property L, we also have $\det(zI - xA' - B') = \prod_{i=1}^{m_1} (z - \beta_i) \prod_{i=m_1+1}^{n} (z - x\alpha_i - \beta_i)$, so the coefficient of x^{n-m_1} is $\prod_{i=1}^{m_1} (z - \beta_i) \prod_{i=m_1+1}^{n} (-\alpha_i)$. But $\det(-A_{22}) = \prod_{i=m_1+1}^{n} (-\alpha_i)$, and this is nonzero, so we must have $\det(zI - B_{11}) = \prod_{i=1}^{m_1} (z - \beta_i)$. Applying this argument to each of the eigen-values of A yields $\det(zI - B') = \prod_{i=1}^{t} \det(zI - B_{ii})$.

The second part of the theorem is obtained by examining the coefficient of x^{n-2} on both sides of $\det(zI - xA' - B') = \prod_{i=1}^{m_1} (z - \beta_i) \prod_{i=m_1+1}^{n} (z - x\alpha_i - \beta_i)$.

Observe that the conclusions of this theorem would obviously hold for a pair of triangular matrices; we can think of the theorem as telling us that when A and B have property L they retain some of the behavior of a pair of matrices which have property P. From this theorem it is easy to prove the result that Hermitian matrices with property L must commute.

Theorem (Motzkin, Taussky [OTT 56]): If A and B are Hermitian matrices with property L, then $AB = BA$.

Proof. Since A is Hermitian, we can reduce it to Jordan form with a unitary similarity P, so A' is diagonal and B' is still Hermitian. The second part of the previous theorem then tells us that all of the off-diagonal blocks of B' are zero. Since each diagonal block of B' pairs up with a scalar block in A, we see that A' and B' commute. Therefore, $AB = BA$.

References

1. INTRODUCTION

See also OTT: 9, 174, 211, 225 and [7].

1. *Linear Algebra and Its Applications* 13, no. 1/2, ed. David H. Carlson and Richard Varga, dedicated to Olga Taussky Todd (1976).
2. Hans Schneider, "On Olga Taussky-Todd's Influence on Matrix Theory and Matrix Theorists," *Linear and Multilinear Algebra* 5, 197–224 (1977–78).
3. Richard Varga, "Obituary: Olga Taussky Todd," *SIAM News* 29 (January–February 1996), 6.
4. Hershey Kisilevsky, "Olga Taussky-Todd's Work in Class Field Theory," *Pacific Journal of Mathematics* 181, Olga Taussky-Todd memorial issue, 219–224 (December 1997).

5. Phil Hanlon, "To the Latimer-MacDuffee Theorem and Beyond!" *Linear Algebra and Its Applications* 280, 21–37 (1998).
6. Tom Laffey, "Some Aspects of Olga Taussky's Work in Algebra," *Linear Algebra and Its Applications* 280, 51–57 (1998).
7. Special issue in memory of Olga Taussky Todd, *Linear Algebra and Its Applications*, ed Helene Shapiro, 280, no. 1 (1998).

2. INTEGRAL MATRICES, THE LATIMER MACDUFFEE CORRESPONDENCE

See also OTT: 45, 95, 177 and [7].

8. Dennis R. Estes and Robert M. Guralnick, "Representations under Ring Extensions: Latimer-MacDuffee and Taussky Correspondences," *Advances in Mathematics* 54, 302–313 (1984).

3. THE GERSHGORIN CIRCLES

See also OTT: 38, 43, 47, 52, 107, 119 and [7].

9. A. Brauer, "Limits for the Characteristic Roots of a Matrix II," *Duke Math. J.* 14, 21–26 (1947).
10. P. Henrici, "Bounds for Eigenvalues of Certain Tridiagonal Matrices," *J. SIAM* 11, 281–290 (1963).
11. Richard S. Varga, "On Smallest Isolated Gerschgorin Disks for Eigenvalues," *Numerische Mathematik* 6, 366–376 (1964).
12. Richard S. Varga, "Minimal Gerschgorin Sets," *Pac. J. of Math.* 15, 719–729 (1965).
13. John Todd, "On Smallest Isolated Gerschgorin Disks for Eigenvalues," *Numerische Mathematik* 7, 171–175 (1965).
14. Richard S. Varga, "On Recurring Theorems on Diagonal Dominance," *Linear Algebra and Its Applications* 13, 1–9 (1976).
15. Richard A. Brualdi, "Matrices, Eigenvalues, and Directed Graphs," *Linear and Multilinear Algebra* 11, 143–165 (1982).
16. Richard A. Brualdi and Stephen Mellendorf, "Regions in the Complex Plane Containing the Eigenvalues of a Matrix," *Amer. Math. Monthly* 101, 975–985 (1994).

4. THE LYAPUNOV THEOREM

See also OTT: 34, 98, 101, 106, 110, 120, 133 and [7].

17. Alexander Ostrowski and Hans Schneider, "Some Theorems on the Inertia of General Matrices," *J. Math. Anal. Appl.* 4, 72–84 (1962).
18. David Carlson and Hans Schneider, "Inertia Theorems for Matrices: The Semidefinite Case," *J. Math. Anal. Appl.* 6, 430–446 (1963).
19. Hans Schneider, "Positive Operators and an Inertia Theorem," *Numerische Mathematik* 7, 11–17 (1965).

20. Bryan Cain, "Inertia Theory for Operators on a Hilbert Space" (Ph.D. thesis, University of Wisconsin, 1967).
21. David Carlson, "A New Criterion for H-stability of Complex Matrices," *Linear Algebra and Its Applications* 1, 59–64 (1968).
22. Raphael Loewy, "On the Lyapunov Transformation for Stable Matrices" (Ph.D. thesis, California Institute of Technology, 1972).
23. David Carlson and Raphael Loewy, "On Ranges of Lyapunov Transformations," *Linear Algebra and Its Applications* 8, 237–248 (1974).
24. Raphael Loewy, "On Ranges of Lyapunov Transformations II," *Linear and Mulitilinear Algebra* 2, 227–237 (1974–1975).
25. Raphael Loewy, "On Ranges of Lyapunov Transformations III," *SIAM J. Appl. Math.* 30, no. 4, 687–702 (1976).
26. Raphael Loewy, "On Ranges of Lyapunov Transformations IV," *Glasgow Math. J.* 17, no. 2, 112–118 (1976).
27. David Carlson and Richard Hill, "Generalized Controllability and Inertia Theory," *Linear Algebra and Its Applications* 15, 177–187 (1976).
28. David Carlson and Richard Hill, "Controllability and Inertia Theory for Functions of a Matrix," *J. Math. Anal. Appl.* 59, 260–266 (1977).
29. Bryan Cain, "Inertia Theory," *Linear Algebra and Its Applications* 30, 211–240 (1980).
30. Bryan Cain, "Corrections to 'Inertia Theory,'" *Linear Algebra and Its Applications* 42, 285–286 (1982).
31. Biswa Nath Datta, "Stability and Inertia," *Linear Algebra and Its Applications* 302–303 (1999), 563–600.

5. Generalized Commutativity, Commutators, Cramped Matrices

See also OTT: 63, 82, 92, 102, 109, 111, 137 and [7].

32. G. Frobenius, "Über der von L. Bieberbach gefundenen Beweis eines Satzes von C. Jordan," *Sitzungsber. Akad. Wiss. Berlin*, 241–248 (1911).
33. G. Frobenius, "Über unitäre Matrizen," *Sitzungsber. Akad. Wiss. Berlin*, 373–378 (1911).
34. Kenjiro Shoda, "Einige Satze über Matrizen," *Japanese Journal of Mathematics* 13, 361–365 (1937).
35. H. Zassenhaus, "A Remark on a Paper of O. Taussky," *J. Math. Mech.* 10, 179–180 (1961).
36. R. C. Thompson, "Commutators in the Special and General Linear Groups," *Trans. Amer. Math Soc.* 101, 16–33 (1961).
37. Marvin Marcus and Robert C. Thompson, "On a Classical Commutator Result," *Journal of Mathematics and Mechanics* 16, 583–588 (1966).
38. F. J. Gaines, "Kato-Taussky-Wielandt Commutator Relations," *Linear Algebra App.* 1, 127–138 (1968).
39. F. J. Gaines, "Kato-Taussky-Wielandt Commutator Relations and Characteristic Curves," *Pacific Journal of Mathematics* 61, 121–128 (1975).
40. Helene Shapiro, "Commutators Which commute with One Factor," *Pac. J. Math.* 181, Olga Taussky-Todd memorial issue, 323–336 (1997).

6. Generalized Commutativity, Properties P and L

See also OTT: 56, 69, 85, 94, 112, 137, 151, 164, 182 and [7].

41. G. Frobenius, "Über vertauschbare Matrizen," *Sitzungsber. Akad. Wiss. Berlin*, 601–614 (1896).
42. N. H. McCoy, "On the Characteristic Roots of Matrix Polynomials," *Bull. American Math. Soc.* 42, 592–600 (1936).
43. M. P. Drazin, J. W. Dungey, and K. W. Gruenberg, "Some Theorems on Commutative Matrices," *J. London Math Society* 26, 221–228 (1951).
44. N. Wiegmann, "Normal Matrices with Property L," *Proc. Amer. Math. Soc.* 4, 35–36 (1953).
45. H. Wielandt, "Pairs of Normal Matrices with Property L," *J. Res. Nat. Bur. Standards* 51, 89–90 (1953).
46. Harley Flanders, "Methods of Proof in Linear Algebra," *American Mathematical Monthly* 63, 1–15 (1956).
47. M. Gerstenhaber, "On Dominance and Varieties of Commuting Matrices," *Ann. of Math.* 73, 324–348 (1961).
48. F. J. Gaines, "Some Generalizations of Commutativity for Linear Transformations on a Finite Dimensional Vector Space" (Ph.D. thesis, California Institute of Technology, 1966).
49. F. J. Gaines and R. C. Thompson, "Sets of Nearly Triangular Matrices," *Duke Math. Journal* 35, 441–453 (1968).
50. D. B. Wales and H. J. Zassenhaus, "On L-Groups," *Mathematische Annalen* 198, 1–12 (1972).
51. G. P. Barker, L. Q. Eifler, and T. P. Kezlan, "A Non-commutative Spectral Theorem," *Linear Algebra and Its Applications* 20, 95–100 (1978).
52. H. Shapiro, "Simultaneous Block Triangularization and Block Diagonalization of Sets of Matrices," *Linear Algebra and Its Applications* 25, 129–137 (1979).
53. Robert M. Guralnick, "Triangularization of Sets of Matrices," *Linear and Multilinear Algebra* 9, 133–140 (1980).
54. Shmuel Friedland, "A Generalization of the Motzkin-Taussky Theorem," *Linear Algebra and Its Applications* 36, 103–109 (1981).
55. Robert M. Guralnick, "A Note on Commuting Pairs of Matrices," *Linear and Multilinear Algebra* 31, 71–75 (1992).

The Taussky Todd Celebration

BETTYE ANNE CASE, KRYSTYNA KUPERBERG, HELEN MOORE, AND LESLEY A. WARD

Adapted from "The Olga Taussky Todd Celebration of Careers in Mathematics for Women," AWM Newsletter 29(6), 1999, 14–23, and 30(1), 2000, 12–18. Kuperberg is a professor of mathematics, Auburn University. Moore was then a lecturer in mathematics, Stanford University, and is now the associate director of the American Institute of Mathematics Research Conference Center. Ward was then an assistant professor of mathematics, Harvey Mudd College, where she is now an associate professor.

The Taussky Todd Celebration featured the legacy of Olga Taussky Todd (1906–1995) and drew more than 100 mathematicians, both women and men, to MSRI for three days of information, inspiration, and camaraderie. The primary goals of the AWM-organized conference were well met: to assist, encourage, and inspire the beginning mathematician participants; to provide a forum for networking among mathematicians at different career stages; and to promote the achievements of women in mathematics. The early assurance of base support from the NSA facilitated conference planning; NSA staff provided helpful advice at many stages, and two women mathematicians at the agency participated in the celebration. Supplementary funding from the Department of Energy, the ONR, and MSRI (also the hosting institution) made it possible to bring an array of panelists and to allow a good number of senior mathematicians to participate in the conference and, with the speakers and panelists, to act as mentors for beginning mathematicians.

The program was highlighted by ten plenary talks by women in the mathematical sciences. Two were expository papers on Taussky Todd's mathematics, one by her student Helene Shapiro and the other by Christa Binder, a mathematical historian who knew Taussky Todd. The other speakers were leaders inside and outside academia; although most have held academic positions at some point, each described some work in government or industry. Cathleen Synge Morawetz and Evelyn Boyd Granville were the senior speakers. As did Taussky Todd, both worked early in their careers on mathematical problems related to World War II efforts and went on to receive many honors during long and distinguished careers in academia.

Direct and indirect connections abound between Taussky Todd and the speakers: Granville worked at the National Bureau of Standards, where both Olga and John were employed. Fern Hunt, who received a Flemming Award in 2000, described research at the successor agency, NIST. Linda Petzold talked about a career in technology pursued in both industry and the academy, while Lani Wu, Microsoft, and Lisa R. Goldberg, Barra, discussed work in the currently hot technology and financial sectors. Diane Lambert and Margaret Wright (see also "Communications Industry," III.3) described work at Bell Laboratories, Lucent Technologies.

"Olga's Irishman," Taussky Todd's widower John Todd, attended every talk and made touching remarks after the conference dinner speaker, Richard Varga, discussed Taussky Todd's "impact on me and on her many students." A symposium session of short talks on the various types of employment of Taussky Todd and her interactions with other mathematicians was organized by Mary Ann McLoughlin.

Women graduate students and recent women Ph.D.'s, thirty-seven altogether, participated in the conference workshop, along with twenty-five more senior women mathematicians who were mentors. This augmented version of the very successful workshops AWM has held at the JMM and SIAM national meetings gave participants the opportunity to present posters on their research in three lively one-hour poster sessions. There was excitement in the air as students enjoyed the wonderful feeling of finding that mathematicians other than their thesis advisors were interested to hear their results.

Workshoppers were paired with senior women mathematicians (from academia and industry) for a relaxed lunchtime discussion. The small-group format elicited effective conversations, touching on themes such as successful grant writing, job search techniques, child care, time management, careers in industry, and negotiation of the path toward academic tenure, as well as discussion of the mentees' own research. Throughout the rest of the conference, participants and their mentors continued to interact informally.

"Issues and Insider Insights for Women in Mathematics" was organized by Sylvia Wiegand (University of Nebraska–Lincoln). The panelists, from an assortment of career stages and types of institutions, gave useful information and pragmatic advice that the workshop participants found eye-opening. Brief summaries of the panelists' main points are given in the following, along with their affiliations at that time.

Ellen Kirkman (Wake Forest University) cautioned that although information from mentors and others is useful, because of differences among institutions, little advice is universal. She emphasized the need to consider the source of the advice and to talk with a range of people.

Jean Taylor (Rutgers University) suggested that reading someone else's papers and talking to that person about them might lead to joint work. One of her collaborations began when she found a mistake in a paper and pointed it out to the author.

Susan Morey (Southwest Texas State University) suggested eating lunch with others in the department as an informal means of finding out about department politics and policies, as well as letting people know how hard you are working.

Tamara Kolda (Sandia National Laboratories) spoke about networking and promoting one's career. She recommended asking people questions about themselves and their work when you first meet them, and giving them information about yourself (e.g., your advisor's name, your area of research) as well.

Maria Klawe (University of British Columbia) gave advice on getting promoted. She suggested meeting with the department chair regularly to ask how to improve your CV. For example, do you need more conference participation and invited talks? Should you organize a conference, referee or review papers, or try to be the editor of a journal? If so, then you should ask senior mentors how to go about doing whatever is required. Often, an inquiry leads to an opportunity.

Claudia Polini (Hope College) was told that she could get a reduced teaching load at her small college (and thus have more time to do research) if she got grants. So she applied for four and received two—a strategy of many applications to increase your odds of succeeding.

A second job-related discussion was Krystyna Kuperberg's "Finding a Traditional or Nontraditional Job and Growing in It." The panelists relied on their diverse experiences working in academic and nonacademic environments. Some started their careers with positions at universities and later switched to different types of employment; some did the opposite. Most mentioned choice as the reason for career moves, although some element of necessity was often involved. The speakers made it clear that traditional and nontraditional math jobs can be rewarding, challenging, and satisfying. The panelists also discussed how to overcome predictable adversities, avoid stumbling blocks, and take advantage of opportunities. Questions from the audience ranged from addressing specific mathematical issues to practical topics such as how to prepare an application for a nontraditional position. The panelists were Karen M. Brucks, University of Wisconsin, Milwaukee; Barbara S. Deuink and Barbara B. Flinn, National Security Agency; Lisa R. Goldberg, Barra; Sarah Holte, Fred Hutchinson Cancer Research Center; Linda R. Petzold, University of California, Santa Barbara; and Margaret H. Wright, Bell Laboratories, Lucent Technologies.[4]

In their evaluations of the conference, participants gave high praise to the experience. The program participants, mentors, MSRI, and the funding agencies made the planning rewarding. The organizers were Bettye Anne Case, chair (Florida State University); Sue Geller (Texas A&M University); Carolyn Gordon (Dartmouth College); Dianne O'Leary (University of Maryland, College Park); Gail Ratcliff (University of Missouri, St. Louis); Jean Taylor (Rutgers University); and Sylvia M. Wiegand (University of Nebraska, Lincoln). Others not mentioned elsewhere in this report who worked on the planning committees were Isabel Beichl, Lynne Billard, Lenore Blum, Mary Ellen Bock, Sun-Yung Alice Chang, Jane Cullum, Sharon Frechette, Mary W. Gray, Jenny Harrison, Gloria Hewitt, Linda Keen, Edith Luchins, Carolyn R. Mahoney, Linda P. Rothschild, Alice T. Schafer, Bhama Srinivasan, Chuu-Lian Terng, Janice B. Walker, and Carol Wood.

Graduate Student and Recent Ph.D. Participants

Elizabeth S. Allman, University of North Carolina-Asheville
"Subgroup Separability and Hyperbolic 3-Manifolds"

Elizabeth A. Arnold, University of Maryland, College Park
"Using Hilbert Lucky Primes to Compute Gröbner Bases"

---

4. See "Outside the Academy," III.3 (this volume), for the contributions of Flinn, Holte, and Wright and "Inside the Academy," III.2, for that of Brucks.

Lora Billings, University of Delaware
"Newton's Method and Chaotic Attractors"

Andrea Codd, University of Colorado at Boulder
"Elasticity: Fluid Coupled Systems"

Sylvia Cook, The University of Iowa
"Two Star-Operations and Their Induced Lattices"

Sharon Frechette, Wellesley College
"Hecke Structure of Spaces of Modular Forms"

Sarah J. Greenwald, Appalachian State University
"Diameters of Spherical Alexandrov Spaces and Constant Curvature
    One Orbifolds"

Cheryl Grood, Swarthmore College
"Centralizer Algebras of SO(2n, C)"

Weiqing Gu, Harvey Mudd College
"Volume-Preserving Great Circle Flows on the 3-Sphere"

Rachel W. Hall, Pennsylvania State University
"Hecke C*-Algebras"

Deborah Heicklen, University of California, Berkeley
"Discretizing Randomly Perturbed Dynamical Systems"

Sanjukta Hota, Columbia State Community College
"A Mathematical Model for Carbon Dioxide Exchange during
    Mechanical Ventilation with Tracheal Gas Insufflation (TGI)"

Chris Hurlburt, University of New Mexico
"Differential Modular Forms"

Lois Kailhofer, University of Wisconsin–Milwaukee
"A Classification of Inverse Limit Spaces with Periodic Critical Points"

Annela Kelly, Northeast Louisiana University
"Analytic Measures"

Megan Kerr, Wellesley College
"New Homogeneous Einstein Metrics of Negative Ricci Curvature"

Sandra Kingan, Trinity College, D.C.
"Structural Results for Matroids"

Tanya L. Leise, Rose Hulman Institute of Technology
"Dynamically Accelerating Cracks along a Bimaterial Interface"

Jing-Rebecca Li, Massachusetts Institute of Technology
"Vector ADI: A Low Rank Right Hand Side Lyapunov Equation Solver,
    with Applications to Model Reduction"

Moira A. McDermott, Gustavus Adolphus College
"Tight Closure and Singularity Theory"

Gema A. Mercado, University of Arizona
"Formation of Hotspots and Dynamics of the Electric Field in Microwave
    Heating"

Dorina Mitrea, University of Missouri–Columbia
"The Transmission Problem for Multilayered Anisotropic Elastic Bodies
    with Rough Interfaces"

Helen Moore, Stanford University/Bowdoin College
"Gauss Map Omissions of Minimal Surfaces"

Susan Morey, Southwest Texas State University
"Associated Primes and Ideals of Graphs"

Regan E. Murray, University of Arizona
"Modeling Reaction Zone Dynamics of the Bioremediation
    Equations"

Nilima Nigam, Institute for Mathematics and Its Applications, University
    of Minnesota
"Variational Methods for Some Problems Exterior to a Thin Domain"

Ruth Pfeiffer, National Cancer Institute, National Institutes of Health
"Some Problems for Stochastic Processes with Hysteresis"

Claudia Polini, Hope College
"Ideals with Small Deviation"

Julianne Rainbolt, Saint Louis University
"Extensions of Periodic Linear Groups"

Victoria Rayskin, University of Texas at Austin
"Degenerate Homoclinic Crossings"

Vanessa Robins, University of Colorado at Boulder
"Computational Topology with Applications to Fractal Geometry"

Karen L. Shuman, Dartmouth College
"Signal Processing with the Jacobi Group"

Masha Sosonkina, University of Minnesota–Duluth
"Preconditioning Strategies for Linear Systems Arising in Tire Design"

Jenny Switkes, Claremont Graduate University
"Models of Coevolutionary Interaction"

Lesley Ward, Harvey Mudd College
"Brownian Motion and the Shape of a Region's Boundary"

Julia M. Wilson, University of Wisconsin-Milwaukee
"Non-uniqueness of Boundaries of CAT(0) Groups"

Golbon Zakeri, Argonne National Laboratory and the University of
     Wisconsin–Milwaukee
"You Too Can Optimize Using a Metacomputer"

A Mathematician at NIST Today

FERN Y. HUNT

Hunt is in the Mathematics and Computational Sciences Division at the National Institute of Standards and Technology.

Introduction

In these highlights of the work of a mathematician in a government laboratory, I will tell you about the similarities to as well as the differences from a mathematician working in academia. The National Institute of Standards and Technology (NIST) is the oldest government laboratory in the United States. It was founded in 1901 as the National Bureau of Standards; it now employs some 2000 scientists and engineers engaged in basic and applied research on the chemical, physical, and electrical properties of materials that are important to U.S. industry and commercial competitiveness. Much of this research is used to develop standards and measurement techniques that are used by industry to certify and maintain the quality of industrial products and processes. Mathematicians have been a part of NIST from its earliest days, providing mathematical expertise and collaborating with scientists and engineers on NIST projects. In addition, mathematicians in the past conducted pioneering work on various aspects of mathematical and computational science; these included numerical linear algebra and Monte Carlo methods. Research on these topics continues today, and newer topics such as parallel computation and mathematical modeling have become important areas of work as well. The project I am about to describe arose from a collaboration with scientists working at NIST who were interested in the problem of gloss loss in paint. I started out working as a mathematician developing a model of paint weathering and used some basic results from the theory of point processes—a topic that if not new to paint technologists is certainly not widely known. The project took an unusual direction as a result of a conversation with a computer scientist (I will describe later), and my role has evolved so that now I am coordinating a computer graphic rendering effort to develop photorealistic images based on light scattering measurement of surfaces. Paint modeling is not an obvious topic of research for most mathematicians—certainly not one with my training. I started working on this topic because one of my "hobbies" is random numbers. Shortly after I arrived at NIST, I attended a talk on clustering in random number sequences. It was one of the best presentations on randomness that I've ever seen. It combined simple, but vivid visual demonstrations (including a big jar of colored jelly beans and a Bingo card cutout of a random number table) of facts about recurrent events that are not well known outside the mathematical community. What made the presentation so surprising was that the speaker was a paint scientist—Lou Floyd of the Glidden Paint Company of Ohio. I realized then and there, this was a potential application area to watch. A few years later, I did get the opportunity to work with the NIST host for that talk, Jon Martin. As removed from the world of mathematics as might be, the paint industry is of great industrial importance. Its revenues include a significant share of the U.S. GDP, and its importance to the automobile industry, for example, comes from the fact that painting a car is the most expensive part of the automobile manufacturing process. There is a long list of other industries for which the issue of paint or coating appearance is critical—textiles, paper, housing, and traffic safety to name a few.

In the beginning, the purpose of our collaboration was to investigate the role of initial paint coating characteristics on subsequent gloss loss of a painted surface that erodes during weathering and exposure to ultraviolet light, as might take place when a car coating loses its gloss and intensity of color over the years due to exposure to sunlight.

1. Modeling and Simulation

The work reported here was done jointly with Jonathan Martin and Michael Galler of the Building and Fire Research Laboratory at NIST [4]. It is probably useful to first define what we mean by paint. We will take it to be a composite material consisting of a polymeric matrix (not the mathematical kind) called a binder and spherical pigment particles. In the gloss loss study, the spherically shaped pigments were assumed to be titanium dioxide with diameters ranging from 0.1 to 0.3 microns. We assumed that gloss loss was caused by surface roughening due to the fact that paint binder erodes under exposure to ultraviolet light. An experimental approach to this problem would involve determining the surface structure of the paint film, measuring the degree of gloss, and then systematically varying the surface structure by changing the paint composition in order to identify the relationship between surface structure, paint composition, and gloss loss over time. This would be an expensive and time-consuming process that could be facillitated by modeling. By examining a simplified caricature of the weathering process, we hoped to identify relationships between a limited number of coating characteristics and gloss (as measured by surface roughness). By quantifying these relationships, we hoped to organize some of the relevant coating variables that would need to be a part of any experimental effort. This would aid, and possibly shorten, the experimental design process and help to distinguish the important variables from the less important. It's important to note that we simulated changes in surface morphology and the layers of coating that are exposed during weathering, not the chemistry of the weathering process. We followed the changes in a two-dimensional 900×400 array of pixels representing a two-dimensional cross section of a paint film. For our purposes, the important parameters of this coating were:

- PVC = pigment volume concentration
      = volume of pigment particles / total volume
- size distribution of pigment particles
- clustering of pigment particles (degree of flocculation)

What Is Gloss? The mirrorlike appearance of a new car that is popularly known as gloss is called "distinctness of image" (DOI) by paint technologists. Human beings are good at detecting this attribute (until recently, they were better than the best glossometers), and gloss plays a strong (sometimes determining) role in the decision to purchase an automobile. Specular intensity, or mean reflectance, will be used as a quantitative measure of the DOI even though the correlation between the two is not perfect.

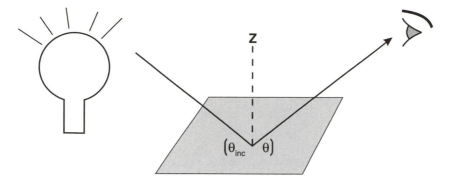

Figure 4. GLOSS. What is it? Specular gloss. $R(\upsilon)$ = ratio of field incident in direction υ to field incident in direction υ_{inc}—the incident direction. R is called the *reflectance*.

Better characterization of DOI is a complex problem involving human psychophysics and is beyond the scope of our discussion. However, we are working on rendering images of surfaces based on good physical models of light reflectance. This should be useful in dealing with some of these difficulties. The reflectance $R(\theta)$ can be described (see figure 4) as the ratio of the light field scattered from the surface in the direction θ to the field incident to the surface with direction θ_{inc} where the angles are measured as depicted in figure 4. To get an expression for the mean reflectance, our version of gloss, we will use physical optics theory as summarized by Beckmann and Spizzichino [1] that, in principle, predicts the angular distribution of light flux reflected off a rough surface as a function of the angle and wavelength of the incident light. After the weathering process has proceeded for a period of time, the top of the two-dimensional cross section is indeed rough in appearance. Weathering is an inherently random process, so the resulting roughened surface may be considered to be random. The theory we use assumes that for each fixed time T, the surface heights $\{z(x, y, \omega, T)\} - \infty \leq x, y, \leq \infty$ are the realization or sample of a Gaussian random process. We simplify the representation of heights even further by assuming they are independent of y, in view of the fact we simulated a two-dimensional cross section. The range of integration is therefore over a single horizontal coordinate x. The mean reflectance is:

$$\mu = \lim_{L \to \infty} \frac{1}{2L} \int_{-L}^{L} \zeta(x, \omega, T)dx \qquad (1)$$

where $z(x, y, \omega, T) = \zeta(x, \omega, T)$ is now the height. The variance of the height is then

$$\sigma^2 = \lim_{L \to \infty} \frac{1}{2L} \int_{-L}^{L} [\zeta(x, \omega, T)]^2 dx - \mu^2 \qquad (2)$$

The parameters μ and σ are limiting values that are independent of x and ω. The existence of these limits is a consequence of the fact that the random process $\{\zeta(x, \omega, T)\}_{x \in (-\infty, \infty)}$ is ergodic and has first and second moments [3].

Figures 5 and 6 show a numerical test of the Gaussian hypothesis (i.e., normally distributed in a single dimension x). For a fixed value of x, a normal probability plot was created by plotting values whose first coordinates were 100 heights at position x created from 100 different runs of the simulation. The second coordinate of these points came from a hundred normally distributed values with mean $\hat{\mu} = \left\langle \sum_{i=1}^{N} z_i / N \right\rangle$ and variance $\hat{\sigma}^2 = \left\langle \sum_{i=1}^{N} (z_i - \mu)^2 / N \right\rangle$ where $\{z_i\}_{i=1}^{N}$ are heights $z_i = \zeta(x_i, \omega, T)$,

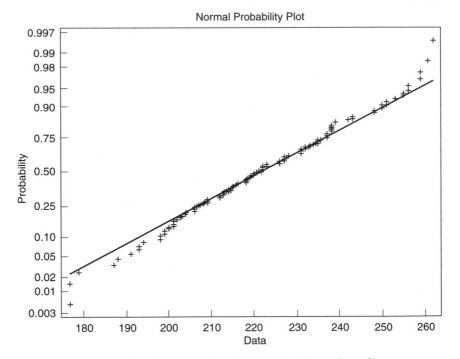

Figure 5. Position $x = 350$; data are heights of profile

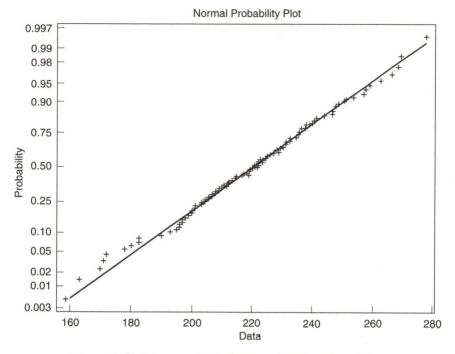

Figure 6. Position $x = 100$; data are heights of profile

$-L \leq x_i \leq L$ and $\langle r(\omega) \rangle$ is the average value of the random variable $r(\omega)$ over the 100 different runs. In the estimates $\hat{\mu}$ and $\hat{\sigma}$, limiting integrals have been replaced by finite sums. The number of points, N, was chosen so that the fluctuation in the values of the sums was relatively small from run to run, thus averaging over 100 runs was not strictly necessary. The horizontal and vertical coordinates were ordered in increasing value and then plotted. If the heights are normally distributed, the points should lie along the 45 degree line. The results of a calculation with $x = 100, 350, T = 1200$ time steps or about 50% of the weathering cycle, and $N = 800$ are shown. We see that agreement with the line is good and therefore consistent with the assumption of normality.

Gloss and Mean Reflectance. Surface roughness as defined by σ is connected with gloss because of the following equation for the mean reflectance:

$$\langle R(\theta) \rangle = \exp\left[-\frac{1}{2}\left(\frac{4\pi\sigma \cos(\theta)}{\lambda} \right)^2 \right] \tag{3}$$

when $\theta = \theta_{inc}$, the incident angle. $\langle R(\theta) \rangle$ is the expected value or average over all sample paths of the process. Note that the reflectance of a ray of light scattered off a randomly rough surface is random, thus we focus on the mean reflectance.

From equation 3 we can see that small roughness is associated with high gloss and vice versa. The derivation uses the assumptions about surface statistics just discussed [1].

Description of Simulation. To simulate photolytic degradation, a collimated uniform "beam" of UV light is projected onto the surface of the paint film, penetrating the binder matrx with a strength that decreases exponentially with depth. As the beam proceeds downward, it damages the binder directly by absorption by binder or indirectly as light is reflected off pigment particles and goes to the binder, where it is absorbed. It is assumed that the damage from the reflected light is uniform over the entire surface and includes binder located in areas that are shielded from direct radiation. Initially, pigment particles shield the binder below them. This eventually leads to the formation of "pedestals" that support single pigment particles as seen in the micrographs of Kampf et al. [5]. Eventually indirect and reflected radiation erodes the pedestals and the now loosened pigment particles are removed from the simulation. More details concerning the simulation can be found in [4].

2. Results of Simulation

The results of the simulations support the contention that paint films, composed of large numbers of small, well-dispersed pigment particles, retain gloss over the long term of the weathering cycle. During the early part of the weathering cycle, however, the opposite conclusions can be drawn.

Effect of PVC on Gloss. Several simulations with PVC values of 15%, 25%, and 35% were performed, each lasting 1500 time steps. Figure 7 shows that after an initial transient, the coatings with the largest PVC had the smallest σ (highest gloss) over the entire course of simulation. Thus, higher PVC leads to higher gloss. It is interesting to note that at the beginning of the simulation, coatings with the lowest PVC have the highest gloss (see figure 8).

Effect of Pigment Size Distribution on Gloss. The simulation used pigment particle of two sizes representing small and large particles, respectively, with the diameter of the large particle (43 pixels) being about three times that of the smaller (15 pixels). With S denoting the percent in the coating of small pigment particles, figure 9 shows the σ values of a set of

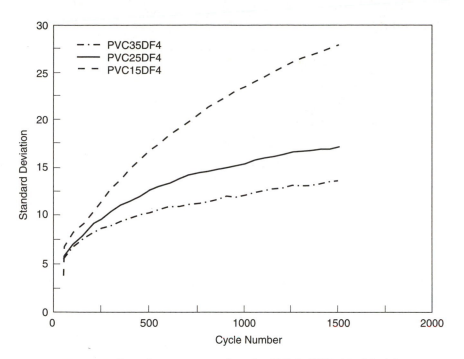

Figure 7. Roughness versus time for DF 4, PVC 15, 25, 35

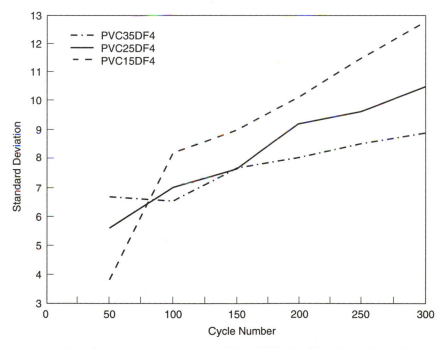

Figure 8. Roughness versus time for DF 4, PVC 15, 25, 35 (early in the simulation cycle)

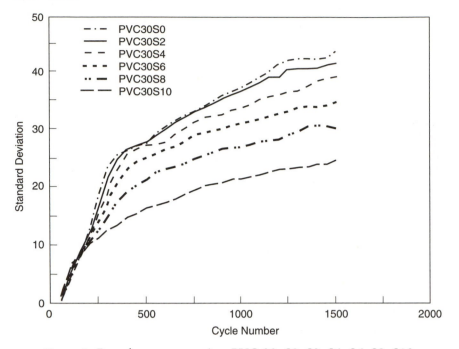

Figure 9. Roughness versus time PVC 30, S0, S2, S4, S6, S8, S10

simulations with varying initial values of $S = 0$, 20, 40, 60, 80, 100%. Except for an initial transient, the coatings with the larger percentage of small pigment particles retained a higher gloss throughout the weathering cycle. Early in the cycle, however (see figure 10), we see, as with PVC, a reversal where, for a period of time, coatings with the larger percentage of small pigment particles have the least gloss. These results suggest that if a manufacturer is interested in coatings that retain their gloss over the long run, then the coatings must contain a high percentage of small pigment particles. However, the kind of gloss that interests consumers is lost well into a weathering cycle [2], so there may be some advantage to using a coating with fewer small pigment particles, since such a coating is (e.g., in the case of titanium dioxide particles) much cheaper.

Effect of Pigment Dispersion on Gloss. At first, we modeled pigment particle dispersion by varying the minimum nearest neighbor distance (dF) beween randomly placed pigment particles in a manner consistent with the PVC. By randomly placed, we mean an approximation of a random close packing. When dF = 0, the pigment particles are allowed to touch each other. If dF is larger, the degree of pigment particle dispersion

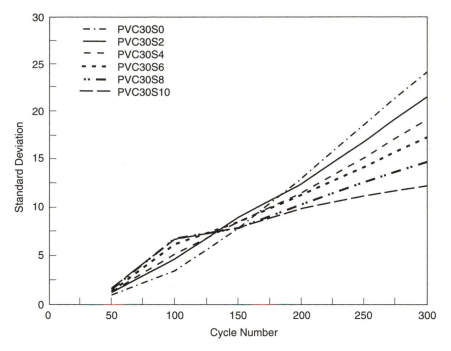

Figure 10. Roughness versus time PVC 30, S0, S2, S4, S6, S8, S10 (early in the simulation cycle)

is larger. In simulations using a number of coatings with several dF values, we found that gloss increases with increase in dF (i.e., better dispersion). This is visually illustrated in figures 11 and 12 showing the cross sections of two simulations at 1200 time steps.

Poisson Point Process. In an effort to derive an analytically tractable way to describe pigment particle dispersion or its opposite (flocculation), we introduced a model of pigment placement based on sampling from the Neyman-Scott point process. To model clustering or flocculation, N uniformly distributed points, (x_i, y_i), $i = 1 \cdots N$, are selected to be the centers of N flocculates or clusters of pigment particles. "Uniform" here means uniformly distributed in the rectangular cross section. However, all these steps can be easily carried out in a three-dimensional setting. Within the ith cluster, we select M_i points uniformly distributed in a circle with center (x_i, y_i) and radius (R). The numbers N and M are discrete Poisson random variables with mean μ and v, respectively. Thus the parameters of this pigment placement scheme are μ, v, and R. It is convenient to introduce the variable $\lambda = \dfrac{\mu}{A}$, the number of clusters per unit area, where A is the area of the cross section. A rough approximation of the pigment volume

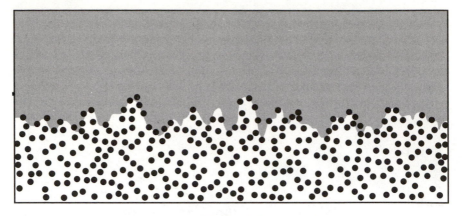

Figure 11. PVC = 35%; $dF = 0$

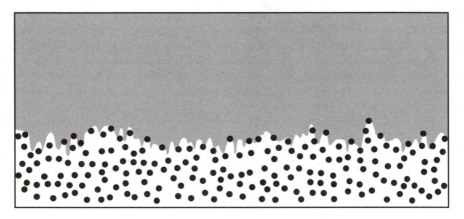

Figure 12. PVC = 35%; $dF = 4$

concentration, the theoretical PVC or TPVC, can be expressed in terms of the model parameters. In the two dimensional setting this is

$$TPVC = \frac{\mu v S}{A} = \lambda v S \qquad (4)$$

where S is the area of a single pigment particle. In three dimensions, areas are replaced by volumes. Suppose Ω is a region of area, $A(\Omega)$. If $n(\Omega)$ is the number of pigment particles in Ω, then we define the degree of flocculation to be the amount of fluctuation in the number of pigment particles as one samples different regions Ω. We express this in terms of the variance of $n(\Omega)$ [6]

$$var[n(\Omega)] = 2\pi p^2 G_1 + pvG_2 + pA(\Omega) - (pA(\Omega))^2 \qquad (5)$$

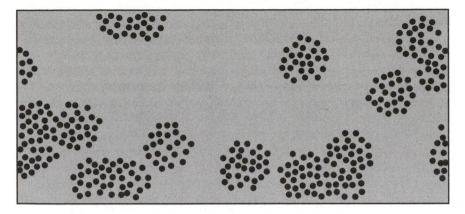

Figure 13.

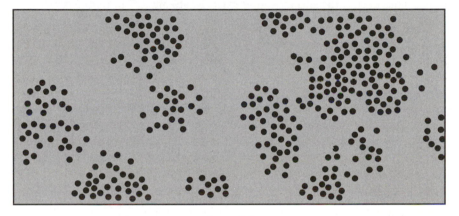

Figure 14.

where $p = \lambda v$, G_1 is a geometric constant depending on the area and shape of Ω and G_2 is another geometric constant that depends only on R which is fixed. The constant G_1 is known for simple geometric shapes Ω. Equation 5 shows that the variability of $n(\Omega)$ can be increased keeping the TPVC fixed by increasing the mean cluster size v but keeping p fixed. Indeed, by equation 4, $TPVC = pS$. We therefore have a way of describing various degrees of dispersion. We illustrate this with figures 13 and 14 that show two realizations of the Neyman-Scott process. The first figure has $\mu = 10$, $v = 30$, and the second has $\mu = 30$, $v = 10$. The TPVC is the same; both figures were chosen so that the actual PVC (in planar terms) differs from the TPVC by no more than 0.1%. However, the pigments in figure 14 appear to be better dispersed than those in figure 13. The figures illustrate a situation that might arise when a surfactant is added to a flocculated paint, because that addition changes the paint dispersion without changing the PVC.

3. Computer Graphic Rendering of Surface Appearance

Members of the more mathematically oriented divisions at NIST often get together for lunch. One day, a year or so before I began to work on the paint project, Holly Rushmeier, a computer scientist in a neighboring division, joined our table. Talk turned to her work. Holly had made fundamental contributions to computer rendering: a new, but very rapidly developing area of computer graphics that used models of light scattering and light source characteristics to develop photorealistic images. This technology had the potential to enable product designers to visualize the appearance of a surface based on information about its light-scattering properties even if the surface didn't exist. Realizing this potential depended first of all on being able to get this information. Optical properties of a surface composed of a given material are summarized in a function known as the BRDF—short for the bidirectional distribution function. Although there is some discussion as to how much, it is clear that the accuracy of a rendered image depends a great deal on how well the BRDF is represented either by data from direct optical measurements, or by mathematical models, or by a combination of the two. Unfortunately, the BRDF is a complicated function, so trade-offs between accuracy and computational efficiency must be made. As Holly, Jon, and I worked to develop ideas for a follow-on project, one of its goals became clear. Since BRDF approximations for rendering programs for computer graphics applications were developed in isolation from the BRDF measurement and modeling community, we would have to build a software interface using the high-precision measurements and modeling done at NIST into formats suitable for input into a rendering program. Just as this new project began, Holly left NIST to take a position at IBM Watson Laboratory. I took over management of this part of the project. Much of what I do involves coordinating the work of optical physicists, material scientists, and computer scientists. The interface work is now completed, and work is proceeding on creating and evaluating images of selected materials. The final figure (figure 15) shows two coated black glass samples with surface roughnesses, 201nm and 805nm, respectively, as measured by the standard deviation we previously discussed. The figures illustrate the decrease in gloss associated with the increased roughness. These images were created by Gary Meyer and Harold Westlund of the University of Oregon from NIST measurements of two coated black glass samples. The samples were 2000 miles away from the

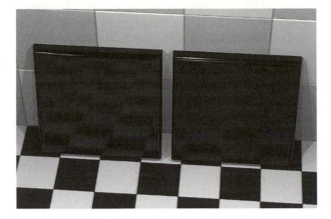

Figure 15. Decrease in gloss associated with increased surface roughness

renderers. A few months later, I flew to Oregon and brought the samples along for visual comparison. Holding up one of them, Meyer said this was the second highlight of his career. It was also a highlight for me—playing a role as part of an interdiscplinary team, using my skills as a mathematician (not in proving theorems but in being able to see the "big" picture), and in communicating and coordinating the work toward its completion.

References

1. P. Beckman, A. Spizzichino, *The Scattering of Electromagnetic Waves from Rough Surfaces* (Norwood, Mass.: Artech House, 1987).
2. J. Braun, "Gloss of Paint Films and the Mechanism of Pigment Involvement," *Journal of Coatings Technology* 63, no. 799, 43, 1991.
3. J. L. Doob, *Stochastic Processes*, Wiley Classic Library Edition (New York: John Wiley and Sons, 1990).
4. F. Y. Hunt, M. A. Galler, and J. W. Martin, "Microstructure of Weathered Paint and Its Relation to Gloss Loss: Computer Simulation and Modeling," *Journal of Coatings Technology* 70, no. 880, 45–54, 1998.
5. G. Kampf, W. Papenroth, and R. Holm, "Degradation Processes in TiO_2 Pigmented Paint Films on Exposure to Weathering," *Journal of Paint Technology* 46, no. 508, 348–356, 1964.
6. D. Stoyan and H. Stoyan, *Fractals, Random Shapes and Point Fields: Methods of Geometrical Statistics* (New York: John Wiley and Son, 1994).

What Use Is Statistics for Massive Data?

DIANE LAMBERT

Lambert is a Bell Labs Fellow and head of the Statistics and Data Mining Research Department at Bell Labs, Lucent Technologies.

1. What Is Statistics?

Most people are first exposed to statistics in a required undergraduate course that is filled with a hundred or more other students who are also required to be there. Partially in response to demands from other departments, the introductory course focuses on traditional methods for the kinds of small experiments, studies, and surveys that are part of the curriculum for those other departments. Typically, nearly all, if not all, the datasets considered have fewer than 100 observations with only a few variables on each, and the objective is to apply simple methods for finding means, variances, confidence intervals, and p-values, and for fitting linear regression models. There is no discussion of computing, probably both for lack of time and because sophisticated computing is not needed. There is some discussion of mathematical properties of techniques, but usually not in ways that are intuitive. Inference centers on confidence intervals and p-values, which are simple to compute but subtle to explain. What most people take away from such a course is that statistics is neither particularly hard nor interesting (except for some peculiar use of language), is only useful for simple data and simple questions (if at all), and anyone can apply it (although it is best avoided), but hardly anyone can understand it.

But statistics is much different, and not just broader or deeper, from what is taught in undergraduate courses. Simply stated, statistics is about extracting information from data that are noisy or uncertain. The unstated position is that all data are noisy. Twenty measurements from a small experiment are noisy, and zillions of transaction records in a data warehouse are noisy. The data may represent a process (like purchasing behavior) that varies across people and from one transaction to the next, even for the same individual. The data may have errors or be incomplete (even in the U.S. Census). Or, the goal may be prediction, so the available data are only a noisy representation of data to come. Statistics, then, is about overcoming the uncertainty and variability (noise) in the data to reveal the information hidden within.

Statisticians take on many roles to analyze data. Some design graphics that bring out the structure of both the random and nonrandom components. Data visualization from the perspective of a statistician is discussed in [24], [10], and [11]. Interactive, dynamic data visualization for the purpose of statistical analysis is discussed in [21]. Statisticians have also designed flexible programming environments for exploring, analyzing, and visualizing data. Two popular languages for statistical computing are S [5] and Lisp-Stat [23]. The 1999 ACM Software Systems Award was given to S because it "has forever altered how we analyze, visualize and manipulate data." Previous winners of this prestigious award include unix, T_EX, and the World Wide Web. Recently, a global collaboration has begun to build a statistical computing environment called Omega that is based on distributed components ([6] and [22]).

Most statisticians, though, spend most of their time working with data, breaking it down into deterministic and random noise components. Usually the first step is informal. The data are plotted in many different ways—with different filters, transformations, and views selected each time—to discover patterns and structure informally. Then those patterns are set aside, often by subtraction, and the residuals of the data from the pattern are examined visually in many different ways to reveal any remaining structure or patterns in the data. Then that structure is subtracted, and the fit and subtract steps are repeated on the new residuals, on and on (with some backstepping, perhaps, to correct missteps in modeling) until no structure can be found. In this way, a statistical model is built empirically.

The basic statistical model is simply

$$data = mean + noise,$$

but both the mean, which is a deterministic representation of the data, and the noise, which represents the variability and noise around the deterministic model, can be complicated. The mean might be linear, nonlinear, additive, multiplicative, nonparametric, a step function (as in a tree), an assignment function (as in classification or clustering), or smooth as a function of a set of explanatory variables. The noise describes the variation in the data, which affects prediction and the reliability of estimates. The noise might be independently normal, correlated, long-tailed, weighted to account for nonrandom sampling, length-biased, spatial, censored, or hierarchically structured, for example. If the noise does not affect the data additively, a more appropriate model is

$$data \sim F_\theta \, mean \, (data) = g(\theta),$$

where F_θ is a distribution that describes the random variability of the data around the deterministic model $g(\theta)$ for the data. F_θ can be a well-known distribution, like the Bernoulli in the case of logistic regression. It can include correlation among observations, mixing to accommodate outliers, weighting to account for nonrandom (but probabilistic) sampling and length biasing, for example. The mean parameter θ may depend on a set of explanatory variables or predictors, and it can be as complicated for an arbitrary distribution F_θ as it is for normal noise. Many kinds of models are possible, each representing both the structure and the variation in the data (see [20], for example). The more complicated the process that generated the data, the more complicated the statistical model may need to be.

A basic premise that drives much of statistics is that modeling the noise component well is as important as modeling the mean structure well. The noise component is needed for prediction, to understand what is likely and what is not. It is also needed to understand how reliable the estimated mean or prediction is. The noisier the data, the less reliable the estimate. A basic tenet of statistics is that using an estimate or prediction without some sense of its reliability is downright dangerous. Although the range and nature of statistical models may seem bewildering to nonstatisticians, a wealth of models is felt to be crucial for representing data when the goal is reliable inference or prediction.

Most statisticians have not dealt with massive data, though. Most do not have the necessary computing or database management skills; their skills lie elsewhere. But it is possible to apply statistics to massive data. Section 2 describes what a statistician sees when looking at massive data and how that differs from what a computer scientist who works with massive data, often called a *data miner*, sees. Section 3 then takes a closer look at applying statistics to fraud detection, which is a problem that has been tackled by many data miners from computer science. Section 4 offers final thoughts on the role of statistics.

2. Massive Data

A Database View. Statisticians find probabilistic models a natural way to think about data, but not everyone looks at data that way. Another view, which is common in computer science, holds that the data in a database are not random or uncertain but an accounting of everything that has happened. Thus, the need is not to make inferences about an underlying process that is observed imprecisely or to estimate deterministic and random models for the data. Instead, the need is to tabulate and process the data.

Statistics might then be used to approximate answers to queries that in principle can be known exactly, given enough computing power and time, but in practice are too costly to obtain (e.g., [18]). How much was spent last month? How many people bought those two products in Florida last weekend? If all the data cannot be processed to answer the questions of interest, then the data can be sampled (not necessarily randomly), and knowledge of the estimating procedure and sampling scheme can be used to provide probabilistic bounds on the difference between the approximate answer and the answer that would have been obtained if all the data had been processed.

The notion that the data are not noisy, or at least noise does not have to be explicitly incorporated into models, is strange to statisticians but not to computer scientists. Many if not most of the data mining tools that computer scientists use were not designed with noise in mind. Instead, a common goal of computer scientists in data mining is to find interesting patterns, associations, and rules from massive sets of data without considering the process that generated the data at hand. For example, an interesting rule might be: "People who buy beer between 7 p.m. and 9 p.m. buy diapers at the same time," or $X \wedge Y \Rightarrow Z$. The value of a rule is measured by its *support* and *confidence*, where support is the fraction of database elements that satisfy conditions X, Y, and Z and confidence is *support(X, Y, Z)/support(X, Y)*. This has the flavor of probability, in the sense that *support* $= P(X \wedge Y \wedge Z)$ and *confidence* $= P(Z \mid X \wedge Y)$ if each database record or item is given equal probability. The resemblance is only superficial, however, because there is no sense of randomness or uncertainty and the goal is to reason only about the data in the database, not about a larger population, judging performance solely by the fraction of a database that satisfies the rule exactly. Finding a good rule then reduces to finding frequent itemsets (which is a hard computational problem).

In contrast, a statistical approach would consider the data in a database as a sample (not necessarily random) from a larger population, such as all people who buy at convenience stores, or a random process, such as disease infection. A statistician would then define "interesting" in terms of a probability model, interpreting "interesting" as current or future observations that are rare under the model, for example. This leads to two important differences with a computer science–based approach. First, what is interesting can be either rare or common. Second, the model can be used to evaluate rules that have not been observed in the data. That is, a statistician uses a model to *estimate* $P(Z \mid X = x, Y = y)$, even at unobserved values of (x, y, z), while the computer scientist relies on partitioning a database and *counting* records with $X \wedge Y$ and $X \wedge Y \wedge Z$.

The view that the data have no noise and extracting information from data amounts to finding frequent itemsets is appropriate in some contexts, but not all. It is not enough when the unit of analysis is small and the answer cannot be computed from the data with certainty, even if all the data are analyzed. Will this customer switch to another wireless service provider? How much will be spent on those two products in that convenience store in Florida next weekend? Noise is also important in dynamic modeling when older data are less relevant than newer data. In that case, a statistical approach allows gradations of uncertainty rather than just binary relevant/irrelevant decisions. Finally, relying solely on counting instances in a database violates the principle of "borrowing strength" or combining many small, different but related problems to make larger inferences. Questions that require going beyond counting instances are the ones that most interest statisticians.

But processing and counting massive data is so difficult that it can overshadow data analysis. The foreword to *Advances in Knowledge Discovery and Data Mining* states:

> Finding new phenomena or enhancing our knowledge about them has a greater long-range value than optimizing production processes or inventories and is second only to tasks that preserve our world and our environment. It is not surprising that it is also one of the most difficult computing challenges to do well. [25]

This quote often takes statisticians by surprise because, from their perspective, the challenges in knowledge discovery go far beyond processing, scaling up algorithms, or using the right database management system. Traditional issues of data visualization and analysis are as challenging as is computing because the data can have complex structure and sources of errors and variability.

A Hard Example: Web Analysis. Suppose the goal is to analyze the useability of a website, and the data consist of the clickstreams of all visitors to the website over a period of time. The simplest approach is to tally the data, counting page hits broken down by time period or geography or referral engine, for example. Counting does not reflect how people use the website, though. Useability requires monitoring how a user traverses pages at the site and how the user changes when the website is revisited. Does a new visitor keep returning to the home page, perhaps because starting over seems the only way to get to the relevant location? Do people look at one page briefly and then leave the site? Do people who are referred from one search engine browse differently from people who are referred by a different search engine? Do people who visit a site frequently differ from those who visit once? How much revenue is generated during a visit to an online store? These questions require thinking of the *visitor* as the unit of measurement, not the entry in a log file. This leads to thorny questions, such as what is a visit? Understanding the bias in the data is also important. Which visitors are of interest? If uninteresting visitors, like robots that access all web pages, dominate the data, then treating all visitors equally introduces bias. The fact that most people look at the home page and then leave may be important, but keeping these shallow visitors in all stages of the analysis may only obscure how the website is used by those who use it at all. Identifying the interesting visitors is difficult not simply because there are many visitors to consider but also because "interesting" is a hidden label that changes as websites evolve and has to be deduced by considering many visitors. Finally, ignoring the tree structure of pages and paths, which are part of the context of the problem, can easily produce misleading results.

The first step in any statistical analysis is to look at the data in multiple ways. Visualizing the behavior of visitors at a website in a way that respects the structure of the data is difficult, but some progress has been made. For example, Mark Hansen and Wim Sweldens of Bell Labs have introduced *synchronized browsing*, which allows an analyst to browse web pages and simultaneously see the results of analyses in the browser, thus tying the structure of the analysis to the structure of the website. As the content provider clicks through the website, results of an analysis of visitors to the page, such as where they came from, what they did on the page, how long they stayed, and where they went next, are displayed. This gives the content provider a sense of the structure of the data that can hint at more formal models of the "average" user, the "typical" way the website is used, and the range of likely departures from the typical and average. Thus, the visualization captures the context, the mean, and the noise of the data.

Although there is a massive amount of data in a web log file, we may still be data-poor when the dynamic tree structure of the data is taken into account. There may be many more ways to traverse a website than there are visitors, and the possible paths change continually as pages are added, deleted, and edited. Still, if a statistical model can be used to describe a visit, then it is possible to make inferences about paths that have never been observed. Even if a complete path has never been observed, visitors may have used parts of the path. With a statistical model, the data can be used to estimate probabilities of the partial paths which can then be combined to estimate the probability of a complete, unobserved path.

Nonstatistical data mining problems, like clustering pages or users, that computer scientists have focused on are also important, but these can be much richer than often assumed. For example, the goal may be not just to segment users into homogeneous groups but to understand how user segments are changing and how the changes relate to changes in the website. A static measure of performance, such as an overall misclassification rate, that gives one summary of performance can then be seriously wrong about current performance. Generally, good performance measures cannot be developed without close collaboration with subject matter experts.

The fact that it is dangerous to analyze data without knowledge of the context has been underscored by many people involved in analysis of data. But it is worth repeating because ignoring it leads to silliness, which can discredit all of statistics and data mining. An April 1998 article in *Forbes* entitled "Diaper-Beer Syndrome" gives a long list of failed attempts to build data warehouses for data mining and an equally long list of unrealistic claims, including some that grew out of the often-cited case that found a correlation between buying beer and diapers in one small convenience store in early evening hours. Several of the papers in [12] also emphasize that a set of general tools applied in a vacuum without taking the context into account is likely to lead to useless results (e.g., [1] and [9]).

In principle, then, statistics does have a role to play with massive data. But has statistics had any success in analyzing massive data? The rest of this paper considers that question in the context of fraud detection, a topic that has previously been discussed in the data mining and knowledge discovery literature (e.g., [3] and [4]). More information on fraud detection is given in [4].

3. Statistical Fraud Detection

There are many kinds of telecommunications fraud. In calling card fraud, a stolen credit card is used to place a call. In wireless fraud, a cellular phone may be cloned. In wireline fraud, a hacker may break into a university's telecommunications network and route calls over the network until the fraud is detected. In subscription fraud, a customer initiates service without intending to pay. Our goal is to find such fraud as quickly as possible by scoring each call for fraud while it is active or as soon as it has ended and then updating a fraud score for the account by the new call score.

If we could predict the next legitimate call on an account precisely, we could score calls for fraud by comparing the observed call with the predicted call. The new call would be scored zero if it matched the predicted call exactly, and scored one if not, and the cumulative call score would measure the severity of the fraud. Of course, calls cannot be predicted without uncertainty, so the natural alternative is to score calls against a probabilistic or predictive model instead of against an exact prediction. The predictive model has to accommodate the variability in the calling pattern, so calls within the usual range of variability do not falsely trigger a fraud alarm. Consequently, fraud detection is as much about describing customer-specific noise as it is about describing a customer-specific average.

Probabilistic call scoring is not easy. First, there are many, many calls, and scoring must be fast enough to keep up with the data flow. So any score must be simple to compute and any prediction model must be simple to maintain. Second, there can be millions of callers, with wildly diverse calling patterns. Some customers make a few calls a week; others make a few thousand calls a day. Some customers never make calls out of business hours; some never make calls in business hours. Some never call China; some make many calls to China. Scoring must be appropriate for *any* caller, so a prediction model must be maintained for *each* customer. Third, the customer base is often volatile, with new customers making calls each day, so the prediction model must be easy to initialize meaningfully for a caller with no previous history. In statistical terms, we are forced to "borrow strength" from our knowledge of previous customers to initialize the distribution for a new customer quickly. Finally, the prediction model has to adapt as more is learned about the customer, without intervention or off-line processing, but the model should adapt only to legitimate behavior, not fraud. Thus, finding a good algorithm for fraud detection involves much more than training a classifier on legitimate and fraudulent calls or legitimate and fraudulent account summaries; it involves modeling legitimate and fraudulent calls probabilistically and scoring calls against both models.

From a statistical perspective, a call can be represented by a random vector $X = (X_1, \ldots, X_k)$, where X_1 might be call duration, X_2 call timing (day-of-week or hour-of-day), X_3 call rate, X_4 geography of the called number (hierarchically organized by country, region, city, exchange, for example), and so on until all the information that can be gleaned from a calling record is included. A legitimate caller i at the time of its call n has a multivariate distribution $C_{i,n}$ on $X_{i,n}$. A fraudster has a, hopefully different, multivariate distribution F. The distributions $C_{i,n}$ and F are high-dimensional, with possibly complex interactions among the variables. For example, call duration may depend on the time of day, so considering only the one-dimensional, marginal distribution of call duration, ignoring the differences in peak and off-peak hours, would be misleading. In any case, the statistical problem is to score a call according to whether it is more likely to be fraud (have come from F) or to be legitimate (have come from $C_{i,n}$), so the first step is to estimate $C_{i,n}$ for each caller i and the fraud distribution F as well as possible, given the constraints on space and processing time.

An ideal estimator of $C_{i,n}$ or F would be

- *nonparametric*, to capture the full set of behaviors over the customer base,
- *short*, to meet constraints on storage space,
- *sufficient*, so the $C_{i,n}$s and F represent all the data relevant to detecting fraud,
- *multivariate*, to capture important dependencies among calling variables,

- *dynamic*, to enable updating $C_{i,n}$ with every unsuspicious call that the customer makes,
- *easily initialized*, so a reasonable starting estimate can be assigned to a new customer, and
- "*optimal*," to predict most customers well and a significant fraction very well.

Histograms (an array of bins and corresponding relative frequencies) are simple nonparametric estimates that have the required properties (see [8] for details). Either fixed-depth histograms with variable endpoints and fixed heights for the bins or fixed-width histograms with variable heights and fixed endpoints can be used. Choosing the set of histograms to monitor amounts to choosing the set of marginal and conditional distributions to be monitored. For example, should durations of peak and nonpeak calls be monitored separately or together? A complication is that to simplify processing the same set of distributions has to be monitored for all customers.

A statistical approach to designing $C_{i,n}$ is described in [19]. They assume that there is a set of "legitimate" historical calls for tens or hundreds of thousands of customers over several months, where legitimate means that the customers experienced no fraud during the period or so little fraud that it was not detected. The set of conditional distributions to track is then chosen by applying χ^2 tests of independence to each customer separately, and combining the resulting p-values across accounts to find the conditioning variables that are needed to model most customers well. Because the procedure is based on statistical testing, it explicitly accommodates uncertainty and limits the structure of the model to one that can be supported by the data. It is also feasible for large databases of customers because each χ^2 test is quick to compute. The resulting structure or set of histograms is called a *signature*. Each customer i at the time of its call n has its own signature $C_{i,n}$.

The next task is to define a way to reestimate or incrementally update a signature with every new unsuspicious call that the customer makes. Call-by-call updating restricts processing to the subset of accounts that have new calls and avoids the need to retrieve past calls from a database, which is inevitably slow. It also enables call-by-call fraud detection. In the statistics literature, incremental updating is called sequential estimation, and it has a long history. Perhaps the most common sequential estimation method, one that is equally familiar to engineers, is exponentially weighted moving averaging (EWMA). If a standard, fixed-width histogram has estimated bin probabilities $p_{i,n}$ after call n, and call $n+1$ is represented by a vector $Z_{i,n+1}$ that is zero in every bin except the one that contains the observed value for call $n + 1$, then the EWMA updated bin probabilities are

$$p_{i,n+1} = (1 - w)p_{i,n} + wZ_{i,n+1},$$

where $0 < w < 1$. The weight w controls how fast old calls are aged out of the signature because the current call has weight w and the kth earlier call has weight $w(1 - w)^k$. With larger w, the estimate adapts more quickly to changes in a customer's calling pattern, but it is also more variable because the effective sample size or number of calls that have nonnegligible weight is smaller. Variants of exponentially weighted moving averages can be used to update fixed depth histograms [7] and top-seller kinds of histograms that try to retain only the bins with the highest frequencies [18].

Sequential estimators require initial values to start, so signatures require initial values. Statistics can be used to find them. Our method assigns each new customer to several customer segments, one for each component of the signature, on the basis of information about its first few calls. Each customer segment has an average histogram that is used to initialize the signature component of any customer assigned to that segment. These customer segments partition the space of all possible customers, but there are two major differences between standard database partitioning and our approach. First, we initialize each component of the signature separately, because the signature is a product of its components by design. This gives a huge number of possible customer segments or products of signature components, some of which were not even observed in the training set. The second difference is that each customer can diverge from its initial assignment and move to its own "segment of one" over time because the initial signatures evolve through exponential updating. Details are given in [8].

The statistical problems in fraud detection do not end with algorithm design. Evaluating a fraud detection algorithm is as challenging as designing one. For example, fraud management centers track the fraction of *investigated* accounts that are mislabeled as fraud, while researchers track the fraction of *all* accounts that are mislabeled as fraud. Thus, the researcher has a larger denominator, a much smaller false alarm probability, and a more optimistic view of performance. Yet changing to the fraud analyst view is hard because that requires modeling the process for deciding which suspicious accounts to investigate first. Moreover, because fraud detection is ongoing, performance has to be tracked over time, not just summarized at the end of a test period. There are also many facets to performance, such as the time to detect fraud, the number of false alarms, and the losses from fraud. Simulating a fraud management center by running the proposed algorithms for scoring calls and accounts and prioritizing high-scoring accounts on a huge set of calls may be the only way to obtain valid estimates of the dynamic performance of a fraud detection system. But then statisticians can again be applied to design the simulation and its evaluation.

4. Final Thoughts

Computing is critical to the analysis of massive data, and statisticians have much to learn from computer scientists about computing with massive data. At the same time, statistics puts new demands on computer science. For example, there is a need for database management systems that allow more than rudimentary ways to explore data. Many statisticians manage data by extracting what they need from a database, applying tools like **awk** and **perl** to the selected data to reduce or aggregate the data further, and then analyzing the reduced or partially analyzed data in a statistical environment like S. This works, but it is not convenient and is probably far from the state of the art. Scaleable algorithms are needed, too, but the ability to explore data and fitted models conveniently is more fundamental.

There are also aspects of data mining in computer science that statisticians should adopt. A statistical model is a principled way to reason about data, but it may not be the best way to start exploring data. Partitioning the data, especially when there is a lot of data, may be better, and it has led to some new powerful techniques of model building, such as boosting ([16] and [17]). Even with procedures like boosting, though, a model helps us to understand why the procedure works, and thus where it works best and where it fails. (See, for example, [15] and [2].)

Massive data raise many questions beyond analysis that could be much better addressed by statisticians and computer scientists working together than by either group working alone. For example, there is growing concern about protecting the confidentiality of the massive data collected in log files on the Internet. Confidentiality has been discussed in the statistics literature for decades, but most of the proposals that statisticians have made are not feasible for large sets of data. Data cleaning is another area of common interest.

Statistics has traditionally had strong ties to mathematics, both for designing new methodology and evaluating the performance of methods. Mathematics continues to be important to the analysis of massive sets of data. Probability is fundamental to every step in statistical modeling. Take away probability, and there is not much left to the approach to fraud detection described in section 3. Probability also plays a role in computing. It is basic to the design and performance of MCMC (Markov Chain Monte Carlo), for example, which is the workhorse of Bayesian model fitting. Other branches of mathematics that are important to understanding massive data include optimization, Bayesian networks, coding, compression, approximation theory, functional analysis, and algebra. Extracting information is not simply computing, nor is it computing combined with statistics. Mathematics continues to provide the principles and foundations for much of statistics. There is much to learn from many sources.

Acknowledgments

Special thanks go to John Chambers, Bill Cleveland, Mark Hansen, José Pinheiro, and Don Sun of Bell Labs and Rick Becker of AT&T Labs.

References

1. R. J. Brachman and T. Anand, "The Process of Knowledge Discovery in Databases," in *Advances in Knowledge Discovery and Data Mining*, ed. U. M. Fayyad, G. Piatetsky-Shapiro, P. Smyth, and R. Uthurusamy (Menlo Park, Calif.: AAAI Press, 1996), 37–57.

2. P. Buhlmann and B. Yu, "Analyzing Bagging," *Annals of Statistics* 30, 2002, 927–961.

3. P. Burge and P. Shawe-Taylor, "Detecting Cellular Fraud Using Adaptive Prototypes," in *AI Approaches to Fraud Detection and Risk Management*, ed. Tom Fawcett, Ira Haimowitz, Foster Provost, and Salvatore Stolfo (Menlo Park, Calif.: AAAI Press, 1997), 9–13.

4. M. Cahill, F. Chen, D. Lambert, J. C. Pinheiro, and D. X. Sun, "Detecting Fraud in the Real World," in *Handbook of Massive Datasets*, ed. J. Abello, P. Pardalos, and M. Resende (New York: Kluwer Press, 2002), 911–933.

5. J. M. Chambers, *Programming with Data* (New York: Springer, 1998).

6. J. M. Chambers, "Users, Programmers and Statistical Software," *J. Comp. and Graph. Statistics* 9, 2000, 404–422.

7. F. Chen, D. Lambert, and J. C. Pinheiro, "Incremental Quantile Estimation for Massive Tracking," *Proceedings of ACM SIGKDD International Conference on Knowledge Discovery and Data Mining* (ACM, 2000), 516–522.

8. F. Chen, D. Lambert, J. C. Pinheiro, and D. X. Sun, "Reducing Transaction Databases, without Lagging Behind the Data or Losing Information," Technical report, Bell Labs, Lucent Technologies, 2000.

9. P. Cheeseman and J. Stutz, "Bayesian Classification (AutoClass): Theory and Results," in *Advances in Knowledge Discovery and Data Mining*, ed. U. M. Fayyad, G. Piatetsky-Shapiro, P. Smyth, and R. Uthurusamy (Menlo Park, Calif.: AAAI Press, 1996), 153–180.

10. W. S. Cleveland, *The Elements of Graphing Data* (Summit, N.J.: Hobart Press, 1985, 1994).

11. W. S. Cleveland, *Visualizing Data* (Summit, N.J.: Hobart Press, 1993).

12. U. M. Fayyad, G. Piatstsky-Shapiro, P. Smyth, and R. Uthurusamy, eds., *Advances in Knowledge Discovery and Data Mining* (Menlo Park, Calif.: AAAI Press, 2000).

13. U. M. Fayyad, S. G. Djorgovski, and N. Weir, "Automating the Analysis and Cataloging of Sky Surveys," in *Advances in Knowledge Discovery and Data*

Mining, ed. U. M. Fayyad, G. Piatetsky-Shapiro, P. Smyth, and R. Uthurusamy (Menlo Park, Calif.: AAAI Press, 1996), 471–494.

14. T. Fawcett and F. Provost, "Adaptive Fraud Detection," *Data Mining and Knowledge Discovery* 1, 1997, 291–316.

15. J. Friedman, T. Hastie, and R. Tibshirani, "Additive Logistic Regression: A Statistical View of Boosting," *Annals of Statistics* 28, 2000, 333–407.

16. Y. Freund, "Boosting a Weak Learning Algorithm by Majority," *Information and Computation* 121, 1995, 256–285.

17. Y. Freund and R. E. Schapire, "Experiments with a New Boosting Algorithm," in *Machine Learning: Proceedings of the Thirteenth International Conference* (San Francisco: Morgan Kaufmann, 1996), 148–156.

18. P. B. Gibbons, Y. Matias, and V. Poosala, "Fast Incremental Maintenance of Approximate Histograms," *Proceedings of the 23rd International Conference on Very Large Databases* (VLDB), Athens, Greece, 1997, 466–475.

19. D. Lambert and J. C. Pinheiro, "Mining a Stream of Transactions for Customer Patterns," *Proceedings of the Seventh ACM SIGKDD International Conference on Knowledge Discovery and Data Mining* (ACM, 2001), 305–310.

20. P. McCullagh and J. A. Nelder, *Generalized Linear Models* (New York: Chapman and Hall, 1989).

21. D. F. Swayne, D. Cook, and A. Buja, "XGobi: Interactive Dynamic Data Visualization in the X Window System," *J. of Comp. and Graph. Statistics* 7, 1998, 113–130.

22. Duncan Temple Lang, "The *Omega* Project: New Possibilities for Statistical Software," *J. of Comp. and Graph. Statistics* 9, 2000, 423–451.

23. L. Tierney, *LISP-STAT: An Object-Oriented Environment for Statistical Computing and Dynamic Graphics* (New York: John Wiley and Sons, 1991).

24. E. R. Tufte, *The Visual Display of Quantitative Information* (Cheshire, Conn.: Graphics Press, 1983).

25. G. Wiederhold, "On the Barriers and Future of Knowledge Discovery," in *Advances in Knowledge Discovery and Data Mining*, ed. U. M. Fayyad, G. Piatetsky-Shapiro, P. Smyth, and R. Uthurusamy (Menlo Park, Calif.: AAAI Press, 1996), vii–xi.

Math, with an Attitude

LINDA PETZOLD

Petzold is professor and chair of the Department of Computer Science and professor of Mechanical and Environmental Engineering at the University of California, Santa Barbara, where she directs the computational science and engineering graduate program.

How is it possible to start out in the mathematical sciences and end up working on a wide variety of problems in engineering and science? How can those problems in turn influence the mathematics? Although I am pleased with what I have managed to accomplish so far, in retrospect it has been mostly unexpected and unplanned. I will try to give some advice based on experience on how, with a more aggressive and well-informed approach, one should be able to do better.

I received my Ph.D. in computer science from the University of Illinois at Urbana-Champaign in 1978. My thesis research was on the development of efficient numerical methods for highly oscillatory ordinary differential equations. I chose to study numerical analysis and computer science because I enjoyed mathematics, but at that time the job market in mathematics was rather bleak; I loved computing, and I was interested in applying my work to the solution of engineering and scientific problems. This seems to have been a good decision, although working in this field is not at all what I thought it would be—in addition to the analysis and the computing that I expected, there is a lot of interaction with engineers and scientists on a wide variety of topics that can also be very rewarding.

For my first job, I went to Sandia National Laboratories in Livermore, California, to work as a numerical analyst in the applied mathematics group. I wasn't sure what I should be working on, and my supervisor's advice was to "make yourself useful." So I set about to talk to the engineers who shared our building. At the time, I found it difficult to approach people and ask what they were working on, and even more difficult to imagine how I could make a worthwhile contribution. But apparently this was my job, so I did it. The people I talked to were mostly doing computational combustion and modeling of fluid flow systems in solar central receiver power plants. These seemed like interesting problems, and there were differential equations that needed to be solved, so I was off to a reasonable start. Both the combustion and solar energy problems could be expressed, after semi-discretization of the spatial derivatives, as systems of differential-algebraic equations (DAEs),

$$F(t, y, y') = 0, \quad y(t_0) = y_0. \tag{1}$$

In both applications the engineers were using, or planning to use, rather primitive methods for the time integration. So, I thought, here is an area where I might be able to contribute, by building a state-of-the-art DAE solver.

I wrote such a solver, and at first it was a huge failure! We tried to use it in simulation of a solar power plant, which by that time was becoming a rather urgent problem. After a few time steps, though, the solver mysteriously failed. I had the feeling then, and still do, that if you build software and your friends and colleagues are kind and courageous enough to try to use the first versions of it for solving their research problems, you should be prepared to give it your full attention when that software fails. I spent the next month doing nothing but looking for the bug that must have caused this failure. After examining and testing everything, including at one point all of the elements of the local Jacobian matrix for this large-scale problem, I was finally convinced that there was no bug in the code that could have caused this failure and began wondering if there was something strange about the problem.

At about the same time, a report from Boeing Computer Services [1] came to my attention, about some new results on the numerical solution of linear constant-coefficient DAEs with applications in circuit analysis. The report showed how a result from linear algebra could be used to decompose the linear constant-coefficient DAE system into simpler sub-systems which could then be analyzed from the point of view of mathematical structure and convergence of numerical methods. Although there were convergence results for some simple numerical methods in the report, the decomposition showed that problems could be written in the form (1) which had properties that were quite different from those of standard-form ordinary differential equations (ODEs) $y' = f(t, y)$. These problems, which were distinguished by their high *index*, a property of the DAEs which is apparent when they are written in the decomposed form, have a mathematical structure which is quite different from that of standard-form ODEs. For example, here is a very simple index-2 DAE:

$$y_1 = g(t)$$
$$y_2 = y_1'. \tag{2}$$

Notice that this differs from a standard-form ODE in some fundamental respects: the solution is completely determined by the right-hand side, and any initial values must be consistent with the right-hand side, the solution is less continuous than the input function $g(t)$, and the DAE has a hidden constraint (that is, the solution must also satisfy $y_2 = g'(t)$).

I wondered whether the solar power plant problem might be a nonlinear version of such a high-index DAE, and if so whether that could somehow be the source of the mysterious failure. After some time pondering the solar power plant problem, I decided that it was indeed a nonlinear version of a high-index DAE system. The high-index problem arises in the simulation of the flow of an incompressible fluid in the pipes in the plant. Roughly speaking, the pressure acts like the "index-two variable" y_2 of (2). Notice that the smoothness of this variable depends directly on the derivative of the input function (or in a more general nonlinear case, on the derivatives of other problem variables), hence one might expect that discretizations (numerical ODE methods) which were designed for smoother problems and variables might not be appropriate for this problem, and in particular for the pressure variable. To try to understand whether this was indeed the source of the problem and get a handle on how to fix it, I studied the discretization schemes applied to the model equation (2). Although the results in [1] indicated that the backward differentiation formula (BDF) methods I was using should converge for this problem, they were only applicable to linear constant-coefficient DAEs and with constant time stepsizes. When I studied the index-two model problem for nonconstant stepsizes, I found that there were problems with the formulas for variable stepsizes, and there were problems with the error estimates which are used to determine the next stepsize and decide whether the current time step has been successful in an ODE code. Indeed, when I tried my solver on (2), it failed in the same way as it did for the solar power problem!

To make a long story shorter, it was easy to see from the model problem that a reasonable strategy for solving the solar power plant problem with my adaptive code would be to remove the index-two variables (pressure at every node) from the stepsize selection and error test decisions. At this point, there was not enough theory to guarantee that this would give accurate solutions to the nonlinear solar power plant problem with the adaptive solver but at least the theory in [1] indicated that for the related index-two model problem (2) solved by the same formulas with a fixed stepsize, there was convergence. So we modified the solver to remove the pressure variables from these decisions. It worked fine! After extensive testing and carefully inspecting the solutions, we were convinced that they were correct except possibly for the pressure variables. Thus we could solve the problem, and in fact went on to develop a tool which could solve many related problems, which was important to our employer.

This success in solving the problem had another important benefit for me. It gave me the chance to go back and study the problem further to really understand it. When I decided to do this, I got some well-intentioned advice from a number of well-known and well-respected colleagues that it was not a good idea to put much effort into this class of problems (DAEs), because it was somewhat "out of the mainstream" of numerical ODEs and perhaps of only limited interest. I thought about this but then decided to go ahead with this work anyway because there were a lot of interesting research issues, and I could see how it could impact the solution of problems at Sandia. That turned out to be a good decision. I spent about fifteen years working intensely on analysis, numerical methods, and software for DAEs. Much of this work is summarized in the two books [2] and [3].

The original software which hadn't worked underwent quite a few modifications and several rewrites while it was solving more and more challenging problems at Sandia. Eventually this code became the DASSL software [2], which is available on the World Wide Web via netlib. I had written DASSL to solve a few problems at Sandia, but to my amazement, even in the days before the World Wide Web, the Sandia engineers told their colleagues about it, and within a few years DASSL had hundreds of users worldwide, solving a wide variety of problems. This was an important development for me because, although in many cases I was not directly involved in the solution of these problems, it brought some challenging problems and interesting new applications to my attention and motivated a lot of my research. In 1985 I left Sandia to become group leader of the numerical mathematics group at Lawrence Livermore National Laboratory, where I worked with Alan Hindmarsh and Peter Brown to develop the code for DASPK [2]. This software uses the basic time-stepping methods and strategies of DASSL, but incorporates methods for efficiently dealing with very large-scale DAE systems that arise, for example, after semi-discretization of the spatial derivatives for two- and three-dimensional time-dependent systems of partial differential equations.

In the meantime, DAEs had gotten "into the mainstream." I believe that the software played a large part in this. If you have developed a good tool, then engineers and scientists will beat a path to your door to get to use it. As a result, more challenging problems will come to your attention, you and others will develop more theory which will allow development of even more powerful software for the solution of more difficult applications. It is a great cycle! But I should caution that software development is a huge investment of your time that can only pay off if there are important problems that could be solved better or more easily with a new software. In this respect, I was really fortunate to be at Sandia and at Lawrence Livermore National Laboratory during the first thirteen years of my career, with relatively easy access to interesting problems and with no need to worry about tenure. Another benefit of working in a nonacademic environment was that when I had my son in 1984, I didn't have to worry about a tenure decision (and I have a nice memory of a baby shower which was given to me by my colleagues, who were all men).

In 1991, I decided to move to academia. I wanted to have students, more direct control over my research funds, a wider range of problems to work on, and a lower overhead rate! I went to be a professor in the Department of Computer Science at the University of Minnesota. The research went well, and I built a strong research group—it was not very difficult to transition the research from the laboratory environment and to obtain grant funding in academia. The transition to teaching was more difficult (and tiring!), but after a while it was successful. In 1997, I moved to the University of California, Santa Barbara, to the Departments of Mechanical and Environmental Engineering and of Computer Science. I am still developing and analyzing numerical methods, writing software (lately for sensitivity analysis and optimal control), and solving engineering problems (lately simulation and optimal control for chemical vapor deposition of superconducting thin films, trajectory control of spacecraft, and tissue engineering—optimizing the fabrication of a bioartificial artery).

At this point in writing the lecture, I decided to try to write down some advice based on what I think I have learned, sometimes by trial and error, sometimes by luck, and sometimes by good advice, regarding career development and research. I cannot guarantee that this is good advice, but for the most part it has worked for me. If it can help even one person who is reading this, then this writing will have been worthwhile.

In research,

1. Look for problems with *impact* as well as *interest*.
2. Be *flexible* and *open* to new ideas and problems, even if you don't understand everything at first.
3. *Follow your instinct.* Is the problem new and interesting? Is there a need? Keep in mind that the research community can be slow to accept change.
4. *Negative results* can be important—in their absence, a bunch of researchers will continue to work in the wrong directions.
5. Don't hesitate to *write papers.* If it is interesting and/or useful, you have something to say.
6. *Accept all invitations* (within reason). Visibility is extremely important in the research community.
7. *Make your work available and accessible.* Some ways to do this are by writing software, interdisciplinary collaboration, directing some of your papers to a wider audience, and via the World Wide Web.
8. In your *lectures*, emphasize the importance of the problem and your contribution. Keep your audience in mind. Some subjectivity is OK (even expected) in a talk, so give your opinion. A flashy title can sometimes be useful (I use this trick :-)).
9. For your *grant proposals*, believe in what you are doing and communicate why it is important and interesting. Remember that if you don't ask, you don't get.
10. When you are asked, take the opportunity to do some *professional service.* Serve as referee, editor, NSF panelist, . . . (but not to the exclusion of your research!) to see firsthand how the system works.

In managing your career,

1. *Focus* on your personal and professional objectives.
2. Balancing personal and professional life is a big issue, especially when you have children. It can (and should) be done, but you will have to be flexible and resourceful in managing your time.
3. Don't be afraid to ask for advice.

4. In choosing an employer, go where the workplace is interesting and friendly, where there is emphasis on science and recognition for good work, where you will have a chance for visibility, and where your work could make an impact.

5. A mentor can be extremely important. I had several great mentors in the early part of my career, all of whom happened to have been men.

6. Friends and collaborators are important, particularly over the long term.

7. Be confident (or at least project confidence—you might be amazed at how many of us have had to "fake it" in the early part of our careers).

8. Be aggressive. Seek out the people and problems you want to work on, and GO FOR IT!!!!!

References

1. R. F. Sincovec, B. Dembart, M. A. Epton, A. M. Erisman, J. W. Manke, and E. L. Yip, *Solvability of large scale descriptor systems*, Technical Report, Boeing Computer Services Company, Seattle, Wash., June 1979.

2. K. E. Brenan, S. L. Campbell, and L. R. Petzold, *Numerical Solution of Initial-Value Problems in Differential-Algebraic Equations*, 2nd ed. SIAM Classics Series, SIAM, 1996.

3. U. M. Ascher and L. R. Petzold, *Computer Methods for Ordinary Differential Equations and Differential-Algebraic Equations*, SIAM, 1998.

V. INTO A NEW CENTURY

The recent turn of the century (and, indeed, of the millennium) prompts thinking about the future in a special way, pondering how individuals and groups found their way to their current positions and where they are headed. Many mathematical organizations asked prominent members to engage in this kind of thinking. The International Mathematical Union (IMU) commissioned a volume of articles to describe the state of mathematics and mathematicians at the end of the twentieth century. In 2000, there was a special AMS conference of plenary talks with similar purpose, "Mathematical Challenges of the 21st Century." For the IMU collection, Oxford mathematician Frances Kirwan wrote the memoir "Mathematics: The Right Choice?"[1] There she explores themes that recur throughout this volume: How did I get started in math? Why did I choose it for my career? How do I combine pursuing my career with enjoying my family? Kirwan concludes that it takes a certain amount of luck, as well as help from friends, colleagues, and mentors. She confirms the rightness of her choice, as would most of the "valiant women," as Cathleen Morawetz categorizes those who succeed both in mathematical careers and in family life. (See the last paragraph of the first paper of part IV.)

Kirwan says, "I have been wondering what sort of people the next century's mathematicians will be, and what will draw them into mathematics (as it is far beyond my ability even to speculate what their mathematics will be like). . . . Perhaps this process of reflection will help me to persuade students in the future to choose mathematics, although the reasons which influence them will probably be entirely different from mine." After describing her pathway to a mathematics career, she says: "Combining a career and a family is always going to be hard work for both mothers and, these days, fathers of small children, whatever the career, and it is impossible to put as much effort into either career or family as one otherwise could. I have always been very grateful that . . . as an academic mathematician, I have very flexible working hours and can work at home a great deal, in contrast to my husband, who usually only sees our children at weekends." She concludes with her hope for the future, that "both men and women will have as much luck as I have had and will find (probably for very diverse reasons, very different from mine) that mathematics is the right choice for them. For

---

1. Frances Kirwan, "Mathematics: The Right Choice?" in *Mathematics: Frontiers and Perspectives*, ed. Vladimir I. Arnold et al. (Providence, R.I.: AMS, 2000), 117–120.

the future health of mathematics we need to do as much as we can to provide the right environment for that to happen."

The women writing for part V explore themes aligned with Kirwan's. They are a diverse group who have investigated a range of mathematical areas and have a wide variety of work and home life experiences. Their contributions have been arranged by date of award of the writer's doctorate. Leading off, there are two senior mathematicians who are members of the National Academy of Sciences. Part V ends with recent doctorates, early in their careers, including one who was a Schafer Prize winner (see II). It was hoped that a steady upward progression in women's status and comfort in the profession would be apparent; warning signs far along the chronology underscore the necessity for constant vigilance as well as for goodwill in the community.

Two of the contributions here are based on presentations at the AMS conference "Mathematical Challenges of the 21st Century," mentioned earlier. Karen Uhlenbeck, in the introduction to her plenary address, and Carolyn Mahoney, who set the stage for a special AWM discussion, both raised the issue that mathematics must become more inclusive, especially in the face of the upcoming population shifts predicted for the United States. The other essays are autobiographical and were invited (near the dawn of the twenty-first century) to provide updates on the complexities in and rewards of the life of a woman mathematician. The themes explored earlier in this volume clearly continue to resonate in the experiences of contemporary women research mathematicians.

The reader will note many similarities and connections among the reflections of the contributors, as well as relationships among several of the individuals. The voices in this part share a passion for mathematics that shall not be denied. The short memoirs vividly and eloquently give context to the mathematical life.

Biased Random Walk: A Brief Mathematical Biography

NANCY KOPELL

Ph.D. 1967, UC Berkeley

Math genes run in my family: my mother and sister majored in math, and my father was an accountant. But I was the first of my family to go to graduate school, with much apprehension aimed my way by my parents, who had in mind for me a more traditional homemaker role.

Berkeley in the middle sixties was a revelation for someone who had just graduated from Cornell, a place that took seriously its duties in loco parentis for women. Among the things that were hard to miss, other than all the paraphernalia and activities of the start of the sixties, was the lack of women on the math faculty, at least to my knowledge. There were also not many women graduate students, and so all of us got a fair amount of attention. This was not true of the men graduate students; at that time, it was widely said that the math department did not know, within 100, how many graduate students were officially registered.

This attention was not always flattering; some of those who were not doing well were gossiped about in terms that were more personal than mathematical. I wasn't aware of unpleasant gossip about me, but I was acutely and uncomfortably aware of the sexualized atmosphere around me. I wasn't subject to what has since been called sexual harassment; I was simply a young, not unattractive woman in a sphere that was almost entirely male. I was also doing well, which made me more of an enigma. It was hard for me to accept that I really was doing well; the unacknowledged sexual attention (and my own insecurities) made me question any praise or grade that I got. It was to be many years before I could accept that I was actually seen to have talent.

My graduate career did not go smoothly. After passing my qualifying exams with honors, I spent much of the next year engrossed in the Free Speech Movement and playing the guitar. I had an advisor, but I rarely saw him except at parties. I was ostensibly working on a problem in algebraic topology, but neither my advisor nor I was interested in that project, and he soon after changed his own field. I was rescued by Steve Smale, who would see me in the hall and suggest problems in dynamical systems. I rejected many of those out of hand (with some snottiness that I'm now embarrassed to remember) on the grounds that I knew nothing about the field; to my everlasting gratitude, he persevered anyway with more possible projects. Finally I agreed to work on one that seemed absolutely elementary, though hard.

I worked mainly by myself for a few months over a summer and got my first result. I then saw Smale (who had been out of town), and we had our first, and perhaps only, fight: he thought that it would be a thesis, and I insisted it was much too little for a thesis. He said other results would now come, and he was pretty sure I would be finished within the year. So, in September, at almost twenty-four, I was on the job market, with a thesis I hadn't yet written up or even done. Smale was always pretty confident of his opinions, and in this case, as in many others, he was right. I had a year of very hard work to finish and write up my thesis, in addition to learning the field of dynamical systems in which my work was imbedded.

This year was one of the best and most memorable of my academic career. Smale was in the middle of reenergizing dynamical systems with many new ideas and a new program and had gathered around him a tight-knit family of graduate students (his first) and other faculty members, including Moe Hirsch and Charlie Pugh. This group adopted me, and I learned the field from the other students, Mike Shub and Jacob Palis. It was another, and wonderful, culture shock to go from an atmosphere in which grad students were shadows in the halls to one in which I was questioned almost every day about what I had figured out since yesterday.

As Smale predicted, I did indeed come up with more results, but not without some pain and tears. At one point, I was sufficiently discouraged that I was ready to quit, and I went to the office of another faculty friend, Bernard Kripke, to cry on his shoulder. While I was there, he got a call from his then new wife, Margaret, who was a grad student in biology. She was studying for her exams, also very discouraged and ready to quit. So, in an inspired moment, he took me home, mixed up a large jug of margaritas, and left us together to laugh, cry, and get drunk. The next day she went back to her exams, and I to my thesis. She became a very well known biologist, and I finished my thesis, with some mathematical help as well from Ber.

My thesis was good enough (and my advisor had enough pull) that my first job was at MIT, as a C. L. E. Moore Instructor of Mathematics. I was the first woman in that fancy position, and so, once again, I didn't escape special attention. (This is an enormous paradox of my life; I was very shy, with massive anxiety attacks about being scrutinized and fantasies about being invisible, yet somehow steered my life so that what I did always placed me in a public setting within the lens of attention.) This time the attention was puzzlement: if I entered the common room, I was apt to be asked, "Why aren't there any good women in mathematics?" ("Present company excluded" was not said.) But, compared with Berkeley, I was mostly ignored, while given a lot of freedom and a light teaching load. I wanted that, because I was trying to figure out what to do next and wanted as much shadow as possible.

I had loved being in Berkeley and very much enjoyed the mixture of geometry and analysis I had used in my thesis, "On Commuting Diffeomorphisms." But I also knew I didn't want to continue the program Smale had set out to uncover the "generic" properties of diffeomorphisms. I wanted to work on something related to the so-called real world. And I was not very qualified: my training was so abstract that, although I described properties of differential equations in my work, I never had to solve one!

The two years at MIT were hard times for me. I was working by myself to try to redefine what I might do with mathematics and was still unconvinced that I had any talent for it. I had dragged around my various "issues" through graduate school, and after my Ph.D. they insisted on coming front and center. So, in addition to being clueless about how I wanted to proceed professionally, I was dealing with an undiagnosed depression. I spent a lot of time doing nothing.

When I was not doing nothing, I was sometimes trying to read biology (I had taken no bio courses in undergrad or grad school) to look for interesting mathematical issues. I was especially taken with the idea of self-organization, including morphogenesis in biology. There was virtually no math biology theory at that time. I read what was available, which included Thom's catastrophe theory and a body of work by Prigogine. I read stacks of papers and decided that, though I admired the deep mathematical work of Thom, he was no more an applied mathematician than I was at the time, and I needed a master more grounded in the nitty-gritty of physical phenomena. Prigogine's work, in a different way, presented theory I couldn't connect to things I wanted to understand better.

While back at square one, I was approached by a grad student in chemistry who was trying to understand the recently discovered phenomenon of self-organization in an oscillating chemical system, the Belousov-Zhabotinsky reaction. Since, by that time, I had acquired a (somewhat exaggerated) reputation for being a local expert in catastrophe theory, he approached me, believing that if I taught him this theory, he would be able to delve into the secrets of this new science. After many sessions talking with him, I concluded that catastrophe theory would not be of use, but the phenomenon was just what I had been looking for. The grad student was disappointed and dropped it (figuratively speaking—the reaction involves concentrated sulphuric acid!) in my lap and went off to work with Prigogine.

By this time, I was teaching at Northeastern University, relieved that I could find a job (and in Boston) after a highly unproductive postgrad pair of years. Northeastern was psychologically easier for me than MIT: it was not then a high-pressure research university, and opinions about what constitutes good math were less uniform (or uniformly enforced) than in the major research departments. It was also filled with colorful faculty, allowing me to blend in with my own strangeness: being a woman mathematician who wanted to work in an area that didn't yet exist.

The Belousov-Zhabotinsky reaction was the catalyst for my return to mathematical life. But I knew I needed help. The only mathematical literature I then knew of that dealt with self-organization in a scientifically detailed way was fluid mechanics, using bifurcation theory. To get references for this, I decided to go back to MIT and see someone who, I was told, was a master of that field. As it turned out, he was away that day, so I knocked two doors down. Lou Howard, another expert in fluids, was in—and interested. After he found out why I wanted these references, we talked mainly about chemistry and self-organization. He then became my collaborator for the next seven years. I regard that time as my real postdoc, as he gently initiated me into new ways of thinking about science and math (as well as how to solve differential equations!). My own, different, mathematical personality, which I would never have been able to develop without his mentorship, was strong enough to show through in our joint papers. This period was also the only time I did experiments for publication, using an old darkroom (previously a men's room) that Lou had commandeered from the fluid dynamics laboratory.

I stayed at Northeastern for seventeen years. During that time, Lou moved, and our mathematical marriage ended (and, separately, my real one to Gabriel Stolzenberg began). I began working on projects by myself, mostly on issues involving geometric singular perturbation theory. Lou's crowd knew me mostly as Lou's junior colleague. A review of an early NSF proposal of mine said flatly, "She never did anything good before working with Howard, and will never do anything good after." NSF ignored that review, and I got funded. The reviewer was in fact wrong on both counts: my thesis has taken on a life of its own in the topology community because of its unexpected implications for foliation theory, and I managed to muddle though some more good work after Lou, both by myself and with others.

My next chief collaborator was Bard Ermentrout. We met for the first time in Washington; I was all dressed up for a panel meeting, and Bard, who was then a postdoc in DC, sauntered in for our appointment in his hole-filled T-shirt and cutoffs. We almost immediately became a mathematical odd couple, with different skills and approaches but with great mutual admiration (or, at least, mine for him). For many years, all our papers were done jointly. Then one day he said, "I don't want to be married any more; I want to be promiscuous." (He had, in the meantime, actually married.)

After getting over the shock of being mathematically jilted (we remain collaborators and close friends), I realized that his decision was good for both of us and also started collaborating with many people. I had already begun to seek out biologists to work with and was actively collaborating with Karen Sigvardt and Thelma Williams (and Bard) on the mathematics of Central Pattern Generators (CPGs), networks of the nervous system that govern rhythmic motor activity like walking and swimming. That collaboration had arisen from what then seemed to be another step in my professional random walk: Avis Cohen, a biologist working on lamprey swimming, convinced me that the expertise I had in working with oscillating chemical reactions should somehow be useful to people working in CPGs, so I should write a tutorial chapter on oscillators for a book on motor control that she was helping to edit.

Avis, who is about my age and was not as senior academically, became my next major mentor and teacher. The chapter was to be about sixty pages, and I refused to write it unless I knew more about the field, so I could focus on what just might be relevant. She sent me a carton of biological papers, which I tried to pick my way though. (I had a fellowship that year, and I felt a bit crazier than usual.) At the beginning, I understood roughly every third word, including "and" and "the." She didn't hang up on me when I called to ask the meaning of words (this was before email, and we were not at the same university) and even sent me follow-up papers. Best of all, she refused to accept the first couple of stumbling attempts that I made at this chapter, making it clear that it *would not do* to write in mathese to a set of bio readers. She was the one who first tutored me in how to begin to cross the cultural barrier beween math and biology, and I learned that it is a continuous process.

There was another happy accident associated with attempting to learn the literature of CPGs and write for that audience: I realized that work I had recently done with Bard on a completely different subject was relevant to an open question about lamprey coordination. So I actually had something to write about! I met Karen and Thelma at a biology meeting in which they were presenting lamprey work and I was giving a poster on work with Bard about this. I recall them sniffing around my poster late in the session. (I was a wallflower for the first part, as everyone went to see their friends' posters first, and no one but Avis knew me at all.) The reaction was polite but suspicious: What does math have to do with anything? We soon warmed up to one another, and a long and happy collaboration ensued. One of the proudest moments in my professional career was when biologists Karen and Thelma presented a math poster at the annual meeting of the Society for Neurosciences by the four of us (including Bard), explaining features of lamprey swimming. This work also led to collaboration with biologist Eve Marder, who was my mentor in learning about the importance of biophysical detail to network dynamics.

I was very lucky to find such a hospitable group of biologists to draw me into thinking about biological issues. I still find that it is almost impossible to fight one's way into a biological subfield; outsiders tend to be ignored, and mathematicians are *way* outside. But I have found many instances (fast increasing) of biologists who have come across some of my work, seen the potential relevance, and asked me to help in some way with their own work. This has been especially true in my most recent set of obsessions, about rhythms in the nervous system associated with sensory processing and cognition. I now have many collaborations with experimentalists that shape the questions and answers of my work.

The random aspect of my training shaped another part of my career: a desire to help people work at the interface between mathematics and biological applications, without having to do a random walk. I'm now at Boston University, where I spend quite a large fraction of my time working with students and younger colleagues. There is still no road map at this frontier (at least in some areas, such as neuroscience), only communities of people to help each other muddle forward. But these communities make it so much more likely that students will get the mentors they need to find their own voices and make their own contributions.

Challenges

KAREN UHLENBECK

Ph.D. 1968, Brandeis University, (see III.2)

The time scale of millennia is beyond what I can comprehend. My own intelligence, gleaned from experience and from family and friends, is good for about fifty years in the past, and the one lesson of the past is that predictions for the future are problematic at best. I keep in mind that the past fifty years has seen unprecedented progress in mathematics and on many other fronts. However, most of us in cities can no longer see the glories of the night sky, which, if we think in terms of millennia, certainly motivated a great deal of mathematics. This loss can hardly be called progress.

While I believe that if I were to start over again, I would most likely be drawn into the kind of mathematics that might help restore the sight of the stars, or the purity of water, this talk is more mundane. I hope to underline the advances that made the theory of time stationary (technically referred to as "elliptic") partial differential equations and systems so successful in applications in geometry, topology, and applied mathematics. By way of looking to the future, I would like to point out by the wayside that comparatively little is known about wave equations, and identify specific problems as fruitful areas of research.

Hilbert's famous problems, delivered at the 1900 ICM and published in full in 1901, give us today an idea of the great changes that have come about in mathematics in this century. Problems 19, 20, and 23 state quite clearly fundamental questions of elliptic linear and nonlinear equations and boundary values, most of which are reasonably well understood today. They give no hint of the marvelous developments in theory and applications that have made this subject so relevant to core mathematics. Interestingly enough, if one examines as well the list of problems published in 1976 in a volume of the AMS dedicated to reporting progress on Hilbert's problems, questions posed by geometers and topologists are as indicative of the future of partial differential equations as those posed by specialists.

Insofar as I can discover, little appears in these past lists of problems about wave equations despite their importance in physical phenomena. It is true that the related subjects of turbulence, integrability in the form of the Korteweg de Vries Equation (KdV), and general relativity appear in the 1976 volume. We can safely report that KdV has been one of our success stories, but both turbulence and general relativity would gain a great deal from some basic understanding of nonlinear wave phenomena in dimensions above one. This subject has a wealth of open, hard problems. Hilbert mentions in several places the importance of geometric applications, and one lesson from the past is to look to the solution of equations of interest outside the field for hints as to what is profitable. Hence my insistence that the "geometric" or "physical" equations are those to be emphasized, and that perhaps something can be learned from the successes of the elliptic theory.

Changes in society have been even more dramatic than the changes in mathematics, and one of those changes has been most important to me in my career. Up until the 1970s, it was essentially impossible, or at least highly unusual, for a woman to have a mainstream career in science. While many of the standard philosophical objections to including women in scientific endeavors do not apply to mathematics, the successful female mathematicians of earlier generations were as remarkable for their success in battling society as for their legacies of theorems. Women also were called upon to do a great deal of scientific work during World War II, but were generally sent home when male manpower became available. In the United States, the launching of the Russian spaceship *Sputnik* in 1959 resulted in a major effort to educate and train scientists, and women were not excluded from this effort. This began what has been a golden age of mathematics in the United States. Even better for me, in the early 1970s discrimination on the basis of race, and somewhat incidentally sex, became illegal. Most observers hoped (or feared) that the gates of scientific research would be thronged with those previously excluded by legal means.

A look at personnel taking part in any major mathematical conference demonstrates the fallacy of this premise, whether made in hope or fear. In the United States, we have been noticeably unsuccessful in recruiting both women and minority students to academic research mathematics in equal numbers with white or Asian males. Moreover, in contrast to the years after *Sputnik*, we in mathematics have been relatively unsuccessful in recruiting and educating the products of our own public school system into the academic research arena. We know less and have been less successful in this than in studying wave phenomena. The numbers of tenured women and minorities in research mathematics departments are both an embarrassment and a discouragement.

May the next few decades bring progress on all fronts! I'm getting to be an old lady, and hope to see the upward trend before I die.

Mathematics in "My Century"

CAROLYN S. GORDON

Ph.D. 1979, Washington University

Whenever I'm tempted to relate my own experiences as a woman in mathematics to those of women starting out today, I'm reminded of a compliment I received from a woman graduate student (not a native speaker of English) well before the recent turn of the century: "It must have been very difficult to get started as a woman in mathematics; I mean, in your century!" Nonetheless, I'll share some of my experiences in the hope that they may still be relevant in the current century.

Without role models and mentors, I doubt that I ever would have embarked on a career in mathematics. My first role model was my older sister. Her love of mathematics helped me feel that it was okay for girls to enjoy mathematics in spite of all the teasing at school. I knew long before I started college at Purdue University that I would be a mathematics major, but I had no idea what, if anything, I would eventually do with a mathematics degree. Indeed, it wasn't even clear to me whether I would have a career of any type. I only realized that I might be able to pursue an academic career when my algebra professor, Stephen Piper, asked me if I had plans for graduate school. Almost as soon as he posed the question, with its implicit encouragement, I began enthusiastically planning my career.

Since my professors in graduate school were also very encouraging, I wasn't disturbed—or so I thought—by the fact that I had never met a woman professor. However, I will never forget the excitement I felt when, as a graduate student, I walked into my first AWM meeting and saw a room full of women mathematicians. Even greater than my surprise at seeing so many women in mathematics was my amazement at my own reaction.

While a bit of encouragement got me started in an academic career in the first place, my career path has been substantially affected by further encouragement and mentoring. One of the most important things I did for myself early on was to attend every conference and meeting that I could and to visit those mathematicians within a two-hour radius whose work was close to mine. This required me to push myself, as I was very shy, but the effects were invaluable. Wonderful mentoring relationships and research collaborations developed, and my research direction was greatly influenced by the many suggestions I received during informal discussions. For example, since I attended the geometry seminar at the University of Pennsylvania (an hour and a half from my location at Lehigh) so regularly, they turned to me when a last-minute one-semester visiting position became available. A few years later, I was able to return for a full-year visit. Penn's geometry group is not only very strong but also extremely friendly and supportive. I have enjoyed research collaborations with several members of the department, and Herman Gluck has been a wonderful mentor as well as friend throughout my career.

Most dramatic was my experience in giving a contributed ten-minute talk at a regional AMS meeting. Initially it appeared to be a waste of energy. There were almost no geometers at the meeting and no geometers at all in my audience until Rich Millman, a geometer whom I had never met, arrived halfway through my brief talk. Afterward, he asked if I had ever considered Mark Kac's question "Can one hear the shape of a drum?" and its generalizations. He thought that what I was working on might have some relevance. That question completely changed the direction of my research, and as a consequence, my career trajectory. Shortly after focusing my research in this direction, I began receiving support from the NSF and also found myself much more mobile. I was thrilled by the opportunity to return to my graduate institution, Washington University, this time as a tenure-track (and later tenured) faculty member. Some years later, I moved to my current position at Dartmouth College.

Without mentors, I most likely would never have envisioned a career in mathematics; certainly my mentors have influenced every stage of my career development. As I begin my term as president of the AWM, I look forward to the opportunity to encourage more women and girls in mathematics.

Turning to the balance between my professional and personal life, I was fortunate that my two-body problem was solved before I knew that it existed. David Webb joined the faculty at Washington University at the same time as me. We met there and were married a few years later. David was an algebraist with very broad interests and knowledge. I soon found myself turning to him with questions that arose in my work, and, before long, he too was captivated by spectral geometry. We have since collaborated on numerous papers, and his influence has been felt in all my work.

Of course, collaborating with one's spouse does have its downside: it's hard to escape when you feel a need to forget work while your collaborator is still going strong!

Having dreamed of a child for many years before the arrival of our daughter, Annalisa, eight years ago, I am always greedy of my time with her. Until she started school, I almost invariably took her with me to conferences, even when David didn't go. Even now, I much prefer to attend conferences that are scheduled during her school vacations so that I can bring her along. Most often, I've been able to find very good local babysitting services. As an occasional alternative, my sister (now a school teacher with summers off) has traveled with us. When Annalisa's snow days or school breaks occur on my teaching days, I follow a tradition begun by Herman Gluck when his children were small: I bring her to my class and, halfway through, pose a problem that I know no one will be able to answer. Just as I'm about to throw up my hands in despair, a little voice will respond with the correct answer. A stunned silence invariably follows until the students realize that she has been prompted in advance. I realized recently that the effect can go beyond the brief diversion. A calculus student struggling with homesickness said it made her feel much more at home in the class. Annalisa has also assisted me with demonstrations in various colloquium talks. Whereas I was once excessively careful to keep my personal and professional lives completely separate, I have since found that each has enriched the other, even if balancing time remains a challenge.

The growth in the number of women students in mathematics since "my century" is very encouraging. When my sister was in college, she was typically the only woman in her mathematics classes. By the time I entered college seven years later, most mathematics classes did have a few women students. On the other hand, I was the only woman in my entering class in graduate school. It is wonderful to see that roughly half the graduate students at Dartmouth are women. I look forward to the day when women are equally represented at all levels of the profession and students are not surprised to walk into a room filled with women mathematicians.

Thought Problems

INGRID DAUBECHIES

Ph.D. 1980, Vrije Universiteit Amsterdam

"Problems that can be solved by thinking" have held a special fascination for me as long as I can remember. Finding a solution just by exercising one's wits is such a powerful and satisfying experience. I am sure this feeling is shared by many people other than professional mathematicians: how else can one explain the success of puzzles in newspapers and magazines?

Another important thread in my life, also going back to my childhood, is the joy of creating something. This act of creation could be physical and concrete, such as weaving grass carefully dried for a week, painting crazy animals on every available piece of cardboard, or putting together clothes for a favorite doll. Or it could be abstract and all in my mind, such as inventing fairy tales for my younger brother, or thinking up new episodes in the long and complicated epic yarns, in which I figured prominently of course, that I had made up and that I reviewed in my mind before falling asleep at night. The longest-lasting of these yarns was set in medieval times, although the period in my fantasies bore little resemblance to what life in the Middle Ages really must have been like. Another one was inspired by *Robinson Crusoe*, or rather, by the adaptation I read and cherished as a child. The Dutch translator had omitted much of the political discourse of the original. Also, he had realized that Defoe's descriptions of Crusoe's small practical discoveries that enabled his survival on an uninhabited island did not make much sense and had quietly improved them. Only when I had left behind the Dutch translation of my childhood and bought the original version in English did I realize, as an adult, that the wonderful descriptions of exactly how Crusoe made cheese, letting it ripen in special caves, how he fired clay pots, or how he produced yeast to make his bread rise, and so many others that had made the book sparkle for me were not in the original. Fortunately my parents still had my old books, and I recovered this one on my next visit; I treasure it still.

Sometimes both "solving by thinking" and creativity played a role in my favorite childhood activities: for instance, designing clothes (whether for dolls or real people) requires quite a bit of three-dimensional geometric insight, and it delighted my seamstress grandmother that I could "see" how to create a shape by cutting away pieces of fabric and sewing the pieces together again, even if my needlework was not up to par. Gaining insight by the combination of thinking and creating can give such a pure, almost child-like joy, a sense of wonder and of deep satisfaction—feelings I experience again when I have learned or built some new mathematics, even though a lot of hard work and frustration may be required before getting there!

My parents were very strict in some ways; I often felt that many of my school friends were allowed much more (staying up late, going out on weekend nights . . .). Only much later did I realize that in other ways they were much more "open" than many other parents: my brother and I were both encouraged to read widely (which we did, voraciously) and to be creative. My father was a coal mine engineer, but he would really have preferred to be a scientist. His parents were poor and had received only minimal schooling; no one in the family had ever gone to high school. When my father's teachers lobbied for this smart boy to be allowed to go to high school and then college, his parents knew only two models of careers for college-educated people: medical doctors and the engineers who worked in the coal mines that dotted their home region of Belgium. Since my father couldn't stomach the sight of blood, it was naturally decided that he would study to be an engineer. Since you have to pick your major before you enter a Belgian university, he didn't even get a chance to change his mind at college. Because of the costs of sending my father to college, his parents couldn't save for a little house of their own for their retirement—as long as my grandfather worked they lived in a house provided by his employer—so they made a deal with my father: they would invest in his future, and he would buy them a place when they retired. And so it happened. At college, my father discovered how much he liked science, especially physics; later in his career, he would take, at every possible opportunity, continuing education courses to learn more. He always took the questions that his curious daughter asked him very seriously and would explain to me with great patience how things worked, if he knew, or spend a lot of time and effort to think about ways in which we might discover the answer. Only later did I understand that he had sometimes pretended not to know, so that I could discover the solution myself, with some guidance.

From when I was little, until almost the end of high school, I had asserted that I wanted to be an engineer. Nobody ever told me that I couldn't, so it didn't occur to me. It may have helped that I went to an all-girls school (all public schools in Belgium were then single-gender schools); I was never exposed to the attitude that girls might not be as good at mathematics or science as boys, because there were no boys. My parents never made any distinction between my brother and me in their expectations for our education and careers. Later on, I did meet people who felt or even articulated very clearly that women were less "suited" for mathematics or science, but by then I was confident enough to take this as a sign of their narrow-mindedness rather than let it influence me. At some point, I was complimented that I "talked mathematics like a man," which I found very bemusing; only later did I realize that this showed an implicit belief on the part of the speaker that women typically do not produce good math.

My decision to become an engineer changed in my senior year in high school. In preparation for choosing their major prior to going to university (as was still customary in Belgium then), high school seniors were invited to visit different departments at neighboring universities. I first visited the mechanical engineering department at the University of Gent, famous as the birthplace, many decades earlier, of concrete reinforced with steel bars. Their very impressive lab, in which materials were tested to the breaking point by having enormous machines pull or squeeze them, impressed me much less than a later visit to the nonlinear optics lab in Brussels where we learned about the physics and mathematics that make a hologram possible. Seeing how a lens transforms a light intensity distribution in a bundle of parallel rays into a completely different pattern in the focal plane, with intriguing interference phenomena that encode the original information (which I later came to understand constituted the Fourier transform of the original distribution), was much more fascinating than watching a big machine pull on a steel rod until it snapped. After these university visits I decided to study science and not engineering, even though my mother grumbled that scientists were like artists, whereas engineering was a real profession.

Although I had always liked mathematics, it wasn't until college that I started to think I might become a mathematician. I had already selected physics as my major; to make it possible to switch majors I took an extra four classes one summer. Nevertheless I remained a physics major, mostly because of one very inspiring professor, again in nonlinear optics. Making a hologram oneself seemed much more fun than complex analysis. For my Ph.D. I worked on mathematical physics, a discipline that tackles with mathematical rigor problems that come from physics; techniques and approaches I learned in that apprenticeship have served me well ever since, even though I now more commonly look at problems originating in signal analysis or computer science.

My not very standard path toward applied mathematics and especially the problems in electrical engineering that have inspired some of my best work have been an asset rather than a drawback, at least in my experience. It has meant that I have had to learn some advanced mathematics on my own, probably along a more meandering path than if I had learned it in an organized curriculum. (It is ironic that this is often the material in courses at Princeton University, where I now teach!) But it also means that I often have a different "take" on a problem, which, when I am lucky, leads to the solution. When that happens, there is again that joy (*yes!!!*) and also a sense of wonder that the patterns we mathematicians learn to recognize and understand can be useful in so many different contexts.

Outreach and Variety

SUZANNE LENHART

Ph.D. 1981, University of Kentucky

When I look back on my education and career choices, I see individual teachers who had a tremendous impact. I was encouraged at an early age by my teachers at Catholic schools in Louisville, Kentucky. I could do mathematics well and could explain mathematics well to others. Teachers in eighth and ninth grades gave me the opportunity to teach other students. Since I was about twelve years old, I wanted to be a teacher. So I feel that I was born to teach.

At an all-girls high school, two teachers, Sisters Doloretta and Theodora, encouraged me in mathematics and science. I had the chance to work at my own pace. The atmosphere at this school made it easier for me to feel comfortable about excelling in mathematics and science.

When I was taking Calculus I at Bellarmine College, my professor, Ralph Grimaldi, started to encourage me to prepare for studying mathematics in graduate school. He even taught me number theory in an independent study course during the summer after my freshman year to teach me about proofs. Through a cooperative program at my college, I could increase the number of courses available to me by taking courses at the nearby University of Louisville.

When starting graduate school at the University of Kentucky, I did not know what area of research I would pursue. I chose partial differential equations after taking my first PDE course during my second year of graduate school—this after having taken no PDEs as an undergraduate. So now I tell students it is okay to go to graduate school undecided about research area. Graduate school can open new horizons and interests in a variety of directions. My advisor, L. C. Evans, did a good job of training me to do research. My dear friend, David Adams, has mentored me, starting in graduate school and continuing through the present. I became interested in mathematical biology through an interdisciplinary modeling course in graduate school, and that interest continues in my work today.

When I earned my Ph.D. in 1981, there were few postdocs available, so I obtained a tenure-track job at the University of Tennessee (UT). At my first day of work, I was shocked into the realities of the job when I was informed that I should write an NSF proposal that month.

I try to keep an open mind about the variety of research problems to work on. I have always been willing to work on problems outside my area of specialization. I am willing to learn something new and try to apply it. This willingness led to collaborations with Curtis Travis, George Wilson, and Vladimir Protopopescu, all from Oak Ridge National Laboratory. The collaboration with Travis helped me get started early in my career. My collaboration and friendship with Protopopescu has been very positive and fruitful and still continues. Interactions with researchers in the UT Department of Ecology and Evolutionary Biology continue to be very beneficial.

I started working as a part-time researcher at Oak Ridge in 1987; this laboratory is managed by UT-Battelle for the Department of Energy. I work in the Complex Systems Group of the Computer Science and Mathematics

Division. It is great to go there one day a week and talk with Protopopescu about current projects. We have worked on a variety of things over the years, partly directed by the current research goals of our section of the laboratory. The diversity of work has been very interesting, with topics including combat models, robots, bioreactors, and lasers. Working at a national laboratory is quite different from university work. The direction of the research is driven by the availability of funds and the current focus of the Department of Energy (DOE). Researchers spend a great deal of time writing proposals to find funding to cover their time; proposals are written to various government agencies, such as DOE, Department of Defense, and NASA, and sometimes proposals are for cooperative projects with private companies. (Yes, even though ORNL is a DOE laboratory, proposals for the major thrusts of a group's work must be written to DOE.) Of course, publishing in refereed journals is also very important in our section.

I am able to do what is important to me—teaching, research, and service in the form of encouraging students. I have been directing a Research Experiences for Undergraduates (REU) program for many years and hope to continue this work. Service to the mathematics and the local communities is important to me. I have been involved in after-school math/science clubs for middle school students for the past four years, and I currently coach a math competition team at a local high school. I continue to be involved in middle school activities through two workshops each year, called SHADES, SHaring ADventures in Engineering and Science. I feel that I am able to do so much volunteer outreach work because my workload at home is light, since my husband is currently a full-time homemaker. (He quit his computing job when I had the opportunity to spend a semester as the Sonia Kovalevsky Visiting Professor at the University of Kaiserslautern in Germany. We all went to Germany for four months, and when we returned, our son was starting kindergarten, and my husband became a homemaker.)

The REU site program is funded by NSF to give opportunities for undergraduate students. There are about thirty-five sites for such programs for undergraduate math majors across the country. In our program, ten students spend eight weeks in Knoxville working on research projects and attending two short courses and a faculty seminar. Each of our participants is matched with an advisor and works on an individual project. The goal is for the students to learn about the process of doing research and to participate in the process. We also try to educate the students about the breadth of mathematics. Our program has had about 50% female students over the years since its start in 1987. There are a variety of types and sizes of REU programs, giving opportunities and choices to many students.

I am glad to have served as president of AWM. I hope to see more women mathematicians across the spectrum of the mathematics community. I especially look for progress for women in leadership roles in mathematics organizations, in university governance, and in the professoriate at the top research institutions.

Demographic Trends and Challenges

CAROLYN R. MAHONEY

Ph.D. 1983, Ohio State University

What population shifts are projected for the near future? What challenges do these changes pose for the mathematical community, especially for women? And, most important: How do we respond to these challenges? At the summer 2000 AMS 21st Century meeting, an AWM Special Presentation was held in order to discuss these questions. This article is adapted from the introductory remarks I made prior to leading that discussion. I enjoyed the opportunity to hear the opinions of AWM members and friends concerning these issues.

Background

According to *Futurework: Trends and Challenges for Work in the 21st Century* [1], the U.S. population is expected to grow from about 275 million people in the year 2000 to an estimated 394 million people in 2050. For the foreseeable future, each year 820,000 immigrants are expected to arrive in the United States, so that by 2050, immigration will have increased the U.S. population by 80 million people. That is, fully two-thirds of the projected U.S. population increase will result from immigration.

These and other population shifts will bring additional diversity to America. In 1995, the United States was estimated to be 83% white; 13% black; 1% American Indian, Eskimo, and Aleut; and 4% Asian and Pacific Islander. Ten percent of Americans, mostly from those above classified as black or white, were also of Hispanic origin. Nearly one in eleven Americans was foreign born.

The future racial and ethnic makeup of America will be considerably different from today's. Demographic trends show that the percentage of whites in the total population is shrinking, while the percentage of Hispanics is growing, and at a faster rate than that of non-Hispanic blacks. By 2050, minorities are expected to have risen from one in four Americans to almost one in two. The growth rates for both the Hispanic-origin and the Asian and Pacific Islander populations may exceed 2% per year until 2030. Such growth is unprecedented, at least in recent times—even at the peak of the baby boom era, the total population never grew by as much as 2% per year. By 2010, Hispanics are likely to become the largest minority group. In fact, after 2002, the Hispanic population is projected to account for more of the U.S. annual population growth than all other groups combined [1].

The implications of these projected trends raise questions that concern us all. The groups growing the fastest currently produce the fewest mathematicians and scientists per capita. Mathematics education, that is, the education of the general citizenry and the training of mathematics specialists, is a matter of paramount importance. The mathematics community must participate both in building a stronger, more diverse workforce in mathematics-related careers and in ensuring an educated and informed citizenry that is capable of using mathematics in daily life. Meanwhile, educators in general face a changing, less encouraging climate.

Educational Attainment Levels

According to *The Condition of Education 2002* [2], for persons twenty-five or older, nearly 83% have completed high school, while 24% hold a baccalaureate degree. Just thirty years ago, for those twenty-five or older, fewer than 54% had completed high school and fewer than 10% had completed college. The average educational attainment of the population is expected to increase, but the levels expected to be attained vary considerably across racial and ethnic lines and disability status. [2]

According to 1997 data on young adults, Asian Americans had the highest high school graduation rate that year at over 90%. In 1997, for the first time high school graduation rates for young blacks and whites were statistically on par at 86% and 88%, respectively. It is troubling that the high school completion rate for young Hispanics was far lower at 62%; over the 1990s, there was no significant change in this rate. Lower high school completion rates among immigrant Hispanics do not account for this phenomenon: high school completion among all foreign born is 65%, versus 84% among the native born; for Hispanics, these rates were 43% and 69%, respectively. For each racial/ethnic group, the likelihood that a student's family spoke a foreign language in the home decreased when the family had been in the United States for three or more generations. Nonetheless, the rate at which Hispanics spoke only English in the home was consistently lower than that of their Asian counterparts. Not surprisingly, first-generation students in each racial/ethnic group were more likely to live at or below the poverty level than their second- and third-generation counterparts. [3]

The gap between (non-Hispanic) whites and their black and Hispanic peers has narrowed but remains large, according to National Assessment of Educational Performance (NAEP) trend data on science and mathematics achievement of seventeen-year-olds between 1973 and 1999. In mathematics, the gap between blacks and whites appears to be somewhat narrower, but although this gap shrank during the 1980s, there is evidence that it is now widening. The gap in mathematics between whites and Hispanics is somewhat wider, as is that between students from low- and high-income backgrounds.[2] There is a wide range in performance levels for any given grade in school. [3]

The gap between the educational levels of people with disabilities and those without is particularly troubling. The percentage of adults with disabilities not completing high school is more than double that of those without disabilities, while the college graduation rate of those with disabilities is less than 10%, which had been achieved by the general population thirty years ago. Nonetheless, the situation is improving. Laws and policies requiring equal access for people with disabilities, coupled with advances in assistive technologies, should result soon in rising rates of educational attainment for people with disabilities. [3]

An explicit goal of educational standards for mathematics and science is that all students, without regard to gender, race, income, or disability, should participate fully in challenging coursework and achieve at high levels. Course-taking patterns have been analyzed in *Science and Engineering Indicators—2002* [4]. Both female and male students are following a more rigorous curriculum than they were two decades ago, and female graduates in 1998 were more likely than males (58% versus 53%) to have completed the "New Basics" curriculum.[3] In the class of 1998, females were less likely than males to have taken remedial mathematics in high school but at least as likely to have taken upper-level mathematics courses such as Algebra II, Trigonometry, Precalculus, and Calculus. Students in all racial and ethnic groups are taking more advanced mathematics and science courses. From a course-taking perspective at least, all racial and ethnic groups appear to be better prepared for college today than they were in the early 1980s, although black, Hispanics, and American Indians/Alaskan Natives are less prepared than their Asian/Pacific Islander and white peers. [4]

Not only are more Americans graduating from high school, but also more are going to college. In the past quarter century, the number of

2. Measured by eligibility for the National School Lunch Program.

3. Four units of English and three units each of science, social studies, and mathematics, as recommended in "A Nation at Risk: The Imperative for Educational Reform," an open letter to the American people by the National Commission on Excellence in Education, April 1983 [online]. U.S. Department of Education, 1983 [accessed 25 June 2003]. Available from the World Wide Web: http://www.ed.gov/pubs/NatAtRisk/index.html.

bachelor's degrees conferred in all fields has risen from 955,000 to nearly 1.2 million annually; master's degrees, from 278,000 to 420,000; and doctorates, from 33,800 to 45,700 [4]. Nearly two-thirds of the new college students were enrolled in four-year institutions, while the rest attended two-year colleges. Young women enrolled in college at higher rates than young men—69% and 62%, respectively. At the end of the century, although blacks and whites graduated from high school at roughly equal rates, blacks continued to lag behind in both enrollment in and graduation from college. Nearly seven out of ten young white high school graduates went on to college, compared with only six of ten young black high school graduates and five out of ten young Hispanic graduates. Of those twenty-five and over, just 13% of black and 11% of Hispanics were college graduates in 1997, compared with 25% of whites and 42% of Asian and Pacific Islanders. [3]

In the United States, the nearly twenty-year population decline in the size of the college-age cohort reversed in 1997; the number of college-age individuals is projected to increase from 17.5 million to 21.2 million by 2010, with strong growth among minority groups. This increase in the college-age population by more than 13% in the first decade of the twenty-first century signals another wave of expansion in the nation's higher education system and growth in science and engineering degrees at all levels. Long-term trends show that the proportion of women enrolled in all graduate science and engineering fields is increasing. In fact, much of the growth in the number of doctorates earned by U.S. citizens reflected degrees earned by white women and minority students of both sexes. In 1999, women earned 42% of U.S. citizen doctorates in science and engineering (due largely to the social and life sciences), up from 18% twenty-five years earlier; minorities earned 15%, up from below 5%. [3]

Responding to the Challenges

How do we respond to these serious challenges to our progress in mathematics education and to the participation of women in mathematics?

U.S. students appear to be losing ground in mathematics and science to students in many other countries as they progress from elementary to middle to secondary school. There are numerous noteworthy programs aimed at helping states create, implement, and sustain educational improvements in mathematics. For example, Achieve, Inc., an independent, bipartisan, nonprofit organization created by governors and corporate leaders at the 1996 National Education Summit, has launched a Mathematics Achievement Partnership Initiative dedicated to helping states improve middle school students' mathematics performance; this collaboration with fourteen states will base much of its work on lessons from top-performing countries.[4] At the secondary level, there is an emerging cross-national dialogue about mathematics curricula; for example, the Board on International Comparative Studies in Education and the Mathematical Sciences Education Board Center for Education are hosting such a conversation among mathematicians, mathematics educators, and representatives from high-tech industries from several industrialized countries in spring 2003.[5]

Improvement efforts in the quality of U.S. education clearly depend critically on classroom teachers. Expected retirements, enrollment increases, and teacher turnover rates have created a need for special efforts to recruit, train, induct, and retain highly qualified mathematics teachers. Federal funding agencies, including the NSF and the U.S. Department of Education, have created new programs to support exemplary efforts toward addressing the mathematics and science teacher shortage. The mathematics community has recently published a consensus report, *The Mathematical Education of Teachers* [5], calling for a rethinking of the mathematical education of teachers within mathematical sciences departments at the collegiate level. A new field of mathematics, "teachers' mathematics," is emerging, devoted to developing a profound understanding of secondary mathematics.

I have participated in two efforts related to this process: one funded by the Advisory Board of the Stuart Foundation to create the textbook *Mathematics for High School Teachers: An Advanced Perspective* [6], and the other, a Texas workshop "In-Depth Secondary Mathematics," developed by Callahan and Stanley [7]. The textbook is designed for use in upper-division or graduate mathematics content courses for future high school mathematics teachers, while the workshop provides professional development for practicing high school teachers. In both the course and the workshop, teachers gain a mathematically sophisticated perspective on the very mathematics high school students are learning through "extended analyses" of problems or concepts found in high school mathematics textbooks. Consistent with one of the *MET* recommendations, my campus is planning to use these materials in a six-unit yearlong capstone course sequence for prospective secondary teachers and other mathematics majors.

It will be important to monitor the extent to which individuals and groups have access to educational opportunities and how they progress through the various levels ([4], [10]).

The mathematical community has a history of creating and implementing summer and academic year programs aimed at increasing the number

---

4. See www.achieve.org/achieve.nsf/MAP?openform [accessed 25 June 2003].

5. See www7.nationalacademies.org/mseb/Webcasts%20and%20Workshops.html [accessed 29 January 2004] for a webcast of and other materials for "Talking It Through: Cross National Conversation about Secondary Mathematics Curriculum Workshop."

and preparedness of minority and female students who study mathematics. For example, I received NSF funding in the late 1980s to operate a mathematics and science Young Scholars Program (YSP) at the Ohio State University (OSU) that served middle and high school students [8]. When I moved to California State University San Marcos in 1990, I was funded by the MAA's Strengthening Underrepresented Minority Mathematics Achievement program to adapt and implement the OSU YSP program in my new southern California setting [9]. And joyfully, upon arrival at Elizabeth City State University (ECSU) in fall 2000,[6] I was privileged to join a more than eight-year tradition of hosting one of AWM's Sonia Kovalevsky Days on the ECSU campus. These programs are but a small sample of the many excellent programs created and supported by professional societies and government agencies that serve to attract, expose, encourage, and support young persons to seek out and succeed in mathematics-based careers.

Conclusion

This paper has presented only a glimpse of the serious consequences, opportunities, and concerns for the mathematical community and the nation, implied by current U.S. demographics.[7] Nevertheless, this glimpse makes a clear and compelling case that the mathematical community must join the national and global effort to increase the number of well-trained mathematicians, scientists, and engineers.

References

1. U.S. Department of Labor, "The Workforce," chapter 1 in *Futurework: Trends and Challenges for Work in the 21st Century* [online]. Labor Day, 1999 [accessed 25 June 2003]. Available from the World Wide Web: http://www.dol.gov/asp/programs/history/herman/reports/futurework/report.htm.
2. U.S. Department of Education, National Center for Educational Statistics, *The Condition of Education 2002*. NCES 2002-025, Washington, D.C.: U.S. Government Printing Office, 2002. Also available from the World Wide Web [accessed 29 January 2004]: http://www.nces.ed.gov/pubs2002/2002025.pdf.
3. National Science Board, *Science and Engineering Indicators—2000*. Arlington, Va.: National Science Foundation (NSB-00-1). Also available from the World Wide Web [accessed 29 January 2004]: http://www.nsf.gov/sbe/srs/seind00/pdfstart.htm.
4. ———, *Science and Engineering Indicators—2002*. Arlington, Va.: National Science Foundation (NSB-02-1). Also available from the World Wide Web [accessed 25 June 2003]: http://www.nsf.gov/sbe/srs/seind02/pdfstart.htm.

6. Mahoney is now vice chancellor of academic affairs and provost at ECSU.
7. See also "Pathways in Mathematics," III.

5. Conference Board of the Mathematical Sciences, *The Mathematical Education of Teachers*. Issues in Mathematics Education, vol. 11 (Providence, R.I.: American Mathematical Society, 2001).

6. Dick Stanley, Zalman Usiskin, Elana Marchisotto, and Anthony L. Peressini, *Mathematics for High School Teachers: An Advanced Perspective* (Upper Saddle River, N.J.: Prentice-Hall, 2003).

7. Dick Stanley and Patrick Callahan, *Unpublished working papers*. The In-Depth Secondary Mathematics Institute, Texas Education Agency and the Texas Statewide Systemic Initiative of the Charles A. Dana Center at the University of Texas at Austin. 2002. See the World Wide Web [accessed 29 January 2004]: http://www.utdanacenter.org/texteams/pdf/leaders/IDSMU.pdf.

8. Carolyn Mahoney, "The Ohio State University Young Scholars Program," in *Mathematicians and Education Reform 1990–1991*, ed. Naomi D. Fisher, Harvey B. Keynes, and Philip D. Wagreich, CBMS Issues in Mathematics Education, vol. 3 (Providence, R.I.: American Mathematical Society in cooperation with Mathematical Association of America, 1993), 3–12.

9. ———, "Middle School Mathematics and Science at California State University, San Marcos," in *SUMMAC Forum* 1(1993).

10. David J. Lutzer, James W. Maxwell, and Stephen B. Rodi. *Statistical Abstract of Undergraduate Programs in the Mathematical Sciences in the United States: Fall 2000 CBMS Survey* (Providence, R.I.: American Mathematical Society, 2002). Also available from the World Wide Web: www.ams.org/cbms/cbmssurvey.html [accessed 2 December 2003].

Me, a Mathematician?

CATHERINE A. ROBERTS

Ph.D. 1992, Northwestern University

I've always wanted to be a teacher. In fourth grade, inspired by Laura Ingalls Wilder, I set up a school in an unused horse stall. There, I drilled my younger sister and other neighborhood recruits in arithmetic and spelling. As an undergraduate at Bowdoin College, I embraced the concept of a liberal arts education to major in both art history and mathematics. Deep thanks are due to professors who invited me to consider majoring in mathematics, pointing out to me that I did, indeed, have what it takes. My plan throughout college was to become a secondary school teacher, until my mentors suggested graduate school. Me, a mathematician? Interestingly, although my college math grades were exceptional, I didn't land a fellowship for graduate school until I'd successfully completed a year of graduate-level coursework. It was suggested to me by the chair of one graduate program that I wasn't perceived as "serious," since I'd spent a semester abroad in an art program. Indeed, throughout my career I can point to several occasions where my multiple interests, which I consider vital to who I am, led to suggestions that I wasn't serious enough to be an authentic mathematician. It's taken some personal stamina to reject these insinuations and to maintain a positive attitude.

A Little Bit about My Research. I was awarded a Ph.D. in applied mathematics, specializing in nonlinear integral equations, from Northwestern University in 1992. More recently, my research has branched out to consider the challenge of modeling the complex interactions between humans and the environment. It's been invigorating to work in a new direction. This work combines mathematical modeling, statistical analysis, artificial intelligence, and computer programming. The result has been a simulation model of white-water rafting traffic within the Grand Canyon National Park. This research has landed me squarely at home with who I am as an applied mathematician. In addition to the multifaceted nature of this applied project, I've also been called upon to use my communication skills to discuss this work with government officials, science writers, politicians, and all sorts of constituent groups interested in the outcomes of my model. I've thrice been able to spend two weeks inside the Grand Canyon conducting field research. Aspects of this work have provided undergraduate students with real research experiences. Finally, I have a way to define myself as a mathematician while keeping alive various other interests!

Yep, I'm a Feminist! What I have found most rewarding personally has to do with the special aspect of my career as a woman mathematician. It is crucially important to encourage all students to consider nontraditional careers as viable and rewarding alternatives. To this end, I go out of my way to mentor female students and to invite them to engage in mathematics research. I am forever grateful to the generations of women mathematicians before me, who helped carve out a place where women could succeed professionally and personally. I characterize my generation of women mathematicians as wanting to have it all: a great job and a balanced personal life, perhaps with children. I was the first woman professor in my department to have a baby—and my timing was terrible, right smack in the middle of the semester. I think more and more often, a woman who has a child while on the tenure track will find senior female faculty members sitting on review committees and serving as department chairs. Challenges remain, but mechanisms are being set in place that will ease our efforts to balance life and work. When young women ask if they should wait until they get tenure before trying to get pregnant, my response is, "Choose what works for you. You don't have to wait if you don't want to, you can make it work."

The Two-Body Problem. This phrase is used to describe the challenge of finding suitable employment for a Ph.D. and her partner or spouse. I feel that I must say something about the two-body problem. Why? Because I've changed jobs twice in order to address my own family situation. Launching another job search is tough. Cycles in the job market and variations in the hiring calendars for two different fields can make it a particular challenge. Both parties need patience and a willingness to compromise. I'd advise women to be resilient in the face of disappointments and to keep perspective in order to find a life balance. Switching jobs has delayed some facets of my own professional success, but I know my time will come. [Roberts is now tenured at the College of the Holy Cross.]

Service: *Find Time for It.* It is so easy to lose sight of the meaning of our work and life. Service keeps me "on track" (in the deeper meaning of the expression). I strongly encourage involvement in meaningful service. I am deeply committed to helping underprivileged students succeed in mathematics and have taught several summers in the Upward Bound program at Bowdoin College. Mentoring such students as they navigate young adulthood and contributing to service projects in my community are important to me. My advocacy in these regards is a fundamental reflection of who I am as a person.

What Keeps Me Sane. My caring spouse and two young children top the list of things that help me keep perspective. My extended family means a great deal to me, and I spend a lot of time with them. (My younger sister, by the way, survived my tutoring in the horse stall and is now a professor of biochemistry.) I try to carve out personal time for exercise and my favorite hobbies, such as gardening. Continuing education is also important to me; last year I became certified as a lifeguard, and this year I took a pottery class. I also fully acknowledge that I need lots (and lots) of sleep. I point all of this out because it makes me sad to hear my students saying that they no longer read for pleasure, or that they can't find time to do the things that they really love. There's no reason for this, life is much too short. Cherish it and have fun.

My Path toward Mathematics

KAREN E. SMITH

Ph.D. 1993, University of Michigan

The Early Days. I was born in 1965 a few miles south of Manhattan on the Jersey Shore. My grandfather was a New York City policeman, and my father also commuted to "the city" to work, while my mother stayed at home until I was twelve or so. My parents divorced when I was thirteen. I was the oldest of three daughters close in age, and we all attended the local public schools. Both my parents had been the first in their families to go to college, and by senior year I realized I too would be going, but it was never anything I worried about much.

I was good at school and always loved math. I remember showing my first-grade teacher how to subtract 3 from 2 on the number line after she claimed such a subtraction was impossible. In second grade, I was fascinated by the patterns in the multiplication tables: 9, 18, 27, 36, 45, 54, 63, 72, 81—why did they all sum to nine? In third grade, my father taught me a trick to check my multiplication homework. This was a great mystery at the time, though I now realize this trick amounts to checking the answer "modulo 9." Upon learning about different bases in middle school, I struggled to understand why the trick no longer worked and was eventually thrilled to discover how to adapt it to any base. It wasn't until my senior year of high school that I was able to use modular arithmetic to prove it. This was so completely satisfying.

I also had negative experiences in school. In sixth grade I was placed in an accelerated "Unified Mathematics" program, a hot new math idea in the seventies. I was one of about ten sixth graders placed into a seventh-grade classroom. I think I was the only girl, but I don't recall that being a big issue. I remember sitting in the back, not being able to hear or see the teacher very well, feeling intimidated, and having no idea what was going on. The next year I was "dropped back" to my same age cohort. Most of the other ten students continued on with the higher grade, and I always thought of them as so much smarter than me, the "brains."

By senior year, the brains had run out of math courses, so a teacher, Mr. Driscoll, offered a special math course for this small group. He suggested that I also sign up, in addition to calculus. I was nervous, but it seemed a much more interesting elective than typing. We read Underwood Dudley's *Elementary Number Theory* the first semester and Greenberg's *Euclidean and Non-Euclidean Geometries* the second. I loved both! Most amazing was that the brains weren't any smarter than I was; in fact, I found myself helping the boys I'd always thought were so much smarter and solving problems that no one else could. Driscoll seemed awed by my abilities, and I began dreaming of being a real mathematician. Probably Driscoll deserves more credit than anyone for really turning me on to mathematics.

As a girl I was always more comfortable doing "boys' things," like building forts in the woods and playing games like "kill the guy with the ball." Perhaps this has helped me in my career. I certainly don't mean that math is a boys' thing, only that I have a long history of being comfortable around boys, especially rough ones.

College. Princeton University was only thirty miles from where I grew up, but it was a world away. I had only vaguely heard of it when my guidance counselor suggested I apply, assuming instead that I would go to Rutgers, the state university of New Jersey. I applied to and was accepted into Princeton's engineering program, having been told that engineering is the natural (only?) path for a student interested in math and science. Princeton offered me almost a full scholarship; otherwise we could not have afforded it. Even so, I took on student loans and worked numerous part-time jobs to make ends meet.[8]

I had a hard time adjusting to the social scene at Princeton. My problem was not gender, but class. Abused for my "uneducated" accent and grammar, I went back to the Jersey Shore to hang out with my friends nearly every weekend. Eventually, I found a niche of other Princeton misfits with whom I felt comfortable. I never belonged to an eating club or otherwise participated much in Princeton's social culture. Academically, it was easy to maintain an A– average without giving up all-night debates about politics and philosophy. I was not an outstanding student, but I never worried much about grades.

A few professors made a lasting impact on me. Charles Fefferman's freshman honors one-variable calculus class had terrifically challenging take-home exams that I really enjoyed. In our only personal conversation, he informed me that I'd earned an A+ and asked me why I wasn't a math major. I replied that I needed to find a job after graduation, and he countered that he was making a quite satisfactory living as a math major. That five-minute discussion was enough for me to change majors immediately, despite the protestations of my family. Another Princeton highlight was Nick Katz's number theory course, with its long problem sets that took me step-by-step through the development of ideas or the proof of a theorem. I try to emulate this style of teaching today.

My only female professor at Princeton was Suzanne Keller, a sociologist and the first woman tenured at Princeton, who was teaching a women's studies course in which I enrolled (somewhat reluctantly) to fulfill a distribution requirement. The course turned out to be perhaps the most important one I took: Like a blind person suddenly able to see, the source of my life's struggles and frustrations suddenly seemed plainly obvious. Without this epiphany, I doubt very much that I would be where I am today; I may not have even finished the math major.

My math advisor never suggested graduate school and steered me toward education courses despite my interest in pure math. He wouldn't allow me to drop a required junior seminar, after I told him of repeated

---

8. Some of my jobs were pizza delivery person, lifeguard, deli meat slicer, hotel maid, assembly line worker in a computer recycling facility, math tutor, and SAT prep course instructor.

unpleasant encounters with the openly sexist professor running the seminar, explaining that I would "have to get used to this sort of thing." A sympathetic dean eventually rescued me, forcing the department to find an alternate course for me. I am also grateful to my (male) classmates who offered their unsolicited support after witnessing the verbal abuse I endured in the seminar.

I graduated in 1987 with a B.A. in mathematics and a teaching certificate. At graduation, I was amazed to learn that almost all my fellow math majors would actually be getting paid to study mathematics at fancy places like MIT and Harvard. It had never even occurred to me to apply to graduate school; I'd barely heard of the idea, and I wasn't a shining student and had huge student loans to pay off. Instead, I found a position teaching mathematics at a public school in New Jersey, thereby fulfilling my parents' prophecy that my math major would lead to employment earning less per year than my annual expenses at Princeton.

Applying to Graduate School. My career as a high school teacher was brief: one year teaching grades nine through twelve at a public high school in a small industrial town on the Raritan River near New Brunswick. I taught Calculus, Algebra II, and Geometry, and it was a hard job. The school had a lot of problems like teenage pregnancy and drugs. There was little support from administrators. I liked the students and enjoyed teaching them; it was my colleagues' cynical attitudes and lack of interest in mathematics that drove me out of the profession. I probably would have stayed in high school teaching if I'd wound up in a strong school district with enthusiastic teachers like my own Mr. Driscoll or those I'd seen in the Princeton public schools during my practice teaching.

It didn't take me long to start thinking of applying to graduate school. However, I had no idea how or where to apply. Eventually, I cobbled together applications to Berkeley and Michigan (because I had friends from those areas), as well as Columbia and Yale (because they were nearby and I had heard that they also had graduate programs).

My visits to Yale and Berkeley were discouraging. Although I'd been accepted, no one I met seemed to have much interest in interacting with me, and I was sure I'd never fit in at either place. I'd just about given up on the idea of grad school altogether, but fortunately, my Michigan experience was quite different. I got a personal phone call from a professor inviting me to visit at their expense. I stayed with a finishing Ph.D. student, sat in on some classes, and found several professors eager to talk to me. The commons room was lively and full of students, including some who seemed not unlike me. Michigan had also offered me a generous fellowship, and I learned that my student loan payments could be deferred. I accepted the Michigan offer without even waiting to hear from Columbia.

Graduate School. My family seemed vaguely disapproving of my decision to "be a perpetual student," wasting my Princeton degree working for peanuts in the Midwest. Nonetheless, I arrived in Ann Arbor in 1988, happy to be getting paid to study math and ready to work hard. Although I didn't envision finishing my degree, I looked forward to a few years of deferred student loans, a small amount of (college!) teaching, and finally doing real mathematics as I had dreamed of in high school.

I was surprised to discover that graduate school was much easier than anticipated. A Princeton math education is a valuable thing, if it doesn't kill you: I'd seen (if not understood) much of the material before, often from the same textbooks. I also worked harder and found the boldness to ask professors to clarify points I didn't understand. I studied hard for the qualifying exams and passed them within the first year.

Although I knew I wanted to study some kind of algebra, when it came time to choose a thesis advisor I chose the *person*, not the subject. I had really enjoyed Mel Hochster's commutative algebra class, and he seemed to respect me. Halfway through my second year I asked him to be my thesis advisor; his enthusiastic "Yes!" was very encouraging. It began to dawn on me that I might actually finish a Ph.D. someday, although I still didn't have very much emotion invested in the idea.

I began research on the subject of tight closure with Hochster in my third year. Our meetings were strictly business: I had forty-five minutes to ask all the questions I'd come up with that week and to show him what I'd done. At the time, I was a bit intimidated by him, but I always felt reenergized after our meetings. Hochster was a fantastic advisor for me, seeming to assume that I would someday write a thesis. My only frustration was that I wanted to move toward more geometric subjects, but he resisted this strongly, saying, "I can't advise you on this topic." Eventually, I wrote a thesis called "Tight Closure of Parameter Ideals and F-Rationality" that managed to have some connection to geometry. My thesis received the department's Sumner B. Myers Thesis Prize, as well as a Distinguished Dissertation Award from the graduate school.

In graduate school, I also met and married my husband, Juha Heinonen, a young Finnish mathematician who had also arrived in Ann Arbor in 1988 for a postdoctoral position at Michigan. Some senior women mathematicians worry that female graduate students jeopardize their careers by marrying mathematicians. However, for me, just the opposite was true. Juha has been very supportive, never doubting my seriousness as a mathematician, even before I had any thesis results. He has made career compromises in order for us to be together. I have also benefited from the wonderful mentoring relationship he has had with Fred Gehring, learning the mechanics of the profession vicariously through Fred.

There were many small things people did that made a difference. The chair, Don Lewis, occasionally asked me how things were going, giving me the feeling I was noticed. When Karen Uhlenbeck visited, the department organized a luncheon for the female students; just meeting her was inspiring. Carolyn Dean, a new tenure-track assistant professor and the department's only female professor at the time, took me out to lunch a couple of times to chat.

Of course, there were negative experiences as well. When I married, the department tried to cut my fellowship, claiming there was no longer a financial need, although they had never asked any of my male cohort for information about their wives' incomes. The constant refrain of "You'll have no problem getting a job because you're a woman," even from two former chairs of the department, was a disheartening confirmation of my worst fears. I now see that in fact quite the opposite is true: despite affirmative action policies and plenty of talk about hiring women, women mathematicians are still judged less worthy than comparable male colleagues.

The Postdoc Years. Finishing graduate school, I was overwhelmed by the opportunities available to me. Princeton offered me an instructorship, which I turned down partially because of my previous experience there but mostly because I heard it was a special "women's program." I wanted to be someplace where I was wanted for my mathematics, not my gender. In contrast, Steve Kleiman, seeming genuinely interested in my work, called from MIT and invited me to speak in his seminar. I decided to accept a position as a Moore instructor at MIT.

Before going to MIT, I spent one year studying commutative algebra at Purdue with Craig Huneke, supported by an NSF Postdoctoral Fellowship. This was an important step; there I learned that my work is valued by a fairly large group of mathematicians around the country and world. If I'd gone straight to MIT, where fewer people appreciate commutative algebra, I might have developed less confidence. On the other hand, I found myself wanting more than ever to learn more algebraic geometry and was not as excited as I felt I should have been about the problems my postdoctoral supervisor was proposing.

My MIT postdoc years (1994–1996) were the mathematical highlight of my career. Juha had gotten a fellowship from Finland, so he was able to be with me in Boston. We loved living and working there. I learned a lot from private discussions with Steve Kleiman, from Joe Harris's algebraic geometry seminar at Harvard, and from the many incredibly smart postdocs and students in Boston, including Lars Hesselholt, Michael Thaddeus, Sue Tolman, Sandor Kovács, Sara Billey, Neil Strickland, Brendan Hassett, and Ravi Vakil. At MIT, there was an inspiring buzz all around;

I worked hard and was essentially doing or talking mathematics all the time. My work was evolving toward more and more algebraic geometry.

At this time, I also began to travel to conferences, which I hadn't done at all as a student. At the 1994 ICM, I met Michel van den Bergh, leading to a fruitful collaboration. I was also inspired by Jean-Pierre Demailly's lecture explaining his use of L^2-analysis to solve some hard problems in algebraic geometry. This got me started on trying to find a characteristic p approach to Fujita's Freeness Conjecture, a thread I followed for a few years with some success. At the 1995 Santa Cruz conference in algebraic geometry, I met for the first time János Kollár, a mathematician whose work I admire and a current collaborator. I was thrilled when he asked me to lecture about my work on Fujita's Conjecture.

In 1996, I was promoted to an assistant professorship (tenure track) at MIT. I would have loved to stay there, even though I knew that MIT rarely tenured anyone. But Juha and I found it hard to find two jobs in the Boston area, and Michigan offered excellent tenured positions to both of us. So we moved back to Ann Arbor.

Professorship at Michigan and Motherhood. Being a professor turned out to be a bit different than expected. I was relieved that the warnings about not being taken seriously by my former teachers were overblown. But I was extremely busy. I quickly acquired five Ph.D. students, as well as a visiting Ph.D. student and a few (informal) postdocs. There were committee meetings and REUs and reading courses and tens of letters of recommendation to write. I missed my days as a postdoc whose main obligation was to do research, but soon adapted to the new demands and settled into enjoying my job.

The 1997–1998 academic year was eventful for me. I spent the fall at MIT, where I discovered a connection linking tight closure to multiplier ideals, a new direction in my research. That spring, I achieved my greatest accomplishment: the birth of my daughter, Sanelma Heinonen Smith, in April. That semester, I could barely think of anything else and essentially did nothing but prepare for her arrival.

Before Sanelma was born, I naively assumed I would be going back to work gradually, beginning when she was about six weeks old. In fact, I was so enchanted by her and by motherhood that I had no desire at all to do mathematics that entire summer. I just enjoyed being a mom. In August, when she was four months old, I did take a few weeks to meet with my collaborator Lauri Kahanpää to finish proofreading our book, *An Invitation to Algebraic Geometry*, but this was for only a few hours a day.

I did not return to classroom teaching until January 1999, when Sanelma was nine months old, although I was having individual weekly meetings with my graduate students that fall. When I approached the chair to ask for relief from teaching that semester, I was denied and instead given a lecture about how Europe's generous social benefits were destroying its economy. Fortunately, I had been to a university-wide women's event and knew that it was typical for women professors at Michigan to receive a semester's relief from teaching after the birth of a child. So I went to a dean, who easily arranged this for me.

How much mathematics did I get done when my daughter was an infant? Truthfully, not much. Between nursings, serving on committees, supervising Ph.D. and undergraduate research students, and meeting with postdocs, I was exhausted. I had little time or energy for thinking about new research, although I did write up things I'd figured out before she was born. I returned to my research roots in tight closure, a subject I knew well, and in her second year wrote a few small papers on this topic. I didn't have the stamina to continue my struggle toward algebraic geometry.

I didn't start feeling inspired again in my research until Sanelma was nearly two. The support of my friend and colleague Rob Lazarsfeld, who repeatedly bugged me to meet with him to discuss mathematics that year, was critical. He was also a relatively new parent and didn't seem to mind a slow and scattered pace. With Lawrence Ein, we began a fruitful collaboration into algebraic applications of multiplier ideals. A year later, I learned I won the AMS Ruth Lyttle Satter Prize, providing a needed psychological boost as I was again picking up speed.

My daughter just turned five this year, 2003. I am satisfied with the balance I have found between mathematics and motherhood; while I get less math done, life is much richer. I have adjusted my social life, having more parties at home instead of going out. I have learned how to spend money to make my life easier: we have a regular housekeeper who shops, cooks, cleans, and runs errands for us twice a week. I have begun saying *no* to the excessive committee work that plagued my early years as a professor, as the department struggled to have women represented on committees without enough female faculty to make it feasible. I have learned to talk to deans and colleagues outside the math department for advice, support, and concrete help. We've also found a great child care system, the Au Pair Program (overseen by the U.S. government). We've had three au pairs, all young women around twenty years old from Finland, who have offered us terrific child care for a year each in return for a chance to see the United States while they figured out what to do with their lives.

Of course, I've made compromises, especially in terms of invitations involving travel. My decision has been simply to bring my daughter along with me, or else decline the invitation. I have enjoyed taking Sanelma with me to Japan, Sweden, Italy (three times), and Canada, as well as twelve states and Washington, D.C. (seven times). I've managed this by bringing along a babysitter—my husband, an au pair, or my mother, who has been delighted to spend an occasional weekend with her granddaughter while I attend to business on the East Coast. It has been worth the extra (nonreimbursable) expense.

I have enjoyed the opportunity to live and work in exotic places. In 2001, we spent six months in Jyväskylä, Finland, while I did research and taught algebraic geometry as a Fulbright professor. This year (2002–2003), the Clay Foundation is supporting me to conduct research in Berkeley, while I serve as an organizer for a special year in commutative algebra at MSRI. The support and flexibility of my husband, Juha, have made these opportunities possible. Although they have reduced our income for those years, we have both found them to be an important part of a satisfying life and career.

The Worst Advice I Ever Got. The only respect in which men and women are undeniably different lies in women's ability to bear and nurse children. Other than annoying minor incidents of sexism and discrimination, the only times I have felt my career profoundly affected by my gender have been in issues related to childbirth.

As a graduate student, I worried a lot about when and how I would have children. In confessing this to an influential (male) mentor, I received the following advice: *Whatever you do, just wait until you have tenure before getting pregnant.* At twenty-six, with no thesis in sight, I added up a few more years of grad school, three years for a postdoc, six years for a tenure-track job, and knew I'd be too old. I couldn't follow this advice, and it terrified me. A mathematician friend of mine heard similar advice as a postdoc: it was even suggested that she plan on using the latest reproductive technology to get pregnant later! As it turned out, I was lucky and did get tenure at thirty-one and a baby at thirty-two. Having tenure helped me relax and enjoy motherhood; however, not everyone has this luxury.

The culture surrounding this advice is a huge reason so many women drop out of mathematics, both before and after having children. The message is this: *If you are eager to start (or have already started) a family, you can forget about a serious career in mathematics.* Nonsense! Women mathematicians have successfully had children at all career stages. The four tenured women in my own department, all successful researchers, have a total of seven children: two born while their mothers were in graduate school, three to postdoctoral mothers, one to a tenure-track mom, and one to a tenured mom (me). Each of us has to make her own decision whether the time is right. There is no ideal time.

My generation of American women was encouraged to delay bearing children until our thirties, and I now fear the consequences for some of us who have waited. I have had miscarriages trying to have a second child starting at age thirty-five. I wish I'd educated myself sooner about the real risks and difficulties of carrying a pregnancy in one's later thirties. I probably would have started building a family sooner and not worried about timing births around conferences or semester finishing dates.[9]

It is frustratingly unfair that a woman's prime childbearing years exactly coincide with the time she is supposed to be intensely working on a thesis problem, impressing people in a postdoc, slaving in a new tenure-track job. What a luxury it would be to turn one's attention to building a family after establishing a career! Even very supportive men seem unaware of their own tremendous advantage in this respect. Of course, ours is a system designed by men and for men, many of whom had full-time wives raising their children. As more of us become chairs and deans and politicians, the climate will begin to change.

We are all pioneers: we need to push ahead and do what is right for each of us, forcing the system to comply as best we can. By our example and by our demands, we make the road a little more passable for the next generation.

A Cautionary Tale

HELEN MOORE

Ph.D. 1995, State University of New York at Stony Brook

Beginnings. I am a mathematician. I suppose I might have been a bench scientist if I hadn't gotten in trouble for experiments with candy and toothpaste, or if I had been allowed to use my older brother's microscope or Erector set or been given my own. My projects were inside my head, or made from paper clips, dental floss spools, tape, and boxes.

But I love math, and I really enjoyed math contests when I was in school. One of the first mathematicians I met was a local college professor who had agreed to judge a math competition at my high school. I confided to him that I really loved math contests. I still remember his scornful expression and disparaging tone as he replied, "Math contests are nothing like real mathematics." Crushed by his response to my enthusiasm, I worried that he might be correct. These days, more philosophically, I allow that he may just have been an unhappy person, or that (as is the case for many good mathematicians) he may not have done well on math contests when young, or that he may have believed that contests fail to stimulate original thinking. Whatever his reason, I strongly disagree with his statement. It was the problem solving I enjoyed during math competitions,

---

9. For those of you who have also struggled with fertility issues, I am happy to report that I gave birth to healthy twins in August 2003.

and I found this much more intellectually satisfying than homework problems.

I had another encounter during high school with local math professors, a panel of women mathematicians who spoke about their careers. After the panel, I talked with some of them. I wasn't doing well in calculus, because I thought the homework was boring and never did any of it. But I loved math and did well on math contests. Did they think it was possible for me to major in math in college? Their response was "Absolutely!" They said if I liked math, I should major in math, despite my imperfect math class grade. The memory of that encouragement has sustained me to this day, and I am grateful for it. These professors had the knowledge I needed (that a bad high school calculus grade did not mean I couldn't do well as a math major in college) and were willing to share it and say positive things that gave me courage. Indeed, I did well at my undergraduate institution.

Graduate Career. The graduate program I entered had three women faculty members (two tenured, one postdoc) and was a top-twenty math department. (I turned down another program partly because they had no women faculty.) My first year was good: extremely challenging work, lots of collaboration, and several women in my classes. After that, things went downhill. The other two women in my year dropped out. All six of the women in the year above mine dropped out. There were no women who entered in the year after mine. There were numerous incidents in which women in our graduate program experienced harassment. One woman found a message on her office door memo board: "—, I want to— your brains out." An obscene message with my name in it was inked on the elevator doors. When I asked the chair to have it painted over before an upcoming conference, I was told the department didn't have the money to pay for it. Sexist comments, including remarks on women speakers' appearances, were regularly tolerated at tea. A male graduate student "just knew" that a female faculty member hadn't made any contribution to a work she had coauthored. A male faculty member was notorious for making sexually harassing comments to female students and teaching assistants. In a graduate class where I was the only woman among thirty students, one of my classmates turned around and stared at me for sixty seconds in the middle of each class meeting, no matter where I sat.

These and other incidents contributed to an atmosphere in which the women graduate students felt isolated and undervalued. I went into that program with such a strong desire to do mathematics that it carried me all the way through. I used to think that the women who dropped out didn't like math enough. However, I eventually realized that all of the additional battles I had to face because of such sexism took a serious toll, and I understood these women's decisions better. Later, I thought about leaving mathematics myself, but decided to remain rather than seek another source of the intellectual satisfaction that I need.

During a five-year period in the 1990s, over two-thirds of the women who entered this program as Ph.D. students left without that degree, but only about 15% of the men did so. Most of the women left when they didn't pass the qualifying or comprehensive exam. Although I excelled in problem-solving classes, I struggled with the comprehensive exam myself. I came close to passing on one occasion when I knew I could have completed a crucial question with a little more time. After that, I showed the department a study by Goldstein, Haldane, and Mitchell.[10] The authors found that the male advantage on high-level cognitive tests existed only under certain testing conditions. If the same test was administered under untimed conditions, the average time difference was not very large, but the male advantage vanished. Our comprehensive exam was used to decide which students should advance in the Ph.D. program, and it seemed incongruous to me that the determining factor should be speed. The department agreed and relaxed the time limit somewhat. Only about one-third of the takers passed the next exam, but I was one of them. When I left the department, it was with my Ph.D.

A women in science group that included graduate students and faculty helped me survive graduate school. We met for dinner once a month, courtesy of the dean's office. Since I was often the only woman in my math classes and seminars, the science group gave me a cohort in which I didn't feel out of place. I enjoyed the group very much and made some close friends: bright scientists in a variety of fields, who have done well since that time.

At my first math conference, I commented to a fellow graduate student that I was surprised to see so few women. She responded that she thought there were *many*, compared to previous years! During that conference, I approached a group of women I didn't know and invited them to lunch. We had a wonderful time, talking about our research and getting to know each other. The next year, we organized a dinner for all of the women attending that conference. Some men thought we would be limiting our opportunities by having an all-woman event, but we met many more senior people than we had previously. One of them was Chuu-Lian Terng, later a president of AWM, who was particularly kind to us graduate students. When we attended other conferences, knowing lots of the women attendees added a level of comfort we hadn't had previously.

After Graduate School. My first job after graduate school was a tenure-track position at a small liberal arts college. While there, I experienced

---

10. "Sex Differences in Visual-Spatial Ability: The Role of Performance Factors," *Memory and Cognition* 18(1990), 546–550.

harassment from several male students in class. As part of fraternity hazing, they were apparently supposed to grab their crotch (extra points if they reached *inside* their pants to do this), raise their other hand, and ask me a question. I was shocked out of my wits when it happened and didn't deal with it in the classroom. But I reported it to the president's office, and fraternities were eventually banned on campus, due to numerous complaints about a variety of incidents targeting female faculty and students.

I left the liberal arts college and took a temporary position at a research university for several years. It was hard to resign a tenure-track position, but I was happier at the university, because it was a better fit with my interest in research. Although my first year there I was the only woman on the math faculty, things were generally better. Over a few years, there were several bad incidents of harassment of women by men graduate students. But the department supported my efforts to organize events for the women graduate students and provided food for occasional events for women in math. Some of the senior male faculty attended events such as panels of women mathematicians in industry.

My organizing efforts on behalf of women in math and my experience teaching a course on women and minorities in science and math, as well as a previous position in a university program for women in science, led naturally to my current job. I am now the associate director of the American Institute of Mathematics Research Conference Center, one of the institutes funded by the NSF in 2002. Because of the power associated with this position, I have encountered many fewer negative incidents, the most common one occurring when someone assumes I am a secretary. In addition to helping run small, focused workshops in all areas of research mathematics, I am responsible for our major efforts to increase the numbers of women and underrepresented minorities involved in our workshops.

Looking Forward. A male faculty member at a research university recently asked me what I thought he could do to encourage women in math. Just by asking the question, he contributed to my sense of comfort in the community. I know that he will take a leadership role on this issue in his department. When a department chair attends a session on how to keep women in graduate school, this speaks volumes to all present and all who hear about it. The senior men mathematicians I have seen at discussions on women in math occupy a warm spot in my heart. I consider them some of our strongest male allies and enjoy seeing their numbers increase.

Effective leadership helps make mathematical experiences more positive for women, by setting the tone and giving a strong message to others. When the people in charge of a department make clear statements about the way things run, others listen. Zero tolerance of harassment, acting quickly on complaints, and caring about individuals and how their studies and research are going—all of these are important. The people who have supported me have helped me focus on the mathematics I love, which is something every researcher deserves.

Of course, in the future I hope for, women mathematicians will read this book and say, "That piece by Helen Moore is completely off the mark. Those things just don't happen any more!"

Role Models and Mentors

JUDY L. WALKER

Ph.D. 1996, University of Illinois

I am in my seventh year as a Ph.D. mathematician, and though I have had an interesting and rewarding career thus far, I feel as if it is too early for me to be dispensing pearls of wisdom. But when thinking about what to write, I realized that I do know one truth that I can share: finding worthy role models and caring mentors is the key to creating the life you want. Moreover, as I began to think about my role models and mentors, I discovered that AWM has had a profound effect on me, often stepping in at crucial points and providing exactly the guidance I needed to progress to the next level of my career.

My first role model was my sister, Susan. She is three and a half years older than me, and when we were kids I wanted to be just like her. I walked early and talked early, simply because I wanted to keep up with her. She was also my first mentor. She would come home from school, when I was still too young to go, and show me what she had learned. Sue taught me to read well before I started kindergarten, and the confidence this gave me stuck throughout my educational career.

I found my first mathematical mentors as an undergraduate at the University of Michigan. I knew no women on the faculty there, but I found plenty of male faculty members to learn from. Mort Brown was the first mathematician truly to be a mentor to me. I remember one day, as I asked him questions about real analysis during his office hours, he told me that I could be a good mathematician if I could only gain some confidence in myself. That comment clearly stuck with me, and with it Brown also became a role model for me. I decided that if I was going to be an academic mathematician, I wanted to be able not only to do mathematics, but also to have a positive impact on my students' lives.

Brown also encouraged me to apply for REU programs, and I chose one on computational group theory at Rose-Hulman Institute of Technology. The leader of the program was Gary Sherman, who became my next role model and mentor. Not only did I learn a good deal of mathematics and have my first taste of mathematical research, but also I learned that research mathematics doesn't have to be a solo effort and that mathematicians don't have to be solitary people. In addition to finding interesting yet accessible problems for undergraduates to work on, he knew how important it was that the REU participants form strong bonds with one another, and so he took great care in selecting the participants for the program. This care paid off, at least for me: three years later I married fellow participant Mark Walker, and another participant was a groomsman in our wedding.

My first contact with AWM occurred at the end of my undergraduate career when I received an honorable mention for the Alice T. Schafer Award. I didn't know much about the organization then, or the substantial role it would play in my career. But at the awards ceremony in Columbus, I met several other women mathematicians, both the other students being recognized and the women from AWM who organized the event. It was the first time I had been around so many other women mathematicians, and I took from that experience encouragement to continue my mathematical studies.

After much deliberation, I chose the University of Illinois for graduate school. Mentors were abundant for me: Bruce Reznick, from whom I never actually took a class but who was always willing to talk with me about whatever was on my mind; Dan Grayson, who spent countless hours talking with me about my research even though he wasn't my thesis advisor; Nigel Boston, who was my thesis advisor and who also spent countless hours working with me and giving me general advice; and Dick Blahut, an electrical engineer who shared my interest in algebraic geometry codes.

My second contact with AWM came toward the end of my graduate career, when I presented a poster at the AWM Workshop at the JMM in Orlando. The experience was similar to the one I had enjoyed six years earlier in Columbus, only better. Again I was surrounded by women mathematicians—other graduate students as well as recent Ph.D.'s and those established in their careers. The daylong workshop provided plenty of opportunity to talk with these other women about mathematics and about life. I was in the midst of applying for jobs, and being able to escape the Employment Register for the supportive environment of the workshop was wonderful.

The job market for mathematics Ph.D.'s in 1996 was not great. It was better than it had been in the early 1990s, but my fellow job seekers and I certainly had to face the very real possibility that we would be unemployed. Mark and I had the added hurdle of needing to find two jobs together. We did well, though, and ended up turning down offers, including a pair of postdoctoral positions, to accept positions at the University of Nebraska–Lincoln (UNL). We chose UNL in part because it gave us a solution to the "two-body problem," but mostly because we knew the department was a good match for us.

I found several new mentors and role models at UNL. One reason I wanted to come to UNL was that Sylvia Wiegand, then the president-elect of AWM, is here. I was finally going to have my first female mentor in the mathematics profession! And, as I expected, Sylvia has been wonderful to me. She sets an excellent example for all women mathematicians and is a role model to me in addition to being a mentor. What I didn't expect, however, was that once again my primary mentor would be male. Jim Lewis, our department chair, has had more influence on my career than anyone else. He presents me with opportunities he thinks would benefit me and shields me from commitments he thinks would not. His door is always open, and I truly value his advice.

My research mentors have often been outside of UNL. Dick Blahut introduced me to Vera Pless (University of Illinois at Chicago) at a conference when I was still a graduate student, and she immediately took me under her wing. Accepting the tenure-track position at Nebraska right out of graduate school meant that I never did a postdoc, but Felipe Voloch (University of Texas) approached me soon after I received my degree with an idea for a project that proved to be the start of a highly productive long-term collaboration.

In 2000 I had a second opportunity to participate in an AWM Workshop, this time giving a talk along with other recent Ph.D.'s. That workshop marked a transition for me: I was assigned a graduate student to mentor, and I was assigned an established female mathematician as a mentor of my own. The experience was wonderful. Though I certainly benefited from the mentoring I received there, I also gained self-confidence through my conversations with my mentee. Once again, my contact with AWM came at a pivotal point in my career, and it gave me needed support at just the right time.

I received tenure in 2001 and no longer think of myself as new to the mathematics profession. However, even as I make the transition from being a mentee to a mentor, I find that my own mentors are becoming ever more important to me. For example, when my colleague Wendy Hines and I wanted to start the ALL GIRLS/ALL MATH program to encourage high school girls to pursue their interest in mathematics, Jim Lewis and Sylvia Wiegand provided their advice and encouragement. Moreover, Vera Pless agreed to be the "distinguished visitor" for our first ALL GIRLS/ALL MATH summer camp. After the UNL Department of Mathematics and Statistics won the 1998 Presidential Award for Excellence in Science,

Mathematics, and Engineering Mentoring, Jim Lewis came up with the idea of starting the Nebraska Conference for Undergraduate Women in Mathematics in celebration of the award and asked me to lead the organizing committee. Sylvia Wiegand agreed to be a plenary speaker for the first of these conferences. In 1999, I was the principal lecturer for the undergraduate portion of the Institute for Advanced Study/Park City Mathematics Institute Mentoring Program for Women; Felipe Voloch suggested my name to Karen Uhlenbeck for this role.

With the recent birth of my daughter, I find myself looking ever more to my mentors as I struggle to find my way as a parent. I turn especially to my fellow AWM members for advice on how they've balanced career and family. And I've come full circle, taking my sister once again as a primary role model: if I can do even half the job with my daughter she is doing with her sons, we're all doing pretty well.

In 2002, I became a member of the AWM Executive Committee. It is an honor and a thrill to be part of this organization that has provided so much support to me in my own career development. I hope that in the future other women are able to draw the same strength from AWM that I have. But above all, I hope that other women are lucky enough to find mentors—women and men, mathematicians and not—who can provide the advice and encouragement they need in order to achieve their goals.

An Energetic Career

TAMARA G. KOLDA

Ph.D. 1997, University of Maryland at College Park

I've been interested in mathematics since I was a young child. I remember that my father, a gym teacher, used to make up word problems to keep me busy on long car trips. The only dilemma was that neither of us knew if I had gotten the right answer!

Though I don't think my parents ever envisioned my becoming a mathematician, they did all the right things to get me where I am today. In grade school and junior high, I participated in summer math camps, computer camps, and even a two-week program at NASA Goddard Space Center at age thirteen. The last experience sticks in my memory because I distinctly recall deciding that engineering was not for me—sitting in front of a computer all day just looked too boring.

In high school, my focus was on surviving puberty. Appropriately, I developed an interest in biology. I even spent two years doing independent research—one year studying the mating habits of hamsters and a second studying the environmental sensitivity of copepods.[11] By the end of my senior year of high school, after far too many births among the hamsters and deaths among the copepods, my interest had waned.

Having given up on biology and engineering, I wasn't quite sure what to do with myself when I started college at the University of Maryland, Baltimore County. Since I liked math and had kept up with it even during my short-lived flirtation with biology, I picked that as my major. I was by no means certain I would stick with it, but I couldn't think of anything else to do! My inspiration to persevere in mathematics came in the form of my professor for Real Analysis I and undergraduate advisor, Muddappa Gowda. I became more seriously interested in research after attending an NSF-sponsored REU at the College of William and Mary.

As an undergraduate, I also began actively learning more about the plight of women in science. As far back as junior high, I was aware that women were not as well represented as men in math and science. More and more often, I was one of only a small number of women in my classes, especially in college physics and computer science. Because of this, I became extremely active in a campus group to promote women in science and also joined the AWM.

My undergraduate professors were fully supportive of my graduate school aspirations, even helping me to obtain a full fellowship from the National Physical Sciences Consortium (NPSC). At the time, the NPSC fellowship was restricted to women and minorities. It paid for up to six years of graduate school and included two summer internships. In my case, I went to the University of Maryland at College Park and was an intern at the NSA.

In graduate school, I revived the atrophied women's group and gave it a new name, Women in Mathematics, or WIM for short. With support from the department chair, Raymond Johnson, for pizza parties and other activities, we had a very active group. The group had no fixed structure, other than a chair, yet we had plenty of volunteers to cover the various events we sponsored—including networking lunches and a seminar for women students. My primary role in WIM was one of coordination, and I did this via email and the (then fairly new) World Wide Web.

Research-wise, I developed an interest in parallel computing and optimization. After hunting high and low for a thesis advisor, someone finally pointed me to Dianne O'Leary. I did not purposely seek out a female advisor, but it was a great advantage. She was (and still is) empathetic to the difficult situations I faced as a women mathematician. More important, she is an archetype of success, showing that women can do it all!

11. Copepods are tiny crustaceans abundant in plankton.

I had fully intended to become a math professor, but, over time, the combination of a tight job market and tenure horror stories weakened my resolve. I applied for both academic and nonacademic positions and got offers from both groups. The offer I could not resist was a two-year named postdoc with complete research freedom and no teaching. It just so happened that this postdoc was at Oak Ridge National Lab, and it became my introduction to the Department of Energy (DOE).

Oak Ridge was a great experience, but I missed having lots of women mathematicians around. To stay connected, I became the Web editor for the AWM. With the help of several volunteers, I created the AWM's website and maintained it for five years.

I came to enjoy working within the DOE so much that I decided to stay on once my postdoc was finished. But I opted for a change of scenery and moved out to sunny California to work at a different DOE lab, Sandia National Labs in Livermore (in the San Francisco Bay Area). Working at DOE is an applied mathematician's dream job. There is no shortage of real-life, unsolved problems to motivate my research, and now and then I have been able to contribute to the solution of these problems. The computing facilities are among the world's best. And, most important, I work in a large group including top-notch researchers in applied math. I have been very lucky to have a wonderful mentor during my postpostdoctoral years, Juan Meza.

Another secret to my success is having the world's best husband. He's moved with me twice and is continually supportive.

So now, I sit here in front of my computer, and it's not so boring after all.

For the Love of Mathematics

TASHA R. INNISS

Ph.D. 2000, University of Maryland, College Park

I have loved mathematics for as long as I can remember. Math is like a game or a puzzle in which you put the pieces together to get *one* right answer. That is simply beautiful! Algebra, geometry, calculus—all lay the foundation and provide tools to solve puzzles. Mathematics is not just about numbers; it is about thinking creatively to solve problems.

As I was growing up, I alternated between wanting to be a teacher, a lawyer, a teacher, a pilot, a teacher, an accountant, a teacher. . . . It is no surprise that I kept returning to the idea of teaching. My grandfather, who received his master's degree in education from Harvard in the early 1950s, is a retired sixth-grade teacher. My mother, who has a Ph.D. in sociology, is a college professor; and my uncle, who graduated from my alma mater, Xavier University of Louisiana, is a college mathematics professor. My teachers have had a tremendous positive impact on my life because they fostered my interest in math and demonstrated their confidence in my mathematics ability regardless of my race or gender.

In my freshman year of high school I realized that whatever I did in life, it would involve mathematics. The encouragement and nurturing I received from my family and teachers inspired me to pursue higher degrees and contributed to my overall success. In college, I found that I had a gift for explaining concepts and imparting math knowledge to others and decided that I wanted to be a college professor.

My biggest inspiration is my mother, Leslie B. Inniss. My mom raised my brother and me alone while working a full-time job. My grandfather and my brother were strong, positive male influences in my life. Getting my Ph.D. in 2000 was even more special because that same summer my brother completed his doctorate in environmental engineering.

From kindergarten through college, I excelled in all subjects, but my passion was always mathematics. I never felt that being a woman or an African-American would hamper my pursuit of a career in mathematics. There were several factors that worked against my possible success in mathematics. First, I attended public schools in Louisiana, a state not highly ranked in public education. Second, I am a product of a single-parent household. Last, my family was of low socioeconomic status. I did not let any of those factors or possible negative attitudes of society toward women in math deter me from my goals. Instead, with the solid backing of my family's belief in the importance of education, I persevered.

It was not until I began graduate school that I encountered overt sexism and racism. There were times on my journey to the Ph.D. where I was treated unfairly or not respected because I am a woman of color. In those times, I considered giving up my dream. Through soul-searching, encouragement from family, support from mentors, a sincere belief in my abilities, and my involvement in such professional societies as AWM and NAM,[12] I stayed on track and did not let the negative shadow of the two "isms" (sexism and racism) hinder my progress. Despite the obstacles, I was successful in becoming the college mathematics professor I always dreamed of being.

I am currently a Clare Boothe Luce (assistant) Professor of Mathematics at Trinity College in Washington, D.C.[13] The position involves teaching,

12. The National Association of Mathematicians (NAM) is a professional organization dedicated to the promotion of excellence in the mathematical sciences and of the mathematical development of underrepresented American minorities.

13. Inniss recently accepted an assistant professorship at Spelman College, also a liberal arts college for women students.

scholarship, and service. The Luce Foundation provides funding to purchase equipment for research, to maintain memberships in professional societies, and to travel to professional meetings and conferences.

Trinity is a small liberal arts college for women. One of my goals is to be a good role model for the students. What I find most fulfilling is to instill math confidence in students who have been intimidated by mathematics in the past. I believe that anyone can understand mathematics, if it is taught well. I want to dispel the myth that women cannot do mathematics. I find it gratifying to mentor and encourage the Trinity women to pursue their goals and dreams, as I show them that it is not impossible for a woman of color to get an advanced degree in a scientific field.

I also like to *do* mathematics! As an undergraduate, I did on-campus research in such areas as information theory, optimization, and mathematical programming; I participated in a summer research program at the Ohio State University and an NSF REU program at Florida State University. These experiences exposed me to a part of mathematics that I pursued in my graduate studies. My research areas of interest are in statistics and aviation operations research. While in graduate school, I became a member of the National Center of Excellence for Aviation Operations Research. Commissioned by the Federal Aviation Administration (FAA) to utilize academic expertise to solve problems the FAA encounters, the Center of Excellence is composed of four universities: University of Maryland, MIT, UC Berkeley, and the Virginia Institute of Technology. In my dissertation, I addressed a specific problem in air traffic flow management by developing statistical models to help estimate the distribution of arrival capacity at an airport during inclement weather conditions. These models, when used in conjunction with stochastic ground holding models, could aid the specialists at the FAA's Command Center in effectively planning for flight arrivals and departures and could ultimately help in minimizing airline flight delays. Because of the timeliness of this work, I earned the FAA's Centers of Excellence Student of the Year Award. Due to this honor, the FAA offered me an independent contract to continue my dissertation research. Thus, I have the best of both worlds; I get to teach mathematics and *do* mathematics as a visiting researcher at the FAA.

When I received my Ph.D. in mathematics from Maryland, it was a historic moment: I was one of the first three African-American women to do so. I was excited not just to be *finished*, but to be recognized for finishing. Kimberly Weems, Sherry Scott Joseph, and I received a lot of media attention. We were written about in the *Washington Post*, the *Chronicle of Higher Education*, and *Black Issues in Higher Education*. We were interviewed for Black Entertainment Television News and National Public Radio. We were even honored by a reception hosted by the former lieutenant governor, Kathleen Kennedy Townsend, and the Maryland legislature. I had not

expected so much attention upon completing my degree. The most gratifying consequence of all the media attention was being invited to give talks
to inspire and motivate younger women considering advanced degrees in
mathematics and science.

Here is my advice for a woman considering a career in mathematics:
Have faith in yourself. *Never give up* as you follow your dream. Persever-
ance is the key. And by all means, pursue what you *love* because the journey
is not so bad when you truly enjoy what you are doing. Mathematics has its
own language; as long as you can speak that language, it does not matter
what gender you are.

When I tell people that I am a Ph.D. mathematician, they are amazed.
I have never figured out if this is because I am a woman in math, an
African-American woman in math, or just a mathematician!

Mathematics: Mortals and Morals

KATHERINE SOCHA

Ph.D. 2002, University of Texas

In a September 2002 public lecture at Michigan State University, Nobel
Laureate (chemistry) Roald Hoffmann said, "There is no way more sure
to make people think that science is abnormal, not done by normal
human beings (therefore to be feared) than to claim that it is superrational,
that it is accessible only to very smart people, and that it is ethically neutral." Hoffmann's words articulate the stereotype that ordinary people can't
do science, let alone mathematics, queen of the sciences. I believe many
women struggle with the emotional impact of this assertion, the incorrect-
ness of which is obvious to the intellect but perhaps not so obvious to the
heart. Certainly, as a student I did not worry about being a woman mathematician. Rather, I worried about being good enough, and only after several years of graduate school did I develop the perspective and confidence
to ask "good enough by whose standards?"

As I look back from the viewpoint of a professional mathematician, I
recognize that mathematical ability was present in my childhood and
youth, but no one around me had the training to recognize a sensitivity to
symmetry, order, and pattern as intrinsically mathematical. It was only my
stubborn determination to prove my intellectual mettle that kept me
enrolled in mathematics classes. In my sophomore year in college, I finally
met a professor, John Leadley, who fostered my growth as a student and
mathematician by providing a much-needed (by me, at least) structural

approach to understanding mathematics through category theory. What a difference this made to me! I chose to major in mathematics out of interest in its power to create and describe ideas. Had I continued to believe that a problem-solving ("get the answer in the back of the book") approach *was* mathematics, I would not have continued with the program.

Despite this college-level encouragement, I still found the path to my doctorate difficult and circuitous. After a moderately disastrous (academically and personally) first year in graduate school, I transferred to a small, supportive master's program at another school. My success there and the faculty's positive attitudes toward me (particularly the support of Dorothy Williams and Joyce O'Halloran) encouraged me to try again. I entered the small but very supportive doctoral program at Oregon State University (OSU). I was very glad to be at OSU, where Ronald B. Guenther was a superb mentor who cared deeply about the social culture of mathematical life and the roles of women in that culture. His encouragement led me to apply to the 1995 Park City Mathematics Institute and the associated Women's Mentoring Program, held at the Institute for Advanced Study in Princeton. Participation in these programs was a pivotal career experience (even life experience) for me. The faculty of the Women's Mentoring Program created a very welcoming culture; in particular, Susan Friedlander, Barbara Keyfitz, and Karen Uhlenbeck actively encouraged our desire to be mathematicians while opening our eyes to the practical necessity of making good career choices. Through Friedlander's interest and encouragement, I visited and ultimately transferred to the University of Texas at Austin mathematics program, with the sponsorship of her collaborator Misha Vishik. My successes in that program and in my job search last year, both as a mathematician and as an educator, were greatly fostered by Uhlenbeck. She encouraged me in difficult times, both mathematically by talking with me about my research and professionally by talking with me about education and career options. She took my outreach and education interests seriously, not as a distraction from mathematics, but as an expression of individual talents that may one day enrich our field.

Further, Uhlenbeck brought her social consciousness to her role as professor. Very few graduate students in need have not received some sort of help from her. She created a welcoming community for women graduate students in our department, and I think many of us might have faltered without her. What a fantastic and positive impact she has on us, myself most definitely included. This leads me to wonder why the culture of mathematics at the university level can be so unsupportive that the actions of one (admittedly extraordinary) person like Uhlenbeck can dramatically improve the experience of women students.

A spark that led me to articulate a (perhaps partial) answer to this question appeared in another part of Hoffmann's 2002 speech. He said,

"To say that some action (making a molecule, but, yes, even one as seemingly innocent as writing a poem) is ethically neutral; and we'll let society or others worry about the ethical consequences is, I think, deeply immoral. You, I—the maker—has to accompany the creation with an ethical judgment. The action and the consideration of its consequences ought not, cannot be separated." These words generated in me the same excited recognition as that from seeing a mathematical idea fully and clearly for the first time. Of course we cannot act, create mathematics, teach mathematics, write mathematics, and write about mathematics in a moral and ethical void! I need to believe that what we do, as individuals and as a group, matters—and it does matter, enormously.

Even if our mathematical work does not have broad social ramifications, the fact that we choose this work does. Perhaps understanding the quotient rule in calculus or the fundamental theorem of algebra is not relevant to the daily lives of most people in our society, but having confidence in one's ability to understand, to read logically, and to recognize patterns and mathematical ideas is deeply relevant—relevant to having a literate population that can think for itself without being misled by specious reasoning, by misapplied statistical assertions, or by faith in authority.

I believe there will always be individuals attracted to a serious, deep study of mathematics. However, it is up to us to combat the image of mathematics and mathematical thinking as irrelevant to daily life. It is up to us to show that we are part of the human endeavor. It is up to us to make sure that even non-superhumans who wish it can have a fair chance at joining our profession. And it is up to us to abandon "being an authority" in favor of helping people see that it isn't only the smart ones who can do the math.[14]

14. Socha was an assistant professor at Michigan State University and is now an assistant professor at St. Mary's College of Maryland.

AFTERWORD

After a decade of thinking about it, we have worked actively on this book for more than four years. Life around us has continued in its cycles, as we concentrated on the joys and tasks of writing and editing. Anne's father, Ernest Leggett, who would have been so proud to see this book in print, died in April 2004. Among the family of contributors there were births and deaths and all the milestones of living and of building careers.

During production, we were saddened to learn of the deaths of several people connected with the book. In our communications with Etta Zuber Falconer a few weeks before her death in September 2002, she seemed still vital and strong. She expressed the hope that we would include the powerful words of Vivienne Malone Mayes found in "Black and Female," III.1. We are pleased to have honored her desire. Bettye Anne's department was anticipating a return winter visit and a colloquium on partial differential equations by eminent Russian mathematician Ol'ga Ladyzhenskaya. In January 2004, an Italian colleague wrote sadly that Ol'ga had died, fondly recalling, "She was short and tiny and looked like a fairytale grandmother."

And there were several births! Last spring we received an exuberant email from a contributor, in her second trimester with twins following previous difficulties—we share her family's joy in the birth of twins in August 2003. Another author from part V wrote news of her baby daughter as we were reading the galleys.

We two have worked as professional mathematicians for more than thirty years each. During that time, we have witnessed not only "firsts" but also many positive changes in the status of the profession for women. We have seen women honored with National Medals of Science, women serving as presidents of many of the professional societies. A few women are now members of the National Academy of Sciences and full professors at top ten institutions, and the numbers of women in the profession are growing. We have been energized and supported by the sisterhood of women at mathematics conferences.

In our home departments, we serve as mentors of women and men junior faculty. We take pride in seeing women students take wing as they pursue their fledgling careers and in watching from afar as they happily write us of building their careers in the context of their lives.

There are many young women of promise, both within our acquaintance—students, advisees, new faculty, others who seek us out because we are women—and among women presenting their work at meetings and conferences. We hope that the efforts of our mathematical generation have made their pathways easier to follow, and that they will in turn do the same for those who come after them.

As you have read this book, you have noticed the many connections among both the contributors and the subjects that we mentioned in the preface. These interrelationships run the gamut of loci (family, workplace, professional associations, conferences—even the Internet) and are formed both in spirit and through time. The world of women in mathematics (and in the other sciences and engineering) is still too small a place; please join us there!

Ol'ga Ladyzhenskaya (1922–2004)

Etta Zuber Falconer (1933–2002)

PHOTO CREDITS

INDEX OF NAMES

Numbers in bold refer either to text on unnumbered pages or to photo plates (pl.) in the "*Complexities* Photo Album."

ABOUT THE AUTHORS

Bettye Anne Case is the Olga Larson Professor of Mathematics and Director of Financial Mathematics at Florida State University. She is the editor of and a contributor to *A Century of Mathematical Meetings*, *You're the Professor, What Next?* and *Keys to Improved Instruction*.

Anne M. Leggett is Associate Professor of Mathematics at Loyola University Chicago and editor of the Association for Women in Mathematics *Newsletter*.